LINEAR FUNCTIONS AND MATRIX THEORY

LINEAR FUNCTIONS AND MATRIX THEORY

Bill Jacob
Department of Mathematics
University of California, Santa Barbara

Springer-Verlag
New York Berlin Heidelberg London Paris
Tokyo Hong Kong Barcelona Budapest

Textbooks in Mathematical Sciences

COVER: Paul Klee, *Beware of Red*, 1940. Private collection, Milan, Italy. Used by permission of Erich Lessing/Art Resource, NY.

Grateful acknowledgment is also given for permission to use the following: p. 65: Albrecht Dürer, *Albrectus Durerus Nurembergensis Pictor Nuius*..., Paris, 1532, "Demonstration of perspective" (woodcut), Spencer Collection, The New York Public Library, Astor, Lenox, and Tilden Foundations; p. 70: Albrecht Dürer, *The Nativity* (1504), The Metropolitan Museum of Art, Fletcher Fund, 1917.

Library of Congress Cataloging-in-Publication Data
Jacob, Bill.
Linear functions and matrix theory / Bill Jacob.
p. cm. — (Textbooks in mathematical sciences)
Includes bibliographical references and index.
ISBN 0-387-94451-6 (acid-free)
1. Algebras, Linear. 2. Matrices. I. Title. II. Series.
QA184.J333 1995
512′.5—dc20 95-3756

Printed on acid-free paper.

Production managed by Steven Pisano; manufacturing supervised by Jacqui Ashri.
Photocomposed by Integre Technical Publishing Co., Inc. from the author's LaTeX files.
Printed and bound by R.R. Donnelley & Sons, Harrisonburg, VA.
Printed in the United States of America.

9 8 7 6 5 4 3 2 1

ISBN 0-387-94451-6 Springer-Verlag New York Berlin Heidelberg

PREFACE

Courses that study vectors and elementary matrix theory and introduce linear transformations have proliferated greatly in recent years. Most of these courses are taught at the undergraduate level as part of, or adjacent to, the second-year calculus sequence. Although many students will ultimately find the material in these courses more valuable than calculus, they often experience a class that consists mostly of learning to implement a series of computational algorithms. The objective of this text is to bring a different vision to this course, including many of the key elements called for in current mathematics-teaching reform efforts.

Three of the main components of this current effort are the following:

1. Mathematical ideas should be introduced in meaningful contexts, with formal definitions and procedures developed *after* a clear understanding of practical situations has been achieved.
2. Every topic should be treated from different perspectives, including the numerical, geometric, and symbolic viewpoints.
3. The important ideas need to be visited repeatedly throughout the term, with students' understanding deepening each time.

This text was written with these three objectives in mind. The first two chapters deal with situations requiring linear functions (at times, locally linear functions) or linear ideas in geometry for their understanding. These situations provide the context in which the formal mathematics is developed, and they are returned to with increasing sophistication throughout the text.

In addition, expectations of student work have changed. Computer technology has reduced the need for students to devote large blocks of time learning to implement computational algorithms. Instead, we demand a deeper conceptual understanding. Students need to learn how to communicate mathematics effectively, both orally and in writing. Students also need how to learn to use technology, applying it when appropriate and giving meaningful answers with it. Further, students need to collaborate on mathematical problems, and thus this collaboration often involves mathematical investigations where the final outcome depends on the assumptions they make. This text is designed to provide students with the opportunity to develop their skills in each of these areas. There are ample computational exercises so that students can develop familiarity with all the basic algorithms of the subject. However, many problems are *not* preceded by a worked-out example of a similar-looking problem, so students must spend some time grappling with the concepts rather than mimicking procedures. Each section concludes with problems or projects designed for student collaborative work. There is quite a bit of variation in the nature of these group projects, and most of them require more discussion and struggle than the regular problems. A number of them are open-ended, without single answers, and therefore part of the project is finding how to formulate the question so that the mathematics can be applied.

Throughout the text, as well as in the problems, there are many occasions where technology is needed for calculation. Most college students have access to graphing calculators, and the majority of these calculators are capable of performing *all* of the basic matrix calculations needed for the use of this text. Students should be encouraged to use technology where appropriate, and if a computer laboratory is available it will be useful, too. Some problems explicitly require calculators or use of a computer; others clearly do not; and on some occasions the student needs to take the initiative to make an intelligent use of technology. During the in-class testing of the material (as part of a second-year calculus sequence), instructors found that the use of graphing calculators gave students more time to focus on conceptual aspects of the material. Instructors used to assigning a large volume of algorithmic exercises found they had to reduce the number of problems assigned so that students had more opportunity to explore the ideas in the problems.

A brief outline of how the text is organized follows. The first six chapters constitute a core course covering the material usually taught as part of a second-year sequence. Depending on how the class is paced, these chapters require seven to ten weeks to cover. The remaining three chapters deal with more advanced topics. Although they are designed to be taught in sequence, their order can be varied provided the instructor is willing to explain a few results from the omitted sections. Taken together, all nine chapters have more than enough material for a full-semester introductory course. Sections 1.4, 2.4, 3.5, and 4.5 can be omitted if time is tight, although it would be preferable to avoid this, since their purpose is to give geometric meaning to material that students too often view purely symbolically. Chapter 6 could be covered immediately after Chap. 2 should the instructor prefer. No knowledge of calculus is assumed in the body of the text. However, some group projects are designed for use by students familiar with calculus. Answers to the odd-numbered problems are given at the end of the text.

- **Chapter 1** introduces the concept of a linear function. The main purpose of the chapter is to illustrate the numerical and geometric meaning of linearity and local linearity for functions. In Sec. 1.2, real data are analyzed so that students see why this subject was developed. Matrices arise, initially, for convenience of notation. Linearly constrained maxima and minima problems are introduced because the study of level sets in this context provides one of the best ways to illustrate the geometric meaning of linearity for functions of several variables.

- **Chapter 2** studies the linear geometry of lines and planes in two- and three-dimensional space by considering problems that require their use. A main goal here is to provide a familiar geometric setting for introducing the use of vectors and matrix notation. The last section, covering linear perspective, illustrates how these topics impact our daily lives by showing how three-dimensional objects are represented on the plane.
- **Chapter 3** develops the basic principles of Gaussian and Gauss-Jordan elimination and their use in solving systems of linear equations. Matrix rank is studied in the context of understanding the structure of solutions to systems of equations. Some basic problems in circuit theory motivate the study of systems of equations. The simplex algorithm is introduced in the last section, illustrating how the ideas behind Gaussian elimination can be used to solve the constrained optimization problems introduced geometrically in Chap. 1.
- **Chapter 4** treats basic matrix algebra and its connections with systems of linear equations. The use of matrices in analyzing iterative processes, such as Markov chains or Fibonacci numbers, provides the setting for the development of matrix properties. The determinant is developed using the the Laplace expansion, and applications including the adjoint inversion formula and Cramer's rule are given. The chapter concludes with discussion of the LU-decomposition and its relationship to Gaussian elimination, determinants, and tridiagonal matrices.
- **Chapter 5** develops the basic linear algebra concepts of linear combinations, linear independence, subspaces, span, and dimension. Problems involving network flow and stoichiometry are considered and provide background for why these basic linear algebra concepts are so important. All of these topics are treated in the setting of $\mathbf{R}^n$ only, although the results are formulated in such a way that the proofs apply to general vector spaces.
- **Chapter 6** returns to more vector geometry in two- and three-dimensional space. The emphasis is on applying the dot and cross product in answering geometric questions. The geometry of how carbon atoms fit together in cyclohexane ring systems is studied to help develop three-dimensional visual thinking. As mentioned, this material could be covered immediately after Chap. 2 if the instructor chooses. The author, however, prefers to have his students study this chapter after Chap. 5 in order to remind them of the importance of geometric thinking.
- **Chapter 7** studies eigenvalues and eigenvectors and their role in the problem of diagonalizing matrices. Motivation for considering eigenvec-

tors is provided by the return to the study of iterative processes initiated in Chap. 4. The cases of symmetric and probability matrices are studied in detail. This material is developed from a matrix perspective, prior to the treatment of linear operators (Chap. 8), for instructors who like to get to this topic as early as possible. In fact, this material provides nice motivation for Chap. 8.

- **Chapter 8** develops the theory of linear transformations and matrix representations of linear operators on $\mathbf{R}^n$. A main objective is to show how the point of view of linear transformations unifies many of the matrix-oriented subjects treated earlier. The chapter returns to the examples of electrical networks first studied in Chap. 3, where the cascading of networks provides a basis for understanding the composition of linear transformations. The basic geometric transformations of rotations and reflections are also studied.
- **Chapter 9** returns to the geometry of Euclidean space. The Gram-Schmidt process and orthogonal projections can be found here. Least-squares problems are also studied from the geometric point of view. Some of the data given in Chap. 1 are fit using linear regressions, bringing the course to a close by showing how the concepts developed in the class deepen our understanding of some of the original problems considered.

I would like to thank my numerous students and colleagues for their valuable input during the development of this text, provided both anonymously and in person. I am especially grateful to Juan Estrada, Gustavo Ponce, and Gerald Zaplawa for their detailed comments. I would also like to thank Jerry Lyons of Springer-Verlag for his encouragement and support. Finally, I wish to thank my family for their love throughout this project.

Bill Jacob
Santa Barbara, California

CONTENTS

LINEAR FUNCTIONS AND MATRIX THEORY

CHAPTER 1

LINEAR FUNCTIONS

Linear functions are used throughout mathematics and its applications. We consider some examples in this chapter. Most of this book is devoted to the study of how to analyze and apply linear functions.

1.1 Linear Functions

Proportions: An Example

The concept of *proportion*, or *ratio*, is one of the most fundamental ideas in elementary mathematics. Without it we would have great difficulty organizing our daily life. For example, if we begin our day cooking oatmeal for breakfast, we need to get our proportions right: According to the recipe on one box we need $\frac{3}{4}$ cup of milk mixed with $\frac{1}{3}$ cup of oatmeal for each serving. Using too much milk would produce oatmeal soup, and using too little milk would produce oatmeal glue. For a good oatmeal breakfast, the proportion of milk to oatmeal is the key.

One convenient way to understand proportion is through linear functions. Our formula for good oatmeal can be viewed as follows. Since $3 \cdot \frac{3}{4} = \frac{9}{4}$ cups of milk are required for each cup of oatmeal, the proportion of milk to oatmeal is $\frac{9}{4}$. Therefore, we can write the linear function

$$\text{Cups of milk} = \frac{9}{4}(\text{cups of oatmeal}).$$

In this formula we view milk as a function of oatmeal. For example if we use $\frac{3}{2}$ cup of oatmeal, then according to the formula the right amount of

Fig. 1.1. Oatmeal graph

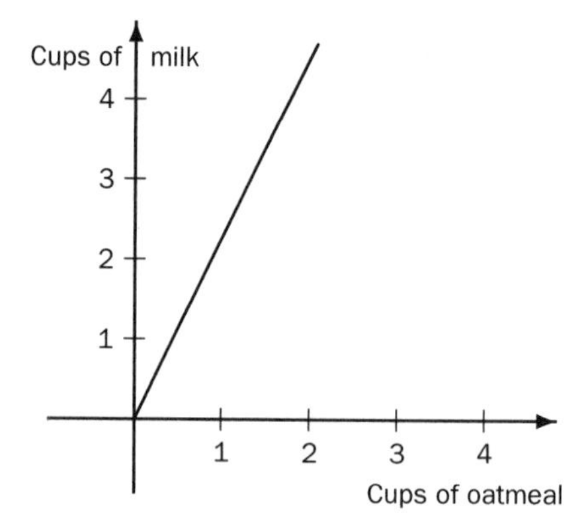

milk is $\frac{9}{4} \cdot \frac{3}{2} = \frac{27}{8}$, or if we are using 16 cups of oatmeal then we need 36 cups of milk. The advantage of this expression is that no matter how much oatmeal we use, the formula tells us the necessary amount of milk for that particular amount of oatmeal.

Linear Functions of One Variable

We say that a variable Y is expressed as a *linear function* of another variable X when we write an equation of the form

$$Y = kX + c,$$

where k and c are real constants. In the oatmeal example above, if Y represents cups of milk and X represents cups of oatmeal, then the real number k is our proportion $\frac{9}{4}$ and the real number c is 0. For any linear function $Y = kX + c$, the constant k is the proportional change of Y as X varies. The constant c is the value of Y when the value of X is 0.[1]

Functions of the form $Y = kX + c$ are called *linear functions* since their graph in the xy-plane is a straight line. Figure 1.1 is the graph of our oatmeal function, where the the cups of oatmeal are plotted on the x-axis and the cups of milk on the y-axis.

Finding the Expression for a Linear Function

Quite often, instead of starting with an explicit formula describing a linear function, we have some information about the values of a function that we believe to be linear. Suppose we want to write an equation for the function. What do we do?

[1] As a notational convenience, in this book we use uppercase letters as variables and lowercase letters as constants in algebraic expressions.

For example, suppose you work in an office and each morning your first task is to make coffee in a large, cylindrical coffee maker. This coffee maker displays the amount (and color) of its contents in a glass tube that runs straight up above its spout. You have learned by experience that once there is only one inch of coffee showing you should throw the contents away because your co-workers complain they get too many grounds in their cups. You have also learned that if you fill the coffee maker to the 6-inch mark then you have the right amount of coffee for ten people.

One morning eight additional coffee-drinking visitors arrive to spend the day working at your office. Your problem is to calculate how high to fill your coffee maker. Instead of guessing, you decide that coffee consumption is a linear function of the number of people drinking coffee. If X denotes the number of coffee drinkers and Y denotes the number of inches you need to fill your pot to satisfy everybody, your problem now is to find the constants k and c in a linear coffee function $Y = kX + c$. You know that the 1-inch mark corresponds to the amount required to satisfy nobody and that when $X = 10$ coffee drinkers, the appropriate Y value is 6 inches. This gives the following input-output chart for your linear function:

(Input) X	Y (Output)
0	1
10	6

The values $X = 0$ and $Y = 1$ show that $c = 1$ in your linear coffee function. Further, substituting $c = 1$, $X = 10$, and $Y = 6$ into the coffee function shows $6 = k \cdot 10 + 1$. Hence, $5 = k \cdot 10$, or $k = \frac{1}{2}$. You have determined that the coffee function is

$$Y = \frac{1}{2}X + 1.$$

This shows that when you have 18 coffee drinkers to satisfy (recall you have 8 visitors for the day) you should fill your coffee maker to the $\frac{1}{2} \cdot 18 + 1 = 10$-inch mark. We have the following interpretation for the constants $c = 1$ and $k = \frac{1}{2}$. The constant $c = 1$ reminds us that there is always 1-inch of cruddy coffee at the bottom of the pot. The proportionality constant $k = \frac{1}{2}$ tells us that each coffee drinker's consumption lowers the pot by an average of $\frac{1}{2}$ inch each morning.

More Oatmeal

There is another way to consider our oatmeal recipe, and it may be better suited for feeding a large family. Suppose in one family there are three big eaters who like to eat one and a half oatmeal servings, three small oatmeal eaters who can eat three fourths of a serving, one normal serving eater, and two who eat no oatmeal. Our problem is to determine how much milk and oatmeal must be cooked. We first compute how many servings must be prepared. We find that

$$\left(3 \cdot 1\frac{1}{2}\right) + \left(3 \cdot \frac{3}{4}\right) + 1 + (2 \cdot 0) = 7\frac{3}{4} = \frac{31}{4}$$

servings are needed.

We next view the quantities of both milk and oatmeal as functions of the number of servings desired. We have the proportionalities: $\frac{1}{3}$ cup of oatmeal per serving and $\frac{3}{4}$ cup of milk per serving. These give the two linear functions

$$\text{Cups of milk} = \frac{3}{4}(\text{number of servings}),$$

$$\text{Cups of oatmeal} = \frac{1}{3}(\text{number of servings}).$$

In the above system of equations we have expressed two outputs (oatmeal and milk quantities) as a linear function of the single input (number of servings).

We find that to make $7\frac{3}{4}$ servings of oatmeal we need $\frac{3}{4} \cdot \frac{31}{4} = \frac{93}{16}$ cups of milk and $\frac{1}{3} \cdot \frac{31}{4} = \frac{31}{12}$ cups of oatmeal. Of course, it is unlikely you will measure your ingredients in this way—with this many people (seven you know eat oatmeal) it would be sensible to plan on 8 servings and measure 6 cups of milk with $2\frac{2}{3}$ cups of oatmeal. The point behind this example is that instead of viewing the amount of milk as a function of oatmeal as we did earlier, it is more natural to view each quantity of milk and oatmeal as a function of the number of servings. These two linear functions are graphed in Fig. 1.2.

Renting an Automobile

As our next example of linear functions, we consider the problem of finding the best deal in a car rental. Two competing rental companies rent the same

Fig. 1.2. More oatmeal

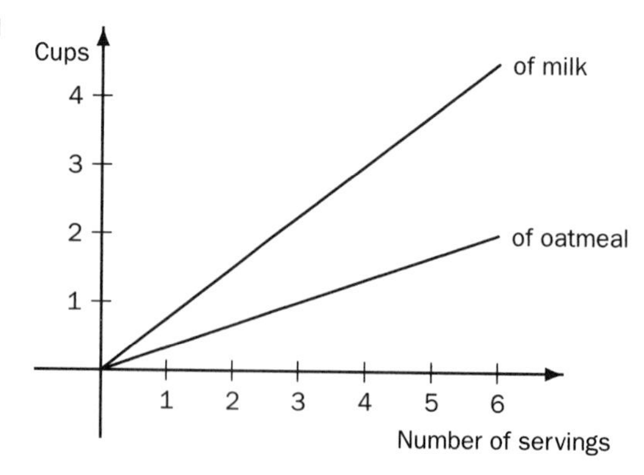

car according to the following pricing. Company A charges \$25 per day plus \$0.25 per mile, while Company B charges \$35 per day plus \$0.10 per mile (in both cases the renter pays for all gas used). In order to decide from whom to rent we need to know how far we plan to drive and how long it will take. The cost of renting from each company is described by the following functions:

$$\text{Price}_A = (\text{days}) \cdot \$25.00 + (\text{miles}) \cdot \$0.25,$$

$$\text{Price}_B = (\text{days}) \cdot \$35.00 + (\text{miles}) \cdot \$0.10.$$

These are called *linear functions of two variables* since they are determined by two quantities and the value of the function is the sum of fixed proportions of these quantities.

For example, if we plan a 200-mile trip for 3 days, we find that the cost of renting from Company A would be $3 \cdot \$25 = \$75$ in daily fees and $200 \cdot \$0.25 = \$50$ in mileage fees, for a total of \$125. Similarly, renting from Company B for the same trip would cost $3 \cdot \$35 = \$105$ in daily fees and $200 \cdot \$0.10 = \$20$ in mileage fees, again for a total of \$125. So the prices are the same for this particular trip. However, if we plan a 1200-mile trip for 5 days, the expense of renting from Company A would be \$425 while the expense of renting from Company B would be \$295. For such a long trip, Company B has the better bargain. On the other hand, if we need a car for two weeks but plan to drive less than 20 miles per day, then we could save by renting from Company A (check this!). We see from this example that in order to make an intelligent decision regarding automobile rental we must take into account two variables: the number of miles and the number of days.

Linear Functions of More Than One Variable

We say that a variable Y is expressed as a *linear function* of other variables $X_1, X_2, \ldots, X_n$ when we write an equation of the form

$$Y = k_1X_1 + k_2X_2 + \cdots + k_nX_n + c,$$

where $k_1, k_2, \ldots, k_n$ and c are real numbers. Both automobile rental pricing functions considered in the previous section were linear functions of two variables, days rented and miles driven.

In this text we will use the symbol $\mathbf{R}$ to denote the set of real numbers, and we write $c \in \mathbf{R}$ to mean "c is a real number." We will use $\mathbf{R}^2$ to denote the set of pairs (a, b) where $a \in \mathbf{R}$ and $b \in \mathbf{R}$, and $\mathbf{R}^3$ will denote the set of triples of real numbers. Using this notation, we will often write $f : \mathbf{R} \to \mathbf{R}$ to indicate that f is a real-valued function of one variable, $f : \mathbf{R}^2 \to \mathbf{R}$ to indicate that f is a real-valued function of two variables, $f : \mathbf{R}^3 \to \mathbf{R}$ to indicate three variables, and so forth.

We noted earlier when considering a linear function of one variable, $Y = kX + c$, that the constant k was the proportional change of Y as X varies. If we look at our expression for Y as a linear function of the variables $X_1, X_2, \ldots, X_n$, we see that each of the constants $k_1, k_2, \ldots, k_n$ has a similar interpretation. The constant k_1 is the proportional change of Y as X_1 varies *while all other variables remain unchanged.* Similarly, the constant k_2 is the proportional change of Y as X_2 varies while all other variables remain unchanged, and so forth. The constant c is the value of Y when all the variables $X_1, X_2, \ldots, X_n$ have value 0.

Finding Expressions for Linear Functions

We previously examined a method for finding the expression of a linear function of one variable given some of its values. Similar ideas work in studying linear functions of more variables, except that additional data about the function are needed. Suppose that the following input-output chart gives some Y values of a linear function of two variables X_1 and X_2.

X_1	X_2	Y
1	0	0
1	2	6
2	2	8
3	1	7

For which constants k_1, k_2 and c can we express the function in the form $Y = k_1X_1 + k_2X_2 + c$?

In order to solve this problem, we might hope to find the constant c first, but we abandon this strategy since we don't know the value of Y when both X_1 and X_2 are zero. Instead we first try to uncover the value of k_1. Recall that k_1 is the proportional change in Y when X_1 changes and X_2 remains fixed. Observe that when $X_2 = 2$ and X_1 increases from 1 to 2, our table shows that Y increases from 6 to 8. This shows that $k_1 = 2$ (because, while X_1 increased by 1, Y increased by 2). Observe that we also know two Y values when $X_1 = 1$. These show that when X_2 increases by 2 (from 0 to 2), the Y value increases by 6. This shows that $k_2 = 3$ since the proportional increase of Y is $\frac{6}{2}$. At this point we know that the linear function must look like

$$Y = 2X_1 + 3X_2 + c$$

for some constant c. We may determine the value of c by substituting any of our function values. When $X_1 = 1$ and $X_2 = 0$ we have $Y = 0$, so substitution gives $0 = 2 \cdot 1 + 3 \cdot 0 + c$. This shows $c = -2$. We have found that our linear function is

$$Y = 2X_1 + 3X_2 - 2.$$

This method for finding the expression of a linear function requires knowledge of enough values to compute the proportional changes in the value of the function when only one variable changes. The procedure is similar to finding the point-slope expression for an equation of a line. Recall that if a line has slope m and passes through the point (a, b) in $\mathbf{R}^2$, it is represented by the equation $(Y - b) = m(X - a)$, known as the *point-slope equation*. For two variables, suppose that the proportional change in Y as X_1 varies is m_1, the the proportional change in Y as X_2 varies is m_2, and the point (a, b, c) in $\mathbf{R}^3$ lies on the graph of the linear function. Then the linear function can be computed using the expression $(Y - c) = m_1(X_1 - a) + m_2(X_2 - b)$.

Problems

1. Suppose you are driving at a constant speed and you travel 220 miles in 5 hours. Write down the linear function that describes how far you have traveled as a function of time during these 5 hours.

2. Find the equation of the linear functions that satisfy the following input-output charts (where X is the input and Y is the output):

(a)

X	Y
-2	0
1	1

(b)

X	Y
1	-0.5
2	-1.5

(c)

X	Y
4	2
2	-1

3. (a) Consider the automobile rental pricing described earlier. If you rent a car for 5 days, how many miles do you have to drive so that the price from either company is the same? What if you rent the car for 7 days? or 8 days?

   (b) Using your answer to part (a), find a linear function $M = kD + c$ that shows that if you rent a car for D days and drive M miles, then the price from either company is the same.

   (c) Show that your answer to (b) can be checked by substituting D for number of days and your expression $kD + c$ for miles into the formulas for Price_A and Price_B given in this section.

4. Find the equations of the linear functions that satisfy the following input-output charts of two variables:

(a)

X_1	X_2	Y
0	4	8
0	3	7
1	3	10

(b)

X_1	X_2	Y
2	1	1
3	1	3
3	2	1
4	3	1

(c)

X_1	X_2	Y
2	0	8
4	0	9
4	1	12

5. Find the equations of the linear functions that satisfy the following input-output charts of three variables:

(a)

X_1	X_2	X_3	Y
1	1	0	−4
1	1	1	−2
0	1	1	0
0	0	0	0

(b)

X_1	X_2	X_3	Y
0	0	0	2
1	0	0	2
1	1	0	3
1	1	1	4

6. Suppose you have \$25 to spend in a candy store. Chocolate costs \$6.80 per pound, suckers cost \$0.10 each, and gumballs cost \$2.50 per quarter pound.
   (a) Suppose you buy C pounds of chocolate and G pounds of gumballs. Express the number of suckers S you can buy as a function of C and G.
   (b) Suppose you need one sucker and one ounce of gumballs for each person attending your birthday party. You want to buy as much chocolate as possible since you don't like suckers and gumballs. If P denotes the number of people who attend your party, express the number of pounds of chocolate C you can buy as a function of P.
7. For each of the following functions, express X as a function of Y:
   (a) $Y = 2X - 4$
   (b) $Y = -3X + 7$
   (c) $Y = 100X + 99$
   (d) Your answer to (a) should be $X = \frac{1}{2}Y + 2$. If we substitute this expression for X into the original function, we obtain $Y = 2(\frac{1}{2}Y + 2) - 4$, which is true after algebraic simplification. What does this mean from the point of view of functions?
   (e) Check your answers to (b) and (c) using the technique given in (d).
   (f) What happens if you try this problem for the constant function $Y = 7$?
8. The Celsius temperature scale was designed so that 0° Celsius is the temperature at which water freezes (which is 32° Fahrenheit) and 100° Celsius is the temperature at which water boils (which is 212° Fahrenheit). Since you may already know the formula for relating Celsius to Fahrenheit, explain how to derive the formula giving degrees Celsius from degrees Fahrenheit using this information. Also, explain what the constants in your formula mean.
9. The following input-output chart does not give enough information to determine Y as a linear function of X_1 and X_2.

X_1	X_2	Y
1	0	3
1	1	3

(a) How much information can you obtain, and what further information would be useful in determining this linear function?

(b) What are the possible linear functions that could fit this chart?

10. An elevator leaves the first floor of a twelve-story building with eight people. Three get off at the third floor, two on the fifth, and one on each of the top three floors. The function giving the elevator height in terms of time is not linear. It is, however, close to what is called *piecewise linear*. The term piecewise linear means what you would guess it means—so try to answer these problems.

    (a) Draw a graph of height vs time for this elevator trip. Make reasonable estimates according to your experiences. Briefly explain in each section of the graph what is happening.

    (b) Suppose the time the elevator stops at each floor is proportional to the number of people getting off. Show how this changes the graph. Do you believe this assumption is reasonable?

Group Project: Polyhedra

Polyhedra can be found everywhere in the world around us. The chances are good that you are sitting on a piece of one right now. Buildings are constructed out of them, soccer balls are made out of them, and they are crucial to understanding many geometric problems. In this project you will explore many of the smaller polyhedra and their basic properties.

We need to recall some terminology so we can talk about polyhedra. Recall that a *polygon* is a flat shape whose edges are line segments. Polygons with three edges are triangles, those with four edges are quadrilaterals, and so forth. The points where the edges of a polygon meet are called *vertices*. Note that if a polygon has three edges then it has three vertices, if it has four edges then it has four vertices, and so forth. *Polyhedra* are solid objects obtained by gluing polygons together at the edges. In the polyhedra we shall study here we will not allow any holes. Along each edge there must be exactly two polygons glued together.

Two famous polyhedra are the *cube* and the *tetrahedron* (a tetrahedron is pictured in Fig. 6.4). The cube is made from six squares and has eight vertices and twelve edges. The tetrahedron is made from four triangles and has four vertices and six edges. The polygons that are used to make a polyhedron are called its *faces*. Of course, the edges of a polyhedron are the edges of its faces, and the vertices of a polyhedron are the vertices of its faces.

(a) Divide your class into groups. Each group will be assigned one (or more) of the numbers 7, 8, 9, or 10. Your group project will be to build as many *different* polyhedra with as many faces as your group number. So, for example, if your group number were 6, you could build a cube as one of your polyhedra. *Be sure to count only the faces on the outside of your polyhedra.* Toothpicks and clay work well for building the polyhedra. Every time you make a new polyhedron you should determine the number of vertices and edges.

(b) Once your group has built at least five different polyhedra with the same number of faces, you next make a polyhedra graph. For the axes of the graph use the number of vertices and the number of edges. For each polyhedron built locate its point on the graph according to its number of vertices and its number of edges. (Make sure that all the polyhedra entered on your graph have the same number of faces!) If there are several polyhedra with the same number of edges and vertices, indicate this on your graph. Do you notice anything about your graph? Can you figure out an equation that relates the number of vertices to the number of edges for your polyhedra with the same number of faces?

(c) Finally, all the groups should get together to compare the information they have generated. Your equations relating vertices to edges should be similar but not quite the same (since each group studied polyhedra with a different number of faces). Can you put your information together and find a single equation that relates vertices, edges, and faces? If you can, you will uncover a famous result known as *Euler's formula.*

1.2 Local Linearity

Many functions that arise in economics, science, and engineering can be studied using linear functions in spite of the fact they may not actually be linear. In this section we consider three different situations and corresponding data that have been obtained either experimentally or from a complicated formula. Our task is to look for patterns in these numbers that resemble the behavior of linear functions. We will then use our observations to find a linear approximation to the function in question.

Storage Battery Capacities

The following table shows how the energy storage capacities (in amp-hours) of a small 12-volt, lead-acid automobile battery are related to the discharge rates (in amps) of the battery. The table shows that the capacity of the battery decreases as the the rate of discharge increases.

Discharge rate (amps)	1	5	10	15	20	25	30	40	50
Amp-hr capacity	70	68	66	62	58	54	50	40	28

This table is based on an actual experiment, not on any particular theory. These numbers were found as follows. The battery was discharged with a constant current rate (electric current rates are measured in *amps*), and the length of time until the battery died was measured. The product of this length of time with the discharge rate gives the *amp-hour capacity* of the battery at that discharge rate. For example, at a 20 amp discharge, it took 2.9 hours before the battery died, giving the $20 \times 2.9 = 58$ amp-hour capacity in the table. The equipment used to make these measurements was not the most accurate available, and for this reason the amp-hour values were rounded to the nearest integer.

We next consider the data in the table. For most practical applications an exact theory of storage battery capacity is not important. What is important is the observation that between 10 and 30 amp discharge rates, the capacity of battery drops about 4 amp-hours for every 5 amp increase. Since the amp-hour capacity of this battery is 58 when the discharge rate is 20 amps, this means that we can write the point-slope equation of the line through $(20, 58)$ and slope $-\frac{4}{5}$ as $C - 58 = -\frac{4}{5}(r - 20)$. Here we are using C for the capacity and r for the rate. In other words, the capacity function in this range is approximated by

$$C \approx 74 - \frac{4}{5}r.$$

A close look at the data reveals that while our linear approximation for C is accurate in the range between 10 and 30 amps, it gives too high a value at 5 amps or below and at 40 amps or more. Nonetheless, since most of the uses for which this battery is designed require 10 to 30 amp discharge rates, this linear approximation provides reasonable values. The expression $C \approx 74 - \frac{4}{5}r$ is what is called a *locally linear approximation* to the capacity function. The nine discharge rates listed in our table and their locally linear approximation in the 10 to 30 amp range are shown in Fig. 1.3.

You may be wondering why the battery capacity drops linearly for a while and then at high currents drops more rapidly. There are several reasons that battery capacity drops with increased current rates. One reason is that the diffusion of compounds required for chemical reaction in the battery

Fig. 1.3. Locally linear battery capacity behavior

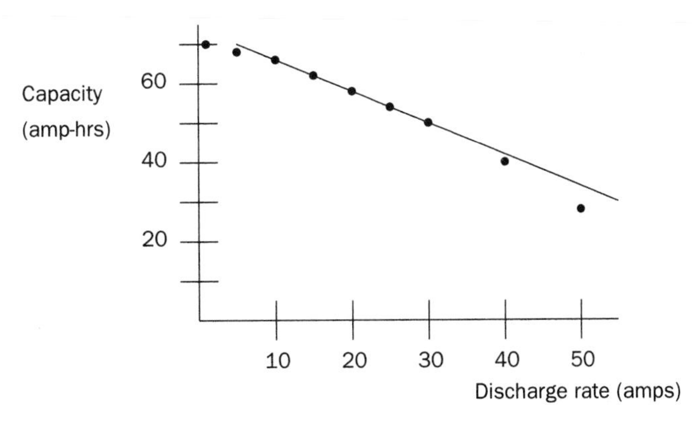

	4%	6%	8%	10%	12%
12 mo.	851.51	860.67	869.88	879.16	888.49
24 mo.	434.25	443.21	452.31	461.45	470.73
36 mo.	295.24	304.22	313.36	322.67	332.14
48 mo.	225.79	234.85	244.13	253.63	263.34
60 mo.	184.17	193.33	202.77	212.47	222.44

Table 1.1. Monthly Payments on a $10,000 Loan.

must occur more quickly at higher current rates. The difficulty of diffusion when the battery is partially discharged becomes significant at higher discharge rates, and this is one possible explanation why the capacity drops so quickly at higher rates. Another reason may be that higher current rates produce more internal heat in the battery, and so energy is lost. In fact, automobile batteries are not made for continuous discharge at high rates. They are designed to discharge at very high rates for short time intervals (when the car is being started), and the rest of the time most of the car's electrical power is supplied by the car's alternator, which also recharges the battery. The considerations of this section are significant, however, in the design of electric cars, where continuous battery discharge is necessary to drive the car.

Amortization

In the typical car or home loan, it is customary to repay the loan over a period of years through monthly payments of equal amounts. During this time period, the borrower is paying interest on the amount still owed. As time passes, the amount owed decreases, and so the amount of the monthly payment applied toward interest decreases and the remainder of the monthly payment reduces the principal. In this plan of repayment, one is said to *amortize* the loan. The amount of a monthly payment in an amortizing loan is determined by three variables: the amount borrowed P, the interest rate i, and the number of payments n. The monthly payment M is not a linear function of these three variables, and usually people look up M in a table once the values of P, i, and n are known. However, for many practical purposes (say, while home and loan shopping) the monthly rate can be estimated using local linearity once its value for several P, i, and n are known.

In Table 1.1 we list the monthly payment required to amortize a $\$10,000$ loan as a function of the number of monthly payments and the annual interest rate. The entries in this table were computed as follows. The values of P, i, n, and M satisfy the equation

$$P = \frac{M\left[1 - \left(1 + \frac{i}{n}\right)^{-nN}\right]}{\frac{i}{n}},$$

and so M can be computed once P, i, and n are known. However, *we ask you to ignore this formula* and instead look closely at the rows of the table. These values for M reveal what is essentially linear behavior. Note that each 2% increase in interest rate increases monthly payments by between \$9 and \$10. This \$9 to \$10 difference represents extra interest that must be paid. On a \$10,000 principal, one month's interest at 2% is $\frac{1}{12} \times .02 \times \$10,000 = \$16.67$. Since monthly interest on an amortizing loan is paid only on the balance (not on the original loan amount), the payment increase due to the extra 2% is only a bit more than half of this.

Our observation shows that the monthly payment is nearly a linear function of the interest rate. For example, on a 36-month repayment schedule, a point-slope calculation shows that the payments are given approximately by

$$M \approx 277 + 4.6 \times i.$$

This approximation gives the values 295, 304, 314, 323, and 332 for the 36-month row of our table, which is a very accurate approximation. Although not an exact determination of a repayment price when a deal is finally negotiated, this linear approximation is more than adequate as a determination of how interest rates will affect the purchaser when the repayment time is fixed.

Note that the columns in Table 1.1 do *not* display linear behavior. As the number of repayment months increases by 12, the payment decreases, but by a significantly smaller amount each time. One reason for this is that the time intervals are increasing rapidly in this chart. If, for example, we were to consider an 8% interest rate over a period of 46, 48, 50, and 52 months, the payment rates would be 253.14, 244.13, 235.84, and 228.20 respectively. Here the drop in payments varies between \$9 and \$7.50 for each increase of two months in amortization time. So it would be more accurate to approximate the payment rate as a function of repayment time as a linear function on this small region of our chart. However, it is not a customary business practice to consider 46- or 50-month repayment schedules, so we won't bother finding this linear approximation.

Airplane Lift and Wing Flaps

Prior to 1915, the design of airplane wing cross-sections consisted mostly of studying existing types followed by the trial use of variations of these to see what happens. However, in the decade that followed, wind tunnel experiments were conducted and the theory of airplane aerodynamics began. Two important numbers that depend on the airplane design and are crucial to determining flight characteristics are the *drag* and *lift coefficients.* Roughly speaking, the drag coefficient measures the air resistance as the plane flies, and the lift coefficient measures the upward pressure (lift) resulting from flight. These coefficients are given in the units (lbs/ft^2)/(mph^2),

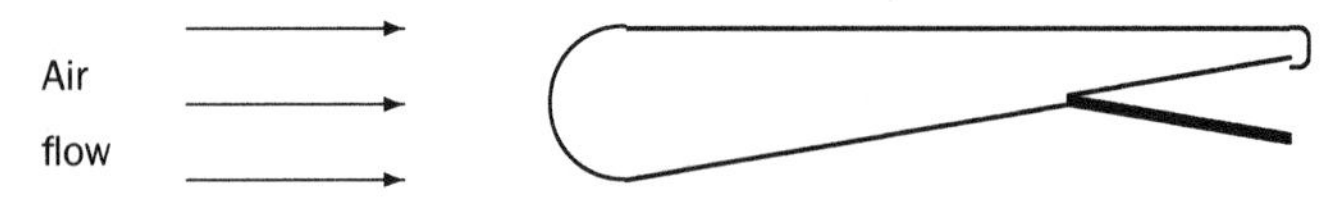

Fig. 1.4. Cross-section of a simple wing flap

which numerically represent the force (in pounds) per square foot of surface area at a given air velocity. Both the drag and lift forces are (essentially) proportional to the square of the velocity (so the force is not a linear function of velocity).

In early planes a lift coefficient of 1.6 was considered good, and for takeoff it was desirable to increase this lift coefficient to the range from 1.8 to 2.5. One device used to increase lift is the wing flap, which lies flush to the wing during flight but is angled down by the pilot during takeoff and landing. Fig. 1.4 illustrates a simple wing flap design.

Some early wind tunnel experiments generated the data given below in Table 1.2,[2] which shows how the *increase* in lift coefficient, ΔC_L, is related to two quantities, the chord ratio E and the flap angle δ. The chord ratio is the ratio between the flap width and that of the wing, and the flap angle is the angle the flap makes with the wing. In Fig. 1.4 the chord ratio is approximately .3 and the flap angle is approximately 20°. Study of this table reveals that each 0.1 increase of E gives an approximate increase of .05 for ΔC_L (except perhaps at the lower right), and each increase of 5° of δ also gives a rough increase of about .05 for ΔC_L. The numbers in the table may not be the most accurate possible (and modern data may differ from that collected in the 1920s), but the approximate increases in ΔC_L just noted show that the table can be reasonably approximated by the linear equation

$$\Delta C_L \approx -.02 + .01\delta + .5E.$$

The corresponding values of this linear approximation are shown in Table 1.3. Although these values are not a perfect match for the data given, it is reasonably close for values of δ near 10° and for values of E near .2. In absence of a precise theory (such as the case in the 1920s), this type of two-variable linear approximation could be quite useful in predicting flight characteristics. Another linear approximation is given in Sec. 9.3 using the technique of linear regression.

	5°	10°	15°	20°
0.1	0.08	0.13	0.19	0.24
0.2	0.12	0.17	0.23	0.30
0.3	0.17	0.22	0.29	0.42

Table 1.2. Experimental Values for ΔC_L.

[2]From Walter S. Diehl, **Engineering Aerodynamics**, Ronald Press Co., New York (1928), pp. 150–152.

	5°	10°	15°	20°
0.1	0.08	0.13	0.18	0.23
0.2	0.13	0.18	0.23	0.28
0.3	0.18	0.23	0.28	0.33

Table 1.3. The Linear Approximation $\Delta C_L \approx -.02 + .01\delta + .5E$.

	5°	10°	20°	30°	40°
0.1	0.005	0.015	0.022	0.035	0.06
0.2	0.01	0.02	0.045	0.075	0.11
0.3	0.02	0.03	0.07	0.12	0.17

Table 1.4. Experimental Values for ΔC_D.

The presence of a wing flap also increases the drag. Table 1.4 shows how the increase in drag coefficient ΔC_D is related to the same two variables, the chord ratio E and the flap angle δ.

These data have characteristics of linearity in each of the variables E and δ. For example, the approximations

$$\Delta C_D \approx .0013\delta \quad \text{when } E = .1,$$

$$\Delta C_D \approx .0027\delta \quad \text{when } E = .2,$$

$$\Delta C_D \approx .0040\delta \quad \text{when } E = .3$$

were suggested by the scientists who obtained this data. Values given by these approximations are listed in Table 1.5 and are fairly close.

It is important to note that the data in Table 1.4 do *not* appear to be that of a linear function of two variables, since the coefficients of δ in the various approximations are quite different. In fact, the text containing this data suggests that the expression

$$\Delta C_D \approx 0.0133E\delta$$

gives a good match to the observed values. Note that this approximation is given by a function that is linear when either variable is fixed *but is not linear itself.* The values given by this approximation are contained in the following table.

	5°	10°	20°	30°	40°
0.1	0.006	0.013	0.026	0.040	0.053
0.2	0.013	0.026	0.053	0.080	0.107
0.3	0.020	0.040	0.080	0.120	0.160

Table 1.5. Values of the Nonlinear Approximation $\Delta C_D \approx 0.0133E\delta$.

Summary

In this section we have seen some linear approximations for various functions. In the cases of battery capacity and the changes in lift and drag coefficients, there was no exact formula for giving these functions, and the linear approximations developed provide one way to study them. In the amortization problem, we had an exact formula, but the linear approximation provided an easier model for understanding how changes in interest rates affect monthly payments.

In each case we estimated proportional changes and then used the point-slope formula to obtain linear approximations. This is all you will be expected to do in the problems and projects in this section. A refinement of this method you can try is to make some small changes in the coefficients and see if that makes the approximation better. This is useful for getting a feel for the problem but is not a sound procedure for serious applications. A powerful method for finding the best linear approximation for a collection of data is given by the method of *least squares*, which is sometimes known as *linear regression*. An introduction to this method is given in Sec. 9.3. In the second group project at the end of this section you will have an opportunity to explore the results of linear regression using a calculator or computer.

All of the calculator projects in this text can be carried out, for example, on a Texas Instrument TI-85 scientific calculator, but many other calculators and computer software packages can do the same thing. You are encouraged to become familiar with the capabilities of whatever system is available. All of the text's instructions will be general in nature, and we will not specify what keys to push. So when the text says "use your calculator to...," it is your responsibility to find out how to accomplish the task on your computer or calculator. When you are learning to use new computer or calculator technology, it is a good idea to work as a team with some classmates and experiment a bit, since instruction booklets can at times be hard to follow. Students should get their instructors involved, too!

Problems

1. Try to find a linear approximation for the first column of Table 1.1, that is, approximate M as a linear function of i for each of 12-, 24-, 48-, and 60-month repayment plans. In the text it was stated that this would not work well, but how close can you get?

2. Find a linear function of two variables that gives a reasonable approximation to the values below. Answers will vary from person to person, so be sure to explain how you found your answer. It might be a good idea to list several possibilities and then explain why your choice is best.

	1	2	3	4	5
−5	0.20	0.46	0.71	0.92	1.20
0	0.05	0.29	0.61	0.74	1.03
5	−0.09	0.15	0.40	0.62	0.89
10	−0.24	0.01	0.24	0.40	0.55

3. Contours on a topographic map describe the elevation function on a region of land. These contours are curves that represent constant elevations, or *level sets*. The elevation function has two input variables: location along the east-west axis and location along the north-south axis. If the elevation function has locally linear behavior in some region, explain in writing and with sketches what the contours look like in that region.
4. Find linear approximations to the change in drag coefficient ΔC_D (given in Table 1.4) as a function of flap angle δ that are more accurate for the angle values between 10° and 20° than the approximations given in this section. Note that answers to this question may vary.
5. Consider the following table of values, which gives the current drain (in amps) for an electric car driving at various velocities on level ground.[3]

Velocity (mph)	10	20	30	40	50
Current (amps)	85	150	235	340	470

   (a) Does it make sense to use a local linear approximation if you want to estimate current drains at 15 mph? Justify your answer. What about at 32 mph?

   (b) Suppose this car was capable of going 60 mph on level ground. (In fact, it wasn't.) How would you estimate the current necessary?
6. Study the experimental data given in Tables 1.2 and 1.4. Is there a linear relationship between ΔC_L and ΔC_D for the values of the chord ratio and flap angle considered there? Part of your answer will be explaining how you make sense of this question and what is meant by a linear relationship in this context.

Group Project: Linearizing Amortization

Suppose that a \$150,000 home loan is to be repaid over a 20- to 30-year period and that the prevailing interest rates are varying between 7% and 9%.

(a) Find a two-variable linear function that approximates the monthly payment depending on the interest rate i and the number of years y of fully amortizing repayment. Set up your function so that it is exact at 25 years and 8%. You will have to use the formula given in the section on amortization (along with a calculator) to compute the amortization payments.

[3] Data collected in 1973 in author's car. The car had a 36-volt battery pack.

(b) How accurate is your approximation for 26 years and 8.25%? For 24 years and 8.25%, and for 30 years and 9%?

(c) Interpret the meaning of the coefficients of the variables in your linear approximation and write a sentence explaining each.

(d) At the end of the section on amortization it was noted *for the chart considered there* that the monthly payment was not close to being a linear function of the number of years of repayment. Is this situation similar or different in this regard?

Group Project: Linear Regression on a Calculator or Computer

The purpose of this project is for you to become familiar with how to use a calculator or computer to find locally linear approximations to data. We will learn more about how the calculator is making these computations (as well as *why* it works) later in Sec. 9.3.

(a) The first step is to learn how to enter data into your calculator. This is usually done by creating and naming a list of points. Enter in the nine pairs given in the storage battery chart.

(b) Next learn how your calculator runs a linear regression on the table of data just entered. (This can be found in the STAT menu of many calculators.) The output of this program usually gives you the coefficients a and b in an approximation of the form $Y = bX + a$ to the data. For the battery capacity example you should find that $a = 73.6$ and $b = -.850$.

(c) Plot your data points on a graph, and graph the linear equation $Y = -.850X + 73.6$ there too. (Better yet, get your calculator or computer to draw this graph.) How does this linear approximation compare to the one considered in Fig. 1.3?

(d) Use your machine to compute linear regressions for the functions given by the rows of Table 1.4 (the ΔC_D values.) How close are these equations to the approximations given in the text?

1.3 Matrices

In this section we continue our study of linear functions by introducing basic matrix algebra. Our first use of matrices will be to simplify the notation. Shortly, however, we shall see that their use is of a much broader scope.

An Assembly Line

In a small assembly line toy cars and trucks are put together from a warehouse full of parts. These parts include car and truck bodies, chassis that both bodies snap onto, wheels, and toy people. In order to assemble a car, you snap a car body onto a chassis, pop on four wheels, and put four people inside. For the truck assembly you snap a truck body onto a chassis, pop on six wheels (four on the rear axle, two on each side), and put two people inside. We record all this assembly information in the following array:

	Cars	Trucks
Car bodies	1	0
Truck bodies	0	1
Chassis	1	1
Wheels	4	6
People	4	2

Each column of this array indicates the number and type of parts needed to assemble your finished product—the first column for cars and the second column for trucks.

We need to assemble 125 cars and 75 trucks in one day, and we want to know how many wheels, people, and so forth are necessary for the job. The array can be used to solve this. If you multiply each entry in the first column by 125, you will obtain a column listing the number of parts needed to build the cars, and if you multiply each entry in the second column by 75, you will obtain a column listing the number of parts needed to build the trucks. This looks like

$$\text{Car parts} = 125 \cdot \begin{pmatrix} 1 \\ 0 \\ 1 \\ 4 \\ 4 \end{pmatrix} = \begin{pmatrix} \text{125 car bodies} \\ \text{0 truck bodies} \\ \text{125 chassis} \\ \text{500 wheels} \\ \text{500 people} \end{pmatrix}$$

and

$$\text{Truck parts} = 75 \cdot \begin{pmatrix} 0 \\ 1 \\ 1 \\ 6 \\ 2 \end{pmatrix} = \begin{pmatrix} \text{0 car bodies} \\ \text{75 truck bodies} \\ \text{75 chassis} \\ \text{450 wheels} \\ \text{150 people} \end{pmatrix}.$$

Finally, adding these columns shows that the total number of parts needed to assemble 125 cars and 75 trucks is

$$\text{Total parts} = \begin{pmatrix} 125 \\ 0 \\ 125 \\ 500 \\ 500 \end{pmatrix} + \begin{pmatrix} 0 \\ 75 \\ 75 \\ 450 \\ 150 \end{pmatrix} = \begin{pmatrix} \text{125 car bodies} \\ \text{75 truck bodies} \\ \text{200 chassis} \\ \text{950 wheels} \\ \text{650 people} \end{pmatrix}.$$

Of course, the answer to the question posed in this example could easily have been found without using arrays of numbers. However, before this section ends we will see problems whose solution would be a notational nightmare without such arrays, or matrices.

Matrix Notation

A *matrix* is a rectangular array of real numbers, where the exact location of each number, or *entry*, is crucial. To be more precise, an $m \times n$ matrix is an array of real numbers with m rows and n columns. The ijth entry of the matrix is located in the ith row and jth column of the array. If the matrix is called A, we often write $A = (a_{ij})$, which means that the a_{ij} are real numbers and the entry in the ith row and jth column of A is a_{ij}. We will also use the notation $A(i, j)$ to denote the ijth entry a_{ij} of A.

For example, the 2×3 matrix

$$A = \begin{pmatrix} 4 & 1 & 7 \\ 2 & 0 & 3 \end{pmatrix}$$

has two rows and three columns. When we write $A = (a_{ij})$, we are specifying the six real numbers $a_{11} = 4$, $a_{12} = 1$, $a_{13} = 7$, $a_{21} = 2$, $a_{22} = 0$, and $a_{23} = 3$. Whenever we describe the location of an entry in a matrix or specify matrix size, the row number is listed first and the column number is listed second. To help you get used to the row and column terminology, we point out that rows go across the page (as does writing in English), and columns go up and down (like the columns on a building).

Matrices are usually denoted by capital letters $A, B, C, \ldots$ to help avoid confusing them with real numbers. *Column matrices* are matrices with one column and are sometimes referred to as *column vectors*. We will often use vector notation, such as $\vec{v}$ to denote column matrices.

We need to know what it means for two matrices to be equal. Two matrices A and B are *equal* if they have the same number of rows, the same number of columns, and precisely the same entries in the same places. In particular, we emphasize that the matrices

$$A = \begin{pmatrix} 1 \\ 2 \end{pmatrix} \qquad \text{and} \qquad B = \begin{pmatrix} 1 & 2 \end{pmatrix}$$

are *not* equal, even though they look similar when rotated by 90°.

Scalar Multiplication and Addition of Matrices

In our assembly-line example we had to add matrices together and multiply matrices by real numbers. The first operation is known as *matrix addition*, and the second operation is known as *scalar multiplication*. Although the ideas behind these two operations are pretty clear (they are called "componentwise operations"), they can also be described using the ij notation just introduced. The definitions read as follows.

Definition (Matrix Addition). If $A = (a_{ij})$ and $B = (b_{ij})$ are both $m \times n$ matrices, then we define the $m \times n$ matrix $A + B$ by $(A + B)(i, j) = A(i, j) + B(i, j)$. Thus to add two matrices of the same shape, one simply adds the corresponding entries of the matrices to obtain another matrix of the same shape.

Definition (Scalar Multiplication). If $A = (a_{ij})$ is an $m \times n$ matrix and k is a real number, then we define the $m \times n$ matrix kA by $(kA)(i, j) = k(A(i, j))$. That is, the ijth entry of kA is ka_{ij}, where a_{ij} is the ijth entry of A. This multiplication by a real number is called scalar multiplication.

For example,

$$\begin{pmatrix} 3 & 3 \\ 2 & -1 \end{pmatrix} + \begin{pmatrix} -2 & 0 \\ 3 & 3 \end{pmatrix} = \begin{pmatrix} 1 & 3 \\ 5 & 2 \end{pmatrix}$$

and

$$3\begin{pmatrix} 1 & 1 & -1 \\ 2 & 3 & -5 \end{pmatrix} = \begin{pmatrix} 3 & 3 & -3 \\ 6 & 9 & -15 \end{pmatrix}.$$

Both matrix addition and scalar multiplication come from the usual multiplication and addition of real numbers applied entrywise to the matrices. Using this we see that the following familiar laws of algebra involving real numbers are also true for these two operations.

Theorem 1. *Suppose that $A, B,$ and C are $m \times n$ matrices and r and s are real numbers. Then the following matrices are each $m \times n$ and*
(a) $A + B = B + A$;
(b) $(A + B) + C = A + (B + C)$;
(c) $r(sA) = (rs)A$;
(d) $r(A + B) = rA + rB$;
(e) $(r + s)A = rA + sA$.

Multiplying Matrices, Part I

When we studied the problem of finding the best automobile rental (see Sec. 1.1), we analyzed the pair of linear functions with two input variables

$$\text{Price}_A = \$25 \cdot (\text{days}) + \$0.25 \cdot (\text{miles}),$$

$$\text{Price}_B = \$35 \cdot (\text{days}) + \$0.10 \cdot (\text{miles}).$$

Matrices provide convenient notation for working with this type of situation. Since we have two rows in the functional expression (one for each variable), we use column matrices to denote the input and output:

$$\begin{pmatrix} \text{Price}_A \\ \text{Price}_B \end{pmatrix} \quad \text{and} \quad \begin{pmatrix} \text{days} \\ \text{miles} \end{pmatrix}.$$

We next consider the 2×2 matrix whose entries are the coefficients that arose in our functional expression

$$\begin{pmatrix} 25 & 0.25 \\ 35 & 0.10 \end{pmatrix}.$$

Matrix multiplication is defined in such a way that the expression

$$\begin{pmatrix} \text{Price}_A \\ \text{Price}_B \end{pmatrix} = \begin{pmatrix} 25 & 0.25 \\ 35 & 0.10 \end{pmatrix} \begin{pmatrix} \text{days} \\ \text{miles} \end{pmatrix} = \begin{pmatrix} \$25 \cdot (\text{days}) + \$0.25 \cdot (\text{miles}) \\ \$35 \cdot (\text{days}) + \$0.10 \cdot (\text{miles}) \end{pmatrix}$$

makes sense and is equivalent to the original expression.

This is accomplished as follows. The first row of our 2×2 matrix of coefficients has entries 25 and 0.25, and if we multiply these entries by the corresponding entries days and miles of our input column and add the result, we obtain $25 \cdot (\text{days}) + 0.25 \cdot (\text{miles})$. This is Price_A. Similarly, multiplying the entries of the second row of our matrix of coefficients (35 and 0.10) with the corresponding entries of our input column and adding the result gives Price_B. This shows how to define matrix multiplication to give our linear pricing function. More precisely, we give the following definition.

Definition. Suppose that $A = (a_{ij})$ is an $m \times n$ matrix and suppose that $\vec{v}$ is the $n \times 1$ column matrix with jth entry b_j. Then we define $A\vec{v}$ to be the $m \times 1$ column matrix whose ith entry is $a_{i1}b_1 + a_{i2}b_2 + \cdots + a_{in}b_n$.

For example, we have

$$\begin{pmatrix} 1 & 4 & 3 \\ 2 & 5 & 1 \end{pmatrix} \begin{pmatrix} 8 \\ 2 \\ 0 \end{pmatrix} = \begin{pmatrix} (1 \cdot 8) + (4 \cdot 2) + (3 \cdot 0) \\ (2 \cdot 8) + (5 \cdot 2) + (1 \cdot 0) \end{pmatrix} = \begin{pmatrix} 16 \\ 26 \end{pmatrix}.$$

Observe that the resulting product has only two rows since the matrix on the left had only two rows. As another example we consider

$$\begin{pmatrix} 4 & 3 \\ 5 & 1 \end{pmatrix} \begin{pmatrix} X \\ Y \end{pmatrix} = \begin{pmatrix} 4X + 3Y \\ 5X + Y \end{pmatrix}.$$

This time we used variables in our column matrix to illustrate the fact that matrix multiplication is defined so that the left-hand matrix acts as the coefficients in a system of equations.

As an additional example, suppose that

$$A = \begin{pmatrix} 2 & 3 \\ 1 & 1 \\ 0 & 0 \end{pmatrix} \quad \text{and} \quad C = \begin{pmatrix} 1 \\ 2 \\ 1 \end{pmatrix}.$$

Then, the matrix equation

$$A\begin{pmatrix} X \\ Y \end{pmatrix} = C$$

represents the system of equations

$$\begin{aligned} 2X + 3Y &= 1 \\ X + Y &= 2 \\ 0X + 0Y &= 1. \end{aligned}$$

Matrix Multiplication, Part II

We next define matrix multiplication in a more general situation than just considered. We need to define the product AB, where A is an $m \times n$ matrix and B is an $n \times p$ matrix. In a natural way this definition extends the case where B is a column matrix. What you do is multiply the left-hand matrix by the columns of the right-hand matrix one at a time, and then string the resulting columns along in order to form the product matrix. For example, since

$$\begin{pmatrix} 1 & 2 \\ 3 & 4 \end{pmatrix}\begin{pmatrix} 5 \\ 6 \end{pmatrix} = \begin{pmatrix} 17 \\ 39 \end{pmatrix} \quad \text{and} \quad \begin{pmatrix} 1 & 2 \\ 3 & 4 \end{pmatrix}\begin{pmatrix} 7 \\ 8 \end{pmatrix} = \begin{pmatrix} 23 \\ 53 \end{pmatrix},$$

we define

$$\begin{pmatrix} 1 & 2 \\ 3 & 4 \end{pmatrix}\begin{pmatrix} 5 & 7 \\ 6 & 8 \end{pmatrix} = \begin{pmatrix} 17 & 23 \\ 39 & 53 \end{pmatrix}.$$

Note in the next definition that it is *not* possible to multiply two matrices of arbitrary size.

Definition (Matrix Multiplication). Suppose that A is an $m \times n$ matrix and B is an $n \times p$ matrix. (Thus the number of columns of A is the same as the number of rows of B.) The *matrix product AB* is defined to be the $m \times p$ matrix given by

$$\begin{aligned} AB(i, k) &= A(i, 1)B(1, k) + A(i, 2)B(2, k) + \cdots + A(i, n)B(n, k) \\ &= \sum_{j=1}^{n} A(i, j)B(j, k). \end{aligned}$$

In the definition, the matrix product AB is the $m \times p$ matrix whose ith row, kth column entry is obtained by adding the products of the corresponding entries of the ith row of A and the kth column of B. Note that since A is $m \times n$ and B is $n \times p$, each row of A and each column of B have precisely n entries. Therefore the phrase "adding the products of the corresponding entries" makes sense.

For example, the product of a 2×2 matrix and a 2×3 matrix is given by

$$\begin{pmatrix} 1 & 4 \\ 2 & 5 \end{pmatrix} \begin{pmatrix} 8 & 9 & 0 \\ 2 & 6 & 0 \end{pmatrix} = \begin{pmatrix} (1 \cdot 8) + (4 \cdot 2) & (1 \cdot 9) + (4 \cdot 6) & (1 \cdot 0) + (4 \cdot 0) \\ (2 \cdot 8) + (5 \cdot 2) & (2 \cdot 9) + (5 \cdot 6) & (2 \cdot 0) + (5 \cdot 0) \end{pmatrix}$$

$$= \begin{pmatrix} 16 & 33 & 0 \\ 26 & 48 & 0 \end{pmatrix}.$$

The product of a 2×3 matrix and a 4×2 matrix,

$$\begin{pmatrix} 1 & 1 & 4 \\ 3 & 1 & 7 \end{pmatrix} \begin{pmatrix} 5 & 1 \\ 8 & 2 \\ 0 & 0 \\ 1 & 0 \end{pmatrix},$$

is *not* defined, although their product in reverse order is

$$\begin{pmatrix} 5 & 1 \\ 8 & 2 \\ 0 & 0 \\ 1 & 0 \end{pmatrix} \begin{pmatrix} 1 & 1 & 4 \\ 3 & 1 & 7 \end{pmatrix} = \begin{pmatrix} 8 & 6 & 27 \\ 14 & 10 & 46 \\ 0 & 0 & 0 \\ 1 & 1 & 4 \end{pmatrix}.$$

This next product is special:

$$\begin{pmatrix} 1 & 0 & 0 \\ 0 & 1 & 0 \\ 0 & 0 & 1 \end{pmatrix} \begin{pmatrix} 1 & 4 \\ 1 & 4 \\ 3 & 2 \end{pmatrix} = \begin{pmatrix} 1 & 4 \\ 1 & 4 \\ 3 & 2 \end{pmatrix}.$$

We emphasize that matrix multiplication is not commutative in general. In other words, even when both AB and BA are defined, *they need not be equal* (nor even of the same size!). For example,

$$(3 \quad 5) \begin{pmatrix} 2 \\ 4 \end{pmatrix} = (3 \cdot 2 + 5 \cdot 4) = (26),$$

but

$$\begin{pmatrix} 2 \\ 4 \end{pmatrix} (3 \quad 5) = \begin{pmatrix} 6 & 10 \\ 12 & 20 \end{pmatrix}.$$

All students should practice matrix multiplication to become used to the process. The matrix product will be used in essentially every section in the rest of this book. It is also possible to multiply matrices on a calculator or computer, and everyone should learn to do this, too. In the first group project in this section it is necessary to use a computer or calculator to carry out the investigation.

Counting the Number of Paths: An Application

In Fig. 1.5 there are four points and four direct paths between them. There is one direct path between A and B, two direct paths between B and C, and one direct path between C and D. We shall say that each of these four

Fig. 1.5. Four points and four paths

direct paths has length 1 (even though they do not have the same geometric length). Since there is no direct path between A and C, the shortest path from A to C has length 2. In fact, you can see there are two such paths, both passing through B. Inspection of the figure also shows that there are two paths of length 3 from A to D.

Suppose our problem is to decide how many paths there are between A and C whose length is at most 4. Retracing the same path will be allowed. Counting these paths is not difficult, but we need to be systematic and make sure that we don't omit any path or count some path twice. Matrices provide a nice tool for carrying out this type of counting process.

For example, let's examine a systematic method for counting the number of paths of length 2 between the points A and C. (We saw the answer was 2, but we examine the counting process carefully.) To count such paths we organize our counting according to the midpoint of each possible path. This means we must add the following four numbers: n_1 = the number of paths with midpoint A; n_2 = the number of paths with midpoint B; n_3 = the number of paths with midpoint C; and n_4 = the number of paths with midpoint D. Since there are no paths of length 1 from A to A, we see that $n_1 = 0$. Since there is one path of length 1 from A to B and two paths of length 1 from B to C, we see that $n_2 = 1 \cdot 2 = 2$. Again, since there no paths of length 1 from A to C or D, we see that $n_3 = 0$ and $n_4 = 0$. Adding, we obtain that there are $0 + 2 + 0 + 0 = 2$ paths of length 2 between A and C.

Our counting procedure for length 2 paths between A and C was the same as the process of computing an entry in a matrix multiplication. This can be written as

$$\begin{pmatrix} 0 & 1 & 0 & 0 \end{pmatrix} \begin{pmatrix} 0 \\ 2 \\ 0 \\ 1 \end{pmatrix} = 0 \cdot 0 + 1 \cdot 2 + 0 \cdot 0 + 0 \cdot 1 = 2,$$

where the first row matrix gives the number of length 1 paths from A to A, B, C, D and the second column matrix gives the number of length 1 paths from A, B, C, D to C. More generally, we consider the following matrix whose entries give the number of paths of length 1 between points in our figure.

	A	B	C	D
A	0	1	0	0
B	1	0	2	0
C	0	2	0	1
D	0	0	1	0

Observe that when we counted the number of paths of length 2 between A and C according to our system, all we did was multiply the first row of our matrix by the third column.

We next set

$$M = \begin{pmatrix} 0 & 1 & 0 & 0 \\ 1 & 0 & 2 & 0 \\ 0 & 2 & 0 & 1 \\ 0 & 0 & 1 & 0 \end{pmatrix} \quad \text{and note} \quad M \cdot M = M^2 = \begin{pmatrix} 1 & 0 & 2 & 0 \\ 0 & 5 & 0 & 2 \\ 2 & 0 & 5 & 0 \\ 0 & 2 & 0 & 1 \end{pmatrix}.$$

The first row, third column entry of M^2 is 2, which is the number of paths of length 2 between A and C. Similarly, the $3, 4$ entry of the matrix M^2 tells us that there are no paths of length 2 between C and D, and the $2, 2$ entry tells us that there are five paths of length 2 from B to B.

We can now use the matrix M to answer our original question. The matrix M^2 tells us the number of paths of length 2 between any two points in our figure. By similar reasoning the matrix M^3 tells us the number of paths of exact length 3, the matrix M^4 tells us the number of paths of exact length 4, and so forth. These matrices are

$$M^3 = \begin{pmatrix} 0 & 5 & 0 & 2 \\ 5 & 0 & 12 & 0 \\ 0 & 12 & 0 & 5 \\ 2 & 0 & 5 & 0 \end{pmatrix} \quad \text{and} \quad M^4 = \begin{pmatrix} 5 & 0 & 12 & 0 \\ 0 & 29 & 0 & 12 \\ 12 & 0 & 29 & 0 \\ 0 & 12 & 0 & 5 \end{pmatrix}.$$

In order to find the number of paths of length at most 4 between any two points of our figure, we add the matrices M, M^2, M^3, and M^4. We find

$$M + M^2 + M^3 + M^4 = \begin{pmatrix} 6 & 6 & 14 & 2 \\ 6 & 34 & 14 & 14 \\ 14 & 14 & 34 & 6 \\ 2 & 14 & 6 & 6 \end{pmatrix}.$$

We can answer our original question. There are 14 paths between A and C of length at most 4. Note that there are 34 paths of length at most 4 between B and itself.

Using a calculator to compute the matrix M^{25}, we can also determine the number of paths of exact length 25 between A and B. The answer is the $1, 2$ entry of M^{25}, which is 1,311,738,121.

Problems

1. Consider the following matrices:

$$M = \begin{pmatrix} 1 & 3 \\ 0 & 2 \end{pmatrix}, \quad N = \begin{pmatrix} 2 & 0 & 6 \\ 2 & -1 & 2 \\ 5 & 4 & 3 \end{pmatrix}, \quad P = \begin{pmatrix} -2 & 3 \\ 2 & -1 \\ 0 & 0 \end{pmatrix},$$

$$Q = \begin{pmatrix} 1 & -3 \\ 2 & 4 \end{pmatrix}, \quad R = \begin{pmatrix} 1 & 0 & 4 \\ -6 & 0 & 7 \\ 3 & -1 & 0 \\ 6 & 0 & -1 \end{pmatrix}, \quad S = (3 \quad -2 \quad 7).$$

Practice the following matrix calculations by hand. If the requested operation does not make sense, write "nonsense."

(a) $M + N$ (b) $M + 4Q$ (c) MP

(d) PM (e) MQ (f) QM

(g) NS (h) RP (i) PR

(j) $PM - P$ (k) Q^3 (l) $PQ + M$

2. Consider the following matrices:

$$A = \begin{pmatrix} 3 & 1 \\ 0 & 2 \end{pmatrix}, \quad B = \begin{pmatrix} .75 & .0833 & .0833 \\ .3333 & 0 & 0 \\ 0 & .75 & 1 \end{pmatrix}, \quad C = \begin{pmatrix} -0.3 & 0 \\ 0 & -0.2 \\ 0 & 0.4 \end{pmatrix}.$$

Use a calculator or a computer to compute A^{10}, B^7C, $A^8 - 5A^7 + 6A^6$, and $B^{17} - B^{16}$.

3. Express each of the following collections of linear functions of several variables in matrix form.

(a) $$\begin{aligned} P_1 &= 2X - 3Y \\ P_2 &= X + Y \end{aligned}$$

(b) $$\begin{aligned} Q_1 &= 3X - Y - Z \\ Q_2 &= X + Y + Z \\ Q_3 &= 2X + Z \end{aligned}$$

(c) $$\begin{aligned} R_1 &= X - Y - Z - W \\ R_2 &= X - Y - Z - W \end{aligned}$$

(d) $$\begin{aligned} S_1 &= 2Y + Z \\ S_2 &= 4X - 5Y - 7Z \\ S_3 &= X - Y + 2Z \\ S_4 &= X + Z \end{aligned}$$

4. For each of the following systems of equations, express the system in matrix form.

(a) $$\begin{aligned} 2X + Y - 3Z &= 3 \\ Z + W &= 5 \\ X - Y - Z + W &= 0 \end{aligned}$$

(b) $$\begin{aligned} X + Y + Z &= 1 \\ X - Y - Z &= 1 \\ 5X + 3Z + 3Y &= 5 \end{aligned}$$

(c) $$\begin{aligned} 2X_1 + X_2 - 2X_3 - 5X_4 + X_5 &= 0 \\ X_1 - X_3 + X_5 &= 0 \end{aligned}$$

(d) $$\begin{aligned} X + Y &= 2 \\ X - Y &= 1 \\ 2X + 3Y &= 4 \end{aligned}$$

5. Suppose that M and N are 2×2 matrices such that

$$MN - NM = \begin{pmatrix} a & b \\ c & d \end{pmatrix}.$$

Show that $a + d = 0$. (Hint: Express M and N as matrices with variables, and expand $MN - NM$.)

6. Find all 2×3 matrices R and S for which

$$R + S = \begin{pmatrix} 1 & 0 & 1 \\ 1 & 0 & 1 \end{pmatrix} \quad \text{and} \quad R - S = \begin{pmatrix} 0 & 1 & 0 \\ 0 & 0 & 1 \end{pmatrix}.$$

7. (a) Use matrices to determine the number of paths of length 3 or 4 between the points A and C in the graph below. How many paths are there of length at most 4?

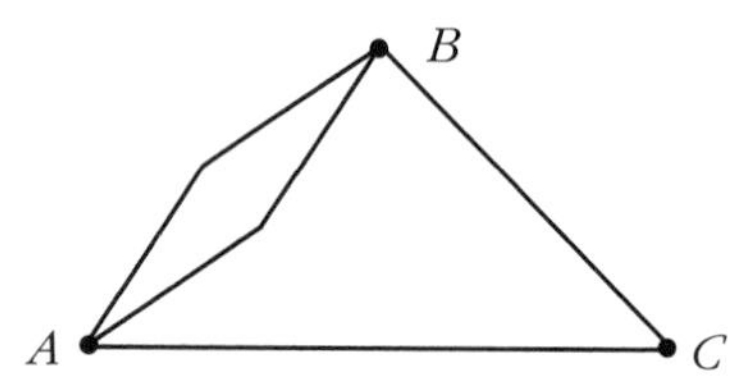

(b) Determine the number of paths of length at most 3 between the points A and C in the graph below. How many between A and A? Between C and C?

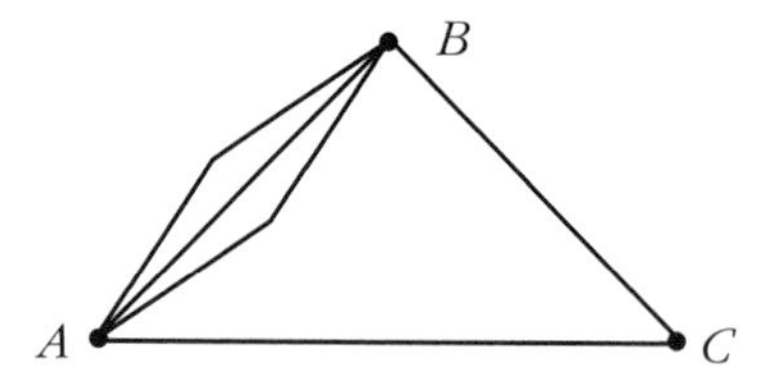

Group Project: Matrix Powers

To begin this project you will first have to learn how to input matrices into a calculator or a computer. Then you will have to learn how to find the result of matrix operations on your machine.

(a) Input the matrix

$$G = \begin{pmatrix} 0.2 & 0.6 \\ 0.8 & 0.4 \end{pmatrix}$$

and find the powers G^2, G^3, G^{50}, G^{51}, and G^{52}. What do you notice?

(b) Input the matrix

$$H = \begin{pmatrix} 0.6 & 0.6 \\ 0.2 & 0.4 \end{pmatrix}$$

and find the powers H^2, H^3, H^{50}, H^{51}, and H^{52}. What do you notice here?

(c) Input the matrix

$$J = \begin{pmatrix} 2 & 1 \\ 1 & 2 \end{pmatrix}$$

and find the powers J^2, J^3, J^{50}, J^{51}, and J^{52}. Now what do you notice?

(d) What are the differences and similarities between the three cases considered in parts (a), (b), and (c)? Explain any theories you have about what is happening.

Group Project: Matrix Games

The type of game described next is known as a *zero sum matrix game.* The idea is somewhat like the elementary school game where two players quickly hold out their hands representing scissors, paper, or a rock. In this game there are two players whom we call X and Y. Each starts with twenty beans. For each play of the game the players have a choice of holding up one or two fingers at the same time. They exchange beans according to the following matrix:

Payment to X	$Y = 1$	$Y = 2$
$X = 1$	-2	4
$X = 2$	1	-3

So if both X and Y hold up one finger, then X gives Y two beans. If X holds up one finger and Y holds up two, then Y gives X four beans.

(a) Play this game for a while to get the idea and keep a record of the payoffs.

(b) Discuss if you think this game is fair. Does one player have an advantage over the other. Why?

(c) What is the best strategy for each player? How do you decide this. (Note: if one player seems to be at a disadvantage, find a strategy that minimizes losses.)

(d) You can represent strategies in matrix form as follows. Suppose X decides to hold up one finger $\frac{2}{3}$ of the time and two fingers $\frac{1}{3}$ of the time. Then player X's average payoff can be represented as the matrix sum

$$\frac{2}{3}\begin{pmatrix} -2 & 4 \end{pmatrix} + \frac{1}{3}\begin{pmatrix} 1 & -3 \end{pmatrix} = \begin{pmatrix} -1 & \frac{5}{3} \end{pmatrix}.$$

(For Y's average payoffs, use columns instead of rows.) We haven't explained what this matrix equation means. As a group try to figure out what this payoff matrix is saying for a given strategy. Represent your strategies for this game using equations of this form.

(e) How could the game be fixed to be fair? How would you represent games using more fingers?

1.4 More Linearity

In the first two sections of this chapter we studied linear functions and in the third we used matrices to organize our work. In this section we investigate other aspects of linearity—namely, problems involving linear constraints. We will continue to use matrix notation whenever it proves useful.

Returning to the Assembly Line

We consider the assembly line for making toy cars and trucks introduced in Sec. 1.3. Our next project is to figure out the best strategy to make some money. A local toy store will pay \$1.40 for each car and \$1.80 for each truck. The store owner will be happy to buy 100 cars and trucks but requires that we sell him at least 25 cars and at least 25 trucks. How many of each should we sell him? Clearly, we cannot answer this question unless we know how much the materials cost. Our suppliers will sell us car bodies for \$0.30, truck bodies for \$0.70, chassis for \$0.10 each, wheels at \$0.05 each, and people for \$0.10 each. In order to calculate expense for parts, we recall our parts matrix

	Cars	Trucks
Car bodies	1	0
Truck bodies	0	1
Chassis	1	1
Wheels	4	6
People	4	2

If we multiply the parts matrix on the left by the row that represents the cost of each part, we obtain a 1×2 matrix whose entries are the total cost of the parts to build a car or truck:

$$(0.30 \quad 0.70 \quad 0.10 \quad 0.05 \quad 0.10)\begin{pmatrix} 1 & 0 \\ 0 & 1 \\ 1 & 1 \\ 4 & 6 \\ 4 & 2 \end{pmatrix} = (1.00 \quad 1.30).$$

Since the toy store owner will pay us \$1.40 for each car, we will make \$0.40 profit on each car, and since he pays \$1.80 for each truck, we will make \$0.50 profit on each truck. Our strategy is now clear! We should sell the toy store owner as many trucks as possible, namely 75 trucks and 25 cars. This would have worked fine, except some difficulties arose. Our wheel supplier just informed us that he now sells wheels only in boxes of 500 (at \$25 per box) and that he will not sell partial boxes. Since we needed

$450 = 75 \cdot 6$ wheels for the trucks and 100 wheels for the 25 cars, we have to reevaluate our plans.

Our supplier's limitations give what is called a *linear constraint* on our problem. We let C denote the number of cars we will assemble, and we let T denote the number of trucks we will assemble. Since we do not want a large number of wheels left over, we decide to buy only one box of 500 wheels. This constraint shows that

$$6T + 4C \leq 500.$$

Assuming we build 25 cars, the constraint means we have $500-4\times 25 = 400$ wheels to use on trucks. Since $400 \div 6 = 66\frac{2}{3}$, we see that we can build 66 trucks. Next we can compute our profit. The cost of assembling a car excluding wheels is \$0.80 and a truck excluding wheels is \$1.00. So, since we spent \$25 on a box of wheels, our material expense can be expressed as $\$(0.80\times C+1.00\times T+25)$. The toy store will pay us $\$(1.40\times C+1.80\times T)$ for our cars and trucks, so we find that our profit P can be expressed as

$$\begin{aligned} P &= \$(1.40 \times C + 1.80 \times T) - \$(0.80 \times C + 1.00 \times T + 25.00) \\ &= \$(0.60 \times C + 0.80 \times T - 25.00). \end{aligned}$$

By selling the toy store owner 25 cars and 66 trucks, we realize $\$(0.60 \times 25 + 0.80 \times 66 - 25) = \$42.80$ in profit.

The plan sounds good, except we realize that we are only selling the toy store owner a total of $25 + 66 = 91$ vehicles, and we could possibly sell him more. Can we make more money by selling more cars and fewer trucks? Our toy store owner will buy up to 100 vehicles with a minimum of 25 each of cars and trucks. In terms of our variables T and C, this means that

$$T + C \leq 100, \;\; T \geq 25, \;\; C \geq 25,$$

which are some new linear constraints to consider. Suppose we try to assemble 100 vehicles without any wheels left over. This would mean finding a solution to the system of equations

$$\begin{aligned} 4C + 6T &= 500 \\ C + \;\; T &= 100. \end{aligned}$$

The second equation is the same as $4C+4T = 400$, and subtracting this from the first shows $2T = 100$ or $T = 50$. In this case $C = 50$ also. If we made 50 cars and 50 trucks, our profit would be $\$(.60\times 50 + .80\times 50 - 25) = \$45.00$. We find that we make an additional \$2.20 following this second scheme.

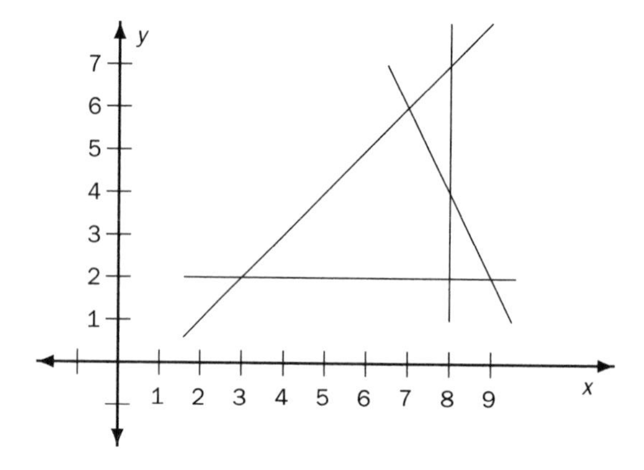

Fig. 1.6. Lines given by equality in linear constraints

We are now left wondering if this second plan is the best possible. It turns out that it is. In order to understand why, we consider this type of problem in a more general setting.

Linear Optimization with Constraints

If we were asked to find the maximum value of the function $Z = 2X+3Y-7$, the correct answer would be to say there is none since the value of Z can be made as large as we like by choosing large X or Y values. However, as we just noted in our toy car and truck assembly problem, most real-life problems have limits on the size of the variables. Suppose now that the values of X and Y are subject to the following constraints:

$$Y \geq 2, \; X - Y \geq 1, \; X \leq 8, \; 2X + Y \leq 20.$$

What can we say about the linear function $Z = 2X + 3Y - 7$? We shall see that Z attains both a maximum and minimum value when X and Y are restricted to the given constraints.

In order to find the maximum and minimum values of our function subject to the constraints, we first must find out which X and Y values are possible. The values of X and Y that satisfy the constraints form what is called the *feasible region*. To find the feasible region we graph the lines given by equality in each constraint inequality. They are $Y = 2$, $X - Y = 1$, $X = 8$, and $2X + Y = 20$ and are shown in Fig. 1.6.

If there is any hope to solve our problem, there must be common solutions to all of the constraint conditions. Each constraint condition has as solutions a half-plane (one side of a line). Therefore, the feasible region is the intersection of half-planes and must be one of the regions bounded by the lines in Fig. 1.6. It turns out that the feasible region is the area inside the quadrilateral shaded in Fig. 1.7. To see that the region pictured is the set of feasible points, we need only check that one point inside does satisfy

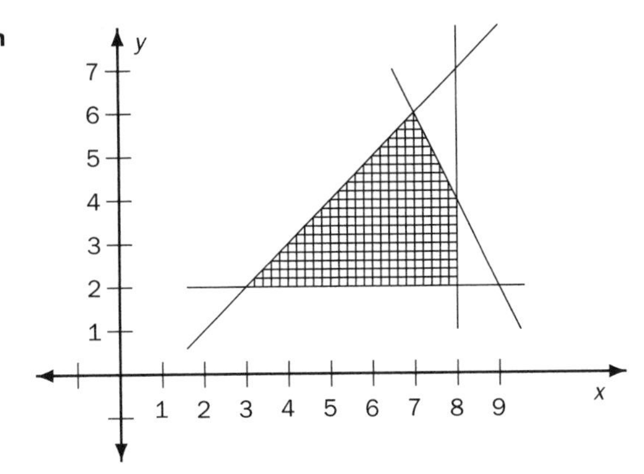

Fig. 1.7. Feasible region

the constraints. For example the point $(5, 3)$ lies in the region pictured and it is quickly checked to satisfy all the constraints.

Next we must decide which points in the feasible region give the maximum and minimum values of our function. For this we use a bit of geometric thinking. The collection of lines $2X + 3Y - 7 = 4$, $2X + 3Y - 7 = 8$, $2X+3Y-7 = 12$, $2X+3Y-7 = 16$, $2X+3Y-7 = 20$, and $2X+3Y-7 = 24$ are drawn on top of the feasible region in Fig. 1.8.

These lines are called *lines of constant value* for the function $Z = 2X + 3Y - 7$ since any two points on the same line give the same output value when used as inputs for the function. Such lines are also called *level sets* for the function. We have pictured only a few of the lines of constant value. All lines of constant value for our function are parallel to these, and the value along any line is always less than the value of a line above it. By visualizing these lines on top of our feasible region, we can see that the minimum value of our function is attained at the lower left corner while the maximum value is attained at the uppermost point.

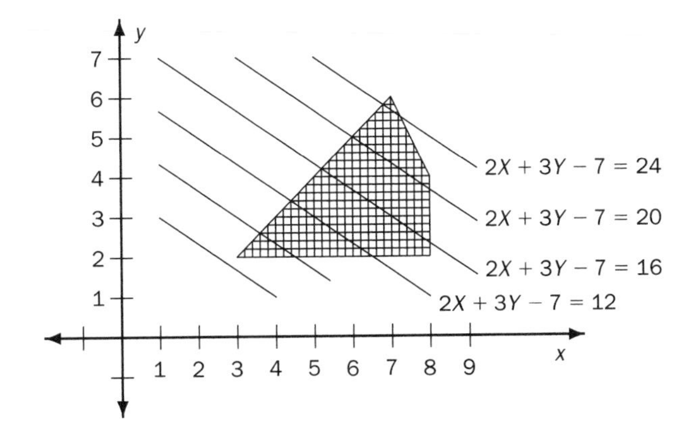

Fig. 1.8. Lines of constant value

The lower left corner of our feasible region is the point obtained by intersecting the boundary lines $Y = 2$ and $X - Y = 1$. It is the point $(3, 2)$, and the minimum value of our function is therefore $Z = 2 \cdot 3 + 3 \cdot 2 - 7 = 5$. The uppermost point of our feasible region is the intersection of the boundary lines $X - Y = 1$ and $2X + Y = 20$. It is $(7, 6)$, and this shows that the maximum value of our function is $Z = 2 \cdot 7 + 3 \cdot 6 - 7 = 25$.

The Principle of Corners

Another way to visualize the optimization problem just considered is to imagine a ruler set on top of the feasible region parallel to the lines of constant value. We slide the ruler upward, keeping it parallel to the constant value lines. As we move it up, it passes on top of larger and larger constant values. If our ruler starts below the feasible region, the point at which it first touches the feasible region will be the point of minimum value. After passing over the feasible region, the point at which the ruler leaves the region will be the point of maximum value. Observe that both these points are vertices of the feasible region.

If we think about linear constraints in two variables X and Y, we can see that feasible regions always have lines for boundaries which meet at vertices. Furthermore, if we are trying to optimize a linear function of two variables, it will always have lines of constant value that are parallel. By the same reasoning as in the above example, our maximum and minimum values will always occur at vertices. We summarize this next.

Theorem 2 (The principle of corners). *If a linear function of two variables subject to linear constraints attains a maximum or minimum value, then these values always occur at the vertices of the feasible region. If the feasible region is a polygon, then a maximum value and a minimum value are attained.*

We remark that it is possible for a linear function subject to constraints to fail to have a maximum or minimum value. This occurs when the feasible regions are unbounded. For example, the function $Z = X + Y$ clearly attains no maximum or minimum when the only constraints are $X \geq 3$ and $Y \leq 4$. More discussion can be found in Prob. 8.

A Few More Examples

The principle of corners shows that one way to solve optimization problems with linear constraints is to determine the locations of the corners of the feasible region and then evaluate the function at these corners. For example, suppose we desire to know the maximum value of $Z = 2X - 3Y$ on the hexagon pictured in Fig. 1.9. The vertices of this hexagon are $(1, 4)$, $(2, 8)$, $(4, 8)$, $(5, 4)$, $(4, 0)$, and $(2, 0)$.

Fig. 1.9. Hexagon

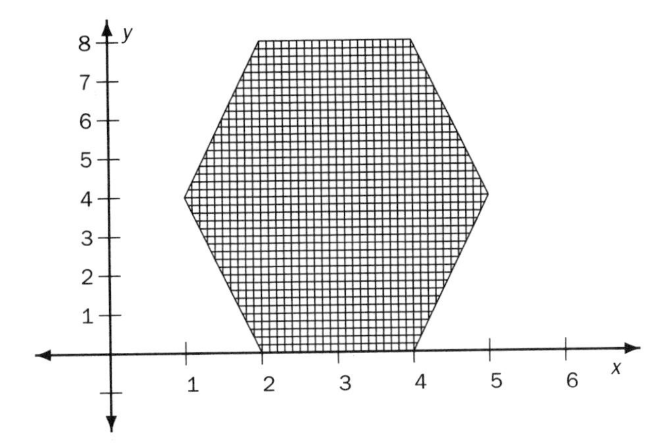

We haven't been given the linear constraints as a set of equations; instead we have a hexagon. But this hexagon could be described by a set of linear constraints. The principle of corners tells us that the maximum value of Z will occur at a vertex, so to find it we need only evaluate our function at the vertices. We find the values are $Z(1,4) = -10$, $Z(2,8) = -20$, $Z(4,8) = -16$, $Z(5,4) = -2$, $Z(4,0) = 8$, and $Z(2,0) = 4$. Hence the maximum value of our function $Z = 2X - 3Y$ on our hexagon is 8.

We might be tempted by the hexagon example to try a shortcut in some problems of this type. For example, consider the linear constraints given in our assembly-line problem described in earlier. These are shown in Fig. 1.10.

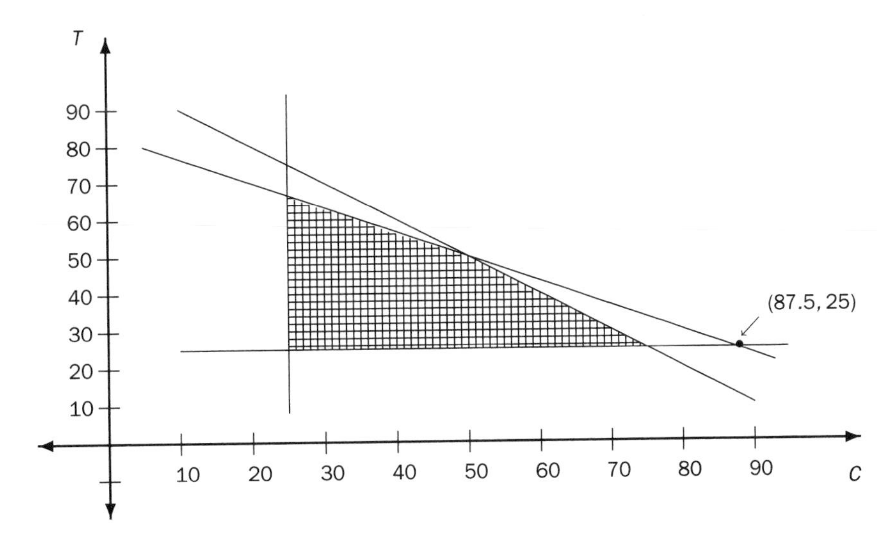

Fig. 1.10. The feasible region for the assembly-line problem

If we consider the four boundary lines, we see by taking them two at a time that there are six possible intersection points. Suppose, for example, we intersect the lines given by $T = 25$ and $6T + 4C = 500$. Solving these equations, we find that $4C = 350$, or $C = 87.5$. If we assembled 87 cars and 25 trucks, our profit function would be $P = \$(.60 \times 87 + .80 \times 25 - 25) = \$47.20$, which is even more profit than obtained earlier. But we forgot about the store owner's constraint that he will only buy 100 vehicles! So we do not have a better solution (unless we can find somebody else who will buy the 12 remaining cars). The lesson here is that when you look for vertices by intersecting boundary lines, be sure to check that your points really lie on the feasible region. Fig. 1.10 illustrates why the intersection point $(87.5, 25)$ just considered lies outside the feasible region.

Problems

1. Suppose we ran into additional difficulties in our toy car and truck business discussed at the beginning of this section. In order to build 50 cars and 50 trucks, our cost for parts is $\$(.80 \times 50 + 1.00 \times 50 + 25) = \$115$. Unfortunately, however, we only have \$100 to invest in parts. So what do we do now? How much profit can we make? Should we go back to our original idea of making only 25 cars and the rest trucks? (Hint: What are the new constraints?)
2. (a) Explain why we know without calculation that there are some linear constraints that have as feasible region the hexagon shown in Fig. 1.9.

   (b) Find some linear constraints that have as their feasible region the hexagon shown in Fig. 1.9.
3. Find the maximum value of the linear function $Z = 2X + 5Y + 1$ subject to the constraints that $X \geq 0$, $Y \geq 0$, $X + Y \leq 6$, and $2X + Y \leq 8$.
4. (a) Graph the feasible region corresponding to the linear constraints $X - 2Y \leq 0$, $-2X + Y \leq 3$, $X \leq 2$, and $Y \leq 3$. What are the vertices?

   (b) What are the maximum and minimum values of $Z = -3X + 3Y + 5$ over the feasible region given in part (a)?

   (c) Suppose the constraint that $X + Y \geq 15$ was added to this problem. How do your answers to (a) and (b) change?
5. (a) Find the maximum and minimum values of $Z = 3X + 4Y$ subject to the constraints that $X \geq 0$, $Y \geq 0$, $X \leq 3$, $Y \leq 3$, and $6X + 8Y \leq 30$.

   (b) Your maximum value in (a) occurred at more than one vertex of your feasible region. Why did this happen? Does your maximum value occur at any other points of your feasible region?

6. (a) A thrift shop received 3200 crayons as a gift. In their storeroom are 100 small boxes that hold 20 crayons and 75 slightly larger boxes that hold 30 crayons. They sell crayons at a price of \$1.50 for a small box and \$2.00 for a larger box. In order to maximize their cash intake, how should they fill the boxes?

   (b) Explain your answer to (a) intuitively without referring to the principle of corners.

   (c) Suppose that the crayons came in 8 colors and that they received 400 of each color. Assuming the shop wants to sell boxes with a balance of colors, how should they fill the boxes in problem (a)?

7. According to a biologist's estimate, there are 500 birds of species A in a small, $100,000$ square yard valley. This number is assumed to be the maximum number the valley can feed. A new species B of birds is introduced to the valley. According to an ecological model, each pair of nesting birds fiercely defends a certain amount of territory and will not nest if other birds intrude. Suppose that a nesting pair of species A defends 100 square yards and that a nesting pair of species B defends 300 square yards near their nests. Suppose also that species A eats twice as much as species B and that they eat the same type of food. What is the maximum number of birds of each type that can coexist in the valley?

In the following two problems the feasible regions are not polygons. The principle of corners still applies to these regions with linear boundaries, but you must interpret it carefully.

8. Consider the feasible region given by the constraints $X + Y \geq 4$, $X \geq 0$, and $Y \geq 0$.

   (a) Graph this feasible region.

   (b) What can you say about the maximum and minimum values of $Z = 2X + Y$ subject to these constraints?

   (c) What can you say about the maximum and minimum values of $Z = -3X - Y + 5$ subject to these constraints?

   (d) What can you say about the maximum and minimum values of $Z = 2X - Y$ subject to these constraints?

   (e) Based on your experience with this example, what theories do you have about when a linear problem with constraints can fail to attain maximum or minimum values?

9. Consider the feasible region given by the constraints $X + Y \geq 4$, $X - Y \leq 4$, and $Y - X \leq 4$. (This is a modification of the feasible region considered in Prob. 8.)

   (a) Graph this feasible region.

(b) Find a linear function that attains a minimum but no maximum on this region.

(c) Find a linear function that attains a maximum but no minimum on this region.

(d) Can you find a linear function that attains neither a maximum or minimum on this region?

(e) How do your answers to this problem affect your answer to part (e) of Prob. 8, if at all?

10. Find the maximum and minimum values of the linear function $Z = X - 2Y + 1$ subject to the constraints that $X \geq 0$, $Y \geq 0$, $3X + Y \leq 8$, and $X + 3Y \leq 8$.

Group Project: Three-Variable Linear Constraints

In this section we studied how to find the maximum and minimum values of a linear function of two variables subject to linear constraints. Suppose the number of variables was increased to three. How are linear optimization problems solved in this case? The next questions should guide your group through some ideas of how to solve such problems.

(a) Suppose you are told that the geometry of this problem in $\mathbf{R}^3$ is analogous to the geometry in $\mathbf{R}^2$ discussed in this section. As a group, discuss what this means and what the principle of corners says in this situation. Why do you believe this?

(b) Using your ideas from (a), find the maximum and minimum values of $W = 2X - Y + 4Z$ subject to the constraints that $X \geq 0$, $Y \geq 0$, $Z \geq 0$, and $X + Y + Z \leq 4$.

(c) In either two or three dimensions, the feasible regions given by linear constraints are known to be *convex*. This means that if two points P and Q lie in the region, then the line segment $\overline{PQ}$ between them also lies in the region. Why are the feasible regions given by linear constraints convex?

(d) Can you draw a convex region in $\mathbf{R}^2$ that is not the feasible region for any finite collection of linear constraints? Where does the linear function $Z = 2X + Y$ attain a maximum on your region?

Group Project: Diet Optimization

You have been asked to design a diet for vacationers visiting the famous Atlas Health and Muscle Spa. For an entire week vacationers are allowed to eat only apples, bananas, carrots, and dates. This diet must be balanced so the following chart is important:

Food	Measure of unit	Sugar (g/unit)	Vitamin C (mg/unit)	Iron (mg/unit)	Cost (cents/unit)
Apples	(large)	1	2	7	8
Bananas	(large)	3	2	4	8
Carrots	1 lb.	3	8	6	16
Dates	2 oz.	2	3	5	7

Health officials have warned the owners of the Atlas Spa not to starve their clients. The law states that each visitor to the spa must consume at least 32 g of sugar each day, at least 28 mg of Vitamin C each day, and at least 35 mg of iron each day. Your task is to find the least expensive way to feed the poor suckers who came to the Atlas Spa for the week.

In this problem, you will need to solve a number of systems of equations in four variables. Now is a good time to learn to do this on your calculator or computer. Also, be sure to check that your solutions do indeed lie within the feasible region.

Historical Note: In 1944 George Stigler in "The Cost of Subsistence," *Journal of Farm Economics*, vol. 27 (1945), pp. 303–314, considered a similar diet problem involving 77 types of foods, their costs, and nutritional content. He constructed a diet for America that satisfied all the basic nutritional requirements and, at 1939 prices, cost only $39.93 per year (less than 11 cents a day!) The diet consisted solely of wheat flour, cabbage, and dried navy beans. Yum yum!

LINEAR GEOMETRY

The purpose of this chapter is to explore some of the uses of matrices and vectors in expressing geometric ideas. We assume you are familiar with the basic properties of plane and space studied in high school geometry. For now, we concentrate on the plane and space; however, later in this text we shall lift the ideas developed here to higher dimensions.

2.1 Linear Geometry in the Plane

In this first section we study the plane (often called the *Euclidean plane*). The Euclidean plane, usually denoted $\mathbf{R}^2$, is the set of ordered pairs (x, y) of real numbers. There is a great variety of geometric objects that can be drawn on the plane—lines, triangles, circles, and rectangles to mention a few. These geometric objects are extremely important in both science and art. Painters communicate their visions of the universe to us on their canvas, which are almost always a piece of a plane. Architects and engineers represent the three-dimensional world around us through their planar drawings. It is important for us to learn how to interpret their work, as well as create our own. This is one good reason for studying geometry in the plane. We shall consider some problems encountered by artists in Sec. 2.4.

Linear Functions Arising in Geometry

Similar triangles can be useful in solving many problems of Euclidean geometry. The reason for this is that the fixed ratio of lengths of sides of similar triangles gives linear functions. We consider an example. In Fig. 2.1 the dot

Fig. 2.1. How far away are the towers?

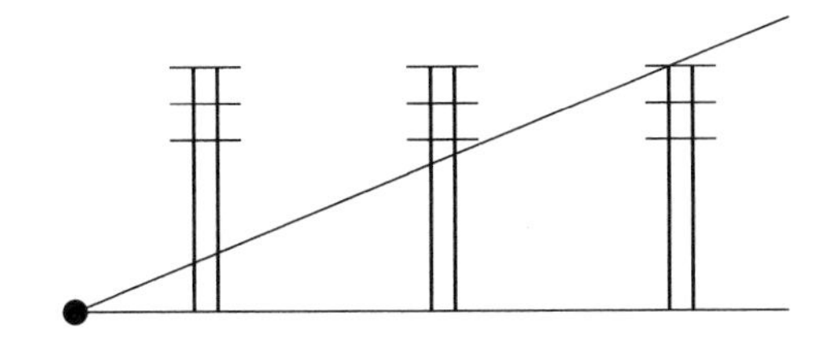

represents the position of a person standing on a flat plane looking at three 100-ft. power towers, which are represented by the vertical segments. This person knows that the closest tower is 500 yds. away and that the towers are 100 ft. tall, and wants to estimate how far away the other towers are. (Note: Fig. 2.1 is not drawn to scale.) The line that angles up toward the top of the third tower represents the observer's line of sight, from which information about the towers can be obtained.

In order to estimate these distances, we notice that there are three similar triangles in the figure, each with the common vertex: our observer. It appears to our observer that the line of sight crosses the first tower about $\frac{1}{5}$ of the way up, or at a height of about 20 ft. If h denotes the height in feet at which this line crosses another pole, and if D denotes the distance in yards of this pole from our observer, then by similar triangles we know

$$\frac{20}{500} = \frac{h}{D}.$$

Rewriting this we obtain a distance function $D(h)$, where the input variable is the height h of the line of sight on the pole. It is the linear function $D(h) = \frac{500}{20}h = 25h$. Since the line of sight appears to cross the second pole about $\frac{3}{5}$ of the way up, or 60 ft., we see that its distance is approximately $25 \cdot 60 = 1500$ yds. We also find that the farthest pole is about $25 \cdot 100 = 2500$ yds. away.

The use of similar triangles, and the linear functions they give, is important in many types of distance estimations. In the second group project in Sec. 2.4 you will have further opportunity to utilize this type of triangular distance estimation.

Quadrilateral Subdivision and Vectors

Consider the two quadrilaterals pictured in Fig. 2.2. No special assumptions were made in constructing them. Note that when new quadrilaterals are formed by connecting the midpoints of each, something special happens. Each of the bold quadrilaterals in Fig. 2.3 is a parallelogram. This will always happen no matter what the starting quadrilateral is, and it can be proved using similar triangles. In order to visualize why this occurs, we will use vectors. Imagine that these quadrilaterals have arrows for edges, as pictured in Fig. 2.4. Two paths start at one vertex of each quadrilateral and end

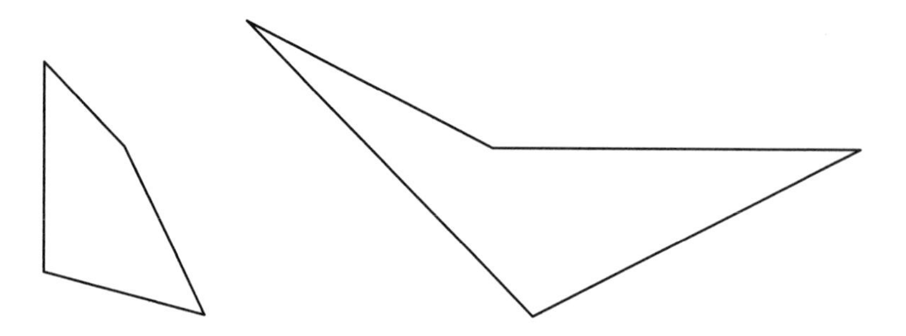

Fig. 2.2. Two quadrilaterals

at the opposite vertex. One path goes to the left and the other to right. Each of these paths "adds" to the direct path going up the middle of the quadrilateral. Next notice the arrows which are thicker in Fig. 2.4 and lie on the sides of the quadrilateral. These arrows are half the length of the sides on which they lie. Note that when they are added, they give paths that are half the length of the central path *and* have the same direction. Further, these paths are two sides of the parallelograms in Fig. 2.3. We have shown, pictorially, why opposite segments connecting the midpoints of sides of a quadrilateral must be parallel and the same length.

The pictorial use of arrows just demonstrated is useful in studying many problems arising not only in geometry, but also in physics, chemistry, engineering, and other subjects. Note that in the problem just considered it was important to consider both the length and direction of each arrow displayed. Whenever both the direction and length of a segment are crucial to consider, we use the notion of a vector. Our next task is to make this idea of a vector more precise. We will return to more applications shortly.

Vectors in the Plane

If P and Q are points on the plane, then we use $\overline{PQ}$ to denote the line segment with P and Q as end points. One often views line segments as a path, from a starting point to an end point. For this reason we have the notion of a directed line segment. A *directed line segment* is a segment $\overline{PQ}$ with a specified starting point P and a specified ending point Q. We will use the notation $\overrightarrow{PQ}$ to denote the directed line segment with starting point P

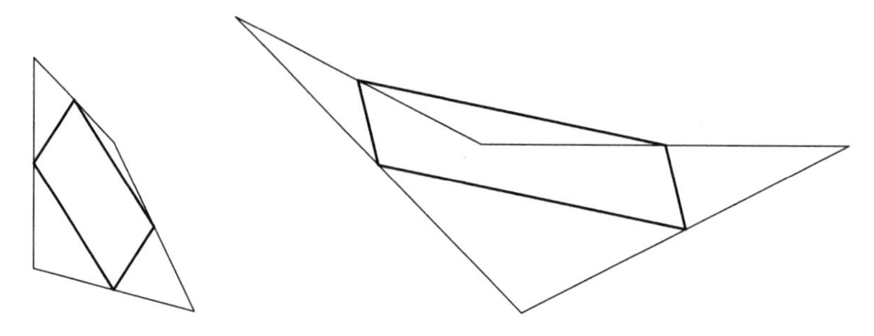

Fig. 2.3. Connecting midpoints of quadrilaterals

Fig. 2.4. Arrows showing why parallelograms arise in Fig. 2.3

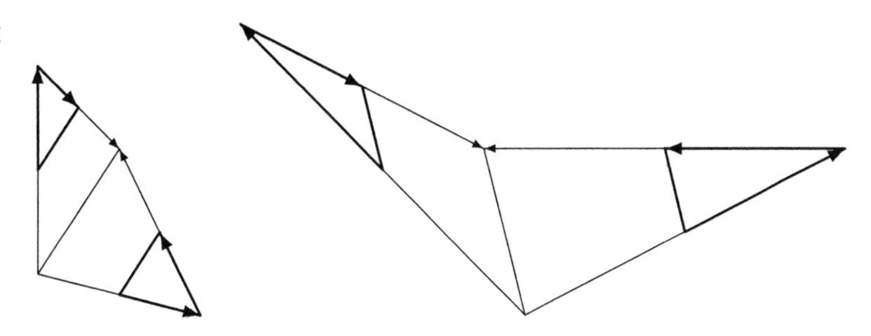

and end point Q. (The reason for specifying starting and ending points will become clear shortly.) Since the phrase "directed line segment" is somewhat long, we refer to directed line segments simply as directed segments.

Often, in geometry as well as in science, it is important to specify both a magnitude and a direction at the same time. Such a combination of information is called a *vector*. We define a vector in $\mathbf{R}^2$ to be a directed line segment with starting point $(0,0) = O$, the origin. The magnitude of the vector $\overrightarrow{OP}$ is the length of the segment $\overline{OP}$, and the direction of the vector $\overrightarrow{OP}$ is the direction of the directed line segment. In case $P = (a, b)$ and $Q = (c, d)$, then by the distance formula in Euclidean geometry we know that the distance between P and Q is $\sqrt{(a-b)^2 + (c-d)^2}$. Since $O = (0,0)$, this says that the magnitude of $\overrightarrow{OP}$ is $\sqrt{a^2 + b^2}$. The magnitude of a vector $\vec{v}$ is denoted $\|\vec{v}\|$.

Observe that a directed segment is essentially a vector with a starting point other than the origin. As such, a directed segment specifies a vector by its length and direction. This is illustrated in Fig. 2.5. We note that any two directed segments that are parallel, point in the same direction, and have the same length specify the same vector (namely, the directed segment "slid"

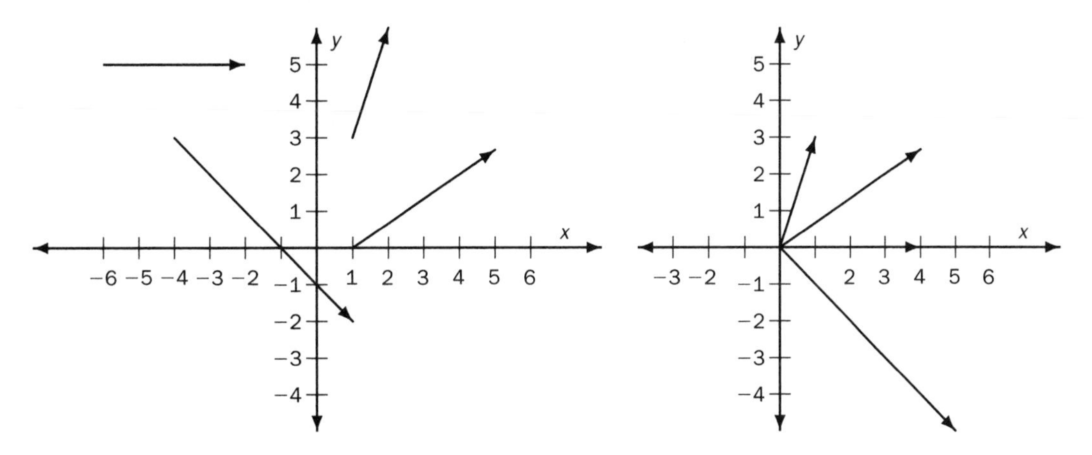

Fig. 2.5. Directed line segments in the plane and the vectors they specify

or "translated" to start at the origin). We shall often view directed segments as vectors that have been translated to have starting points that are possibly different from the origin. We record these ideas in the following definition.

Definition. A *nonzero vector* is a directed line segment whose starting point is at the origin. Any directed line segment $\overrightarrow{PQ}$ specifies a vector, which is the parallel directed segment starting at the origin with the same direction and length.

In case $P = Q$, then it doesn't make sense to talk about the line segment $\overline{PQ}$. However, we often need to talk about vectors of length 0, so we will allow a single vector of length 0, which is often denoted $\vec{0}$. This vector has no magnitude (its length is 0) and consequently has no direction.

We are almost ready to use vectors in the study of geometry. But we need some new notation. Since vectors are directed segments with their starting point at the origin, in order to specify a vector we only need to specify the end point. Hence we give the following.

Definition. Suppose that the point P has coordinates (r, s). Then we use the column matrix

$$\begin{pmatrix} r \\ s \end{pmatrix}$$

to denote the vector $\overrightarrow{OP}$, where O is the origin $(0, 0)$. We shall often use the terminology *column vector* synonymously with column matrix. We shall also say that the vector $\overrightarrow{OP}$ is the *position vector of the point with coordinates* (r, s) since its end point is (r, s). The length of the vector $\overrightarrow{OP}$ is $\sqrt{r^2 + s^2}$ and is denoted $\|\overrightarrow{OP}\|$.

Vector Addition and Scalar Multiplication

In the last chapter we defined matrix addition and scalar multiplication. Since we are using 2×1 matrices to denote vectors in the plane with starting point at the origin, we obtain these two operations for them. It is important to understand the geometric meaning of both these operations.

Whenever

$$\vec{v} = \begin{pmatrix} r \\ s \end{pmatrix}$$

is a vector and t is a real number, the vector

$$t\vec{v} = \begin{pmatrix} tr \\ ts \end{pmatrix}$$

is called the *scalar multiple* of $\vec{v}$ by t. This vector $t\vec{v}$ geometrically results from stretching the vector $\vec{v}$ by length t. Note that in case t is negative, scalar multiplication reverses the direction. Also, if $|t| > 1$, then $t\vec{v}$ is longer than $\vec{v}$, while if $|t| < 1$, then $t\vec{v}$ is shorter than $\vec{v}$.

Consider next two directed segments $\overrightarrow{PQ}$ and $\overrightarrow{QR}$, where the end point of the first coincides with the starting point of the second. Then we define the sum

$$\overrightarrow{PQ} + \overrightarrow{QR} = \overrightarrow{PR}.$$

The sum $\overrightarrow{PR}$ represents the end result of traveling along the directed segment $\overrightarrow{PQ}$ and continuing along $\overrightarrow{QR}$.

In case we are considering two vectors

$$\vec{v} = \begin{pmatrix} r \\ s \end{pmatrix} \quad \text{and} \quad \vec{w} = \begin{pmatrix} t \\ u \end{pmatrix},$$

we define their sum as we do for matrices by

$$\vec{v} + \vec{w} = \begin{pmatrix} r + t \\ s + u \end{pmatrix}.$$

The vector $\vec{v} + \vec{w}$ given by this definition can be understood as follows. The vector $\vec{v}$ is the vector $\overrightarrow{OP}$ where P is the point with coordinates (r, s), and the vector $\vec{w}$ is the vector $\overrightarrow{OQ}$ where Q is the point with coordinates (t, u). If S is the point with coordinates $(r + t, s + u)$, then $\overrightarrow{PS}$ is the directed segment arising by translating the vector $\overrightarrow{OQ}$ to start at P. (The directed segment $\overrightarrow{PS}$ has the same length and direction as $\vec{w} = \overrightarrow{OQ}$, the only difference being that $\vec{w}$ starts at O and $\overrightarrow{PS}$ starts at P.) We see that the sum of directed segments $\overrightarrow{OS} = \overrightarrow{OP} + \overrightarrow{PS}$ is the vector $\vec{v} + \vec{w}$. This is illustrated in Fig. 2.6.

Example. Find (a) $\vec{u} + \vec{v}$, (b) $7\vec{w}$, and (c) the length of $\vec{u} - 3\vec{v} + 2\vec{w}$, where

$$\vec{u} = \begin{pmatrix} 3 \\ 2 \end{pmatrix}, \quad \vec{v} = \begin{pmatrix} -4 \\ 7 \end{pmatrix}, \quad \text{and} \quad \vec{w} = \begin{pmatrix} 0 \\ 9 \end{pmatrix}.$$

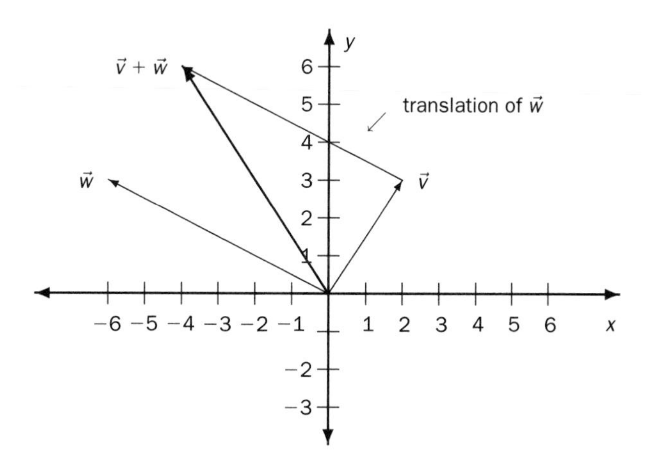

Fig. 2.6. Vector translation and addition

Solution. (a) $\begin{pmatrix}3\\2\end{pmatrix} + \begin{pmatrix}-4\\7\end{pmatrix} = \begin{pmatrix}3-4\\2+7\end{pmatrix} = \begin{pmatrix}-1\\9\end{pmatrix}$

(b) $7\begin{pmatrix}0\\9\end{pmatrix} = \begin{pmatrix}7\cdot 0\\7\cdot 9\end{pmatrix} = \begin{pmatrix}0\\63\end{pmatrix}$

(c) $\begin{pmatrix}3\\2\end{pmatrix} - 3\begin{pmatrix}-4\\7\end{pmatrix} + 2\begin{pmatrix}0\\9\end{pmatrix} = \begin{pmatrix}3+12+0\\2-21+18\end{pmatrix} = \begin{pmatrix}15\\-1\end{pmatrix}$ and so its length is $\sqrt{15^2+(-1)^2} = \sqrt{226}$.

Hexagons, Vectors, and Benzene

Suppose that one vertex of a polygon lies at the origin $(0, 0)$ of the Euclidean plane and that the edge vectors of the polygon are known. Then the other vertices of the polygon can be located by adding the vectors that represent the sides of the polygon. For example, suppose we are interested in determining the vertices of a regular hexagon, one of whose sides has vertices $(0, 0)$ and $(1, 0)$. Then, this edge is represented by the vector $(1, 0)$. We next need to find the vectors that represent the other sides of the hexagon. These vectors are obtained by rotating the vector $(1, 0)$ by multiples of $60°$ around the origin. Adding these vectors in sequence will give the coordinates of the vertices of the hexagon. This process is illustrated in Fig. 2.7.

Using the definition of the trigonometric functions, we know that aside from $(1, 0)$, the five additional end points of the vectors giving our hexagon edges are $(\cos 60°, \sin 60°)$, $(\cos 120°, \sin 120°)$, $(\cos 180°, \sin 180°)$, $(\cos 240°, \sin 240°)$, and $(\cos 300°, \sin 300°)$. In other words, the six vectors that give the edges of our regular hexagon are

$$\begin{pmatrix}1\\0\end{pmatrix}, \begin{pmatrix}\frac{1}{2}\\\frac{\sqrt{3}}{2}\end{pmatrix}, \begin{pmatrix}-\frac{1}{2}\\\frac{\sqrt{3}}{2}\end{pmatrix}, \begin{pmatrix}-1\\0\end{pmatrix}, \begin{pmatrix}-\frac{1}{2}\\-\frac{\sqrt{3}}{2}\end{pmatrix}, \text{ and } \begin{pmatrix}\frac{1}{2}\\-\frac{\sqrt{3}}{2}\end{pmatrix}.$$

Adding these six vectors in sequence gives the coordinates of the six vertices of our regular hexagon,

$$(1, 0), \left(\frac{3}{2}, \frac{\sqrt{3}}{2}\right), (1, \sqrt{3}), (0, \sqrt{3}), \left(-\frac{1}{2}, \frac{\sqrt{3}}{2}\right), \text{ and } (0, 0).$$

Note that the sum of these six vectors gives the final coordinate $(0, 0)$ as it should. One word of caution: Be sure you distinguish between the use

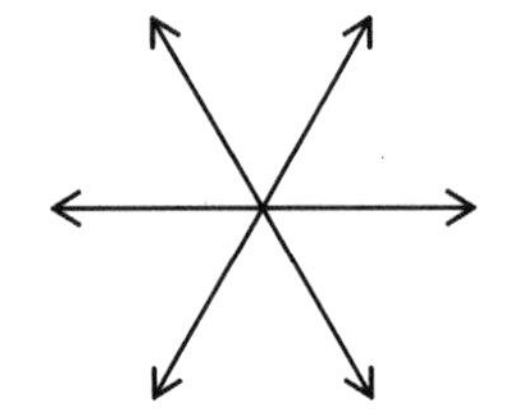

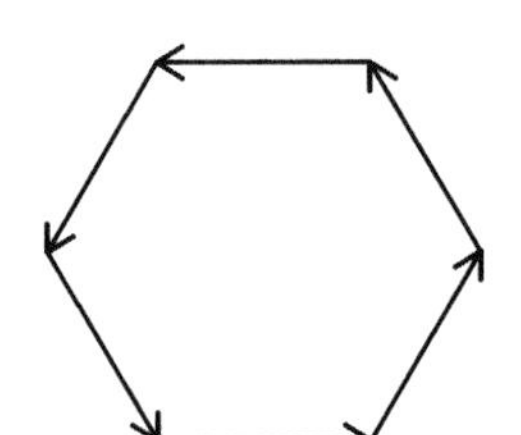

Fig. 2.7. Six vectors, 60° apart, generating a hexagon

of vectors and points in the discussion just considered. It may initially be confusing since similar notation is used, but from context you should always be able to distinguish a vector from the point it represents.

Regular hexagons are important in chemistry because the carbon atoms in a molecule of benzene, C_6H_6, are located on the vertices of a regular hexagon. We can use vector geometry to find the relative locations of the hydrogen atoms as well. The hydrogen atoms in benzene lie on the same plane as the hexagon of carbon atoms, and each carbon atom is bonded to a single hydrogen. The hydrogen atoms are attached to the carbons symmetrically, their bond making a 120° angle with the carbon bonds. The bond lengths between the carbon atoms in benzene is 1.54 Å and the bond length between each hydrogen and carbon is 1.09 Å (the units here are angstroms). This means that if we multiply each of our hexagon vectors by 1.54 Å we can represent our carbon atom hexagon in angstrom units, and if we add an appropriate 1.09 multiples of these vectors to the vertices of the hexagon, we can find coordinates that represent the locations of the hydrogen atoms as well. This representation of the benzene molecule is shown in Fig. 2.8. (The double lines between the carbon atoms indicate double bonds.)

Vector addition will give the coordinates of all the atoms in our benzene model. The coordinates of our six carbon atoms are $(1.54, 0)$, $(2.31, 1.33)$, $(1.54, 2.66)$, $(0, 2.66)$, $(-.77, 1.33)$, and $(0, 0)$ (which are 1.54 times the coordinates of the vertices of the first hexagon). The six coordinates of our hydrogen atoms are obtained by adding 1.09 times the vector with appropriate direction to the coordinate of the carbon atom. They are $(1.54, 0) + 1.09(\frac{1}{2}, -\frac{\sqrt{3}}{2}) = (2.08, -.94)$, $(2.31, 1.33) + 1.09(1, 0) = (3.40, 1.33)$, $(2.08, 3.60)$, $(-.54, 3.60)$, $(-1.86, 1.33)$, and $(-.54, -.94)$. This coordinate information can be useful in determining the distances between atoms in benzene. For example, the distance between two adjacent hydrogen atoms is the distance between $(2.08, -.94)$ and $(3.40, 1.33)$, which using the distance formula in $\mathbf{R}^2$ is $\sqrt{1.32^2 + 2.27^2} = 2.63$ Å.

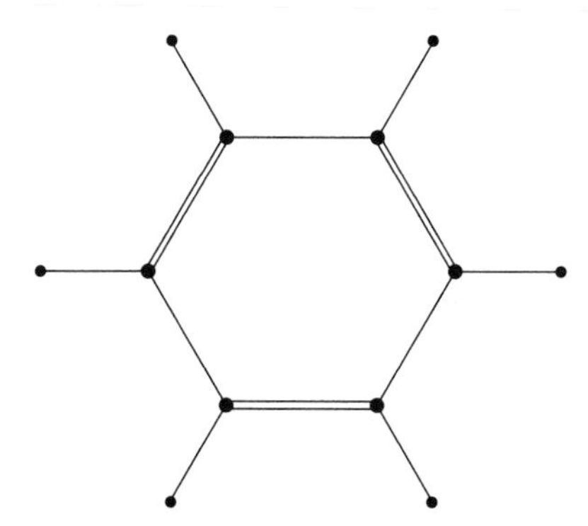

Fig. 2.8. Atom locations in the benzene molecule

Problems

1. Find the sums and scalar multiples of the following vectors as indicated:

$$\vec{u} = \begin{pmatrix} 0 \\ 4 \end{pmatrix}, \ \vec{v} = \begin{pmatrix} 0 \\ 0 \end{pmatrix}, \ \vec{w} = \begin{pmatrix} 2 \\ 6 \end{pmatrix}.$$

   (a) $\vec{u} + \vec{w}$
   (b) $-5\vec{w}$
   (c) $\vec{u} + 3\vec{v} - \pi\vec{w}$

2. Find the sums and scalar multiples of the following vectors as indicated:

$$\vec{p} = \begin{pmatrix} -1 \\ 7 \end{pmatrix}, \ \vec{q} = \begin{pmatrix} 3 \\ -1 \end{pmatrix}, \ \vec{r} = \begin{pmatrix} 2 \\ 6 \end{pmatrix}.$$

   (a) $\vec{p} - 2\vec{q}$
   (b) $-\vec{q} - \vec{r}$
   (c) $\vec{p} + \vec{q} - \vec{r}$

3. Find the coordinates of the vertices of a regular pentagon in the plane if two of the vertices are $(0, 0)$ and $(1, 0)$.
4. You are rowing a boat across a $\frac{1}{2}$-mile wide river that is flowing at 6.5 mph.
   (a) In order to reach a point directly opposite your starting point, you decide to head upstream at an angle of 30°. You are rowing at 8 mph and cannot see the other side through the fog. Use vectors to represent the boat and river velocities (both direction and magnitude) to find out where you end up. If you like, obtain good estimates on graph paper (and avoid messy calculation).
   (b) What angle should you head in order to reach a point exactly opposite you?
   (c) How does your answer to (b) change if you increase your rowing velocity?
5. Find all vectors $\vec{v}$ in the plane of length 2 that satisfy

$$\left\| \vec{v} + \begin{pmatrix} 1 \\ 1 \end{pmatrix} \right\| = 2.$$

   Explain what this condition means geometrically.
6. Find all vectors $\vec{v}$ in the plane that satisfy

$$\left\| \vec{v} - \begin{pmatrix} 1 \\ 1 \end{pmatrix} \right\| = 1 \quad \text{and} \quad \left\| \vec{v} - \begin{pmatrix} 2 \\ 2 \end{pmatrix} \right\| = 2.$$

   Explain what these conditions mean geometrically.

Group Project: Euclidean Geometry Theorems

Use vector geometry to prove the following results from traditional Euclidean geometry. Draw a figure to indicate your use of vectors.

(a) The segments connecting opposite vertices of a parallelogram bisect each other.

(b) The segment that joins the midpoints of two sides of a triangle is parallel and one-half the length of the other side.

(c) The diagonals of an isosceles trapezoid are congruent.

2.2 Vectors and Lines in the Plane

Lines Through the Origin

We next study lines in the plane using vectors. According to Euclidean geometry, a line is completely determined by two points on the line. In other words, there is a unique line through any two distinct points. This basic information can be used to give an algebraic description of lines using vectors.

Example. Consider the two points $O = (0, 0)$ and $P = (-2, 3)$ in $\mathbf{R}^2$. We use vectors to determine all the points on the line $\ell = \overleftrightarrow{OP}$. The idea is this. Consider all the possible vectors that point in the same or opposite direction as the vector

$$\overrightarrow{OP} = \begin{pmatrix} -2 \\ 3 \end{pmatrix}.$$

The end points of all these possible vectors give the line $\overleftrightarrow{OP}$. This is illustrated where $P = (-2, 3)$ in Fig. 2.9.

What are all the possible vectors that start at O and point in the same or opposite direction as the vector $\overrightarrow{OP}$? These vectors are all the multiples

$$\begin{pmatrix} -2t \\ 3t \end{pmatrix}$$

of $\vec{v}$ where t is nonzero. This is because these vectors are obtained geometrically by stretching the vector $\overrightarrow{OP}$ by length t. When $t > 0$ this stretching is in the same direction as $\overrightarrow{OP}$, while when $t < 0$ this stretching includes a direction reversal. Several such vectors are shown in Fig. 2.9. Therefore the line ℓ is the set of end points of all the vectors

$$\begin{pmatrix} -2t \\ 3t \end{pmatrix} \quad \text{where } t \in \mathbf{R}.$$

Fig. 2.9. Vectors in line $\overleftrightarrow{OP}$

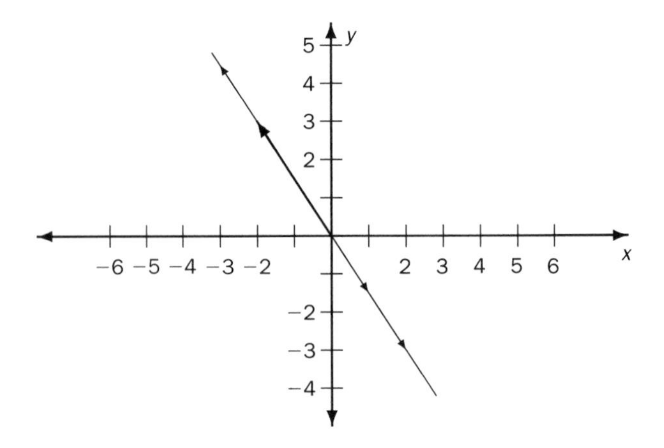

This shows how position vectors can be used to describe lines through $(0, 0)$ in $\mathbf{R}^2$. Using the notation of sets, we can express $\ell = \{(-2t, 3t) \mid t \in \mathbf{R}\}$.[1]

The Parametric Representation of Lines in $\mathbf{R}^2$

We next use vectors to describe lines in $\mathbf{R}^2$ that might not pass through the origin. Consider the line ℓ through the points $P = (0, 2)$ and $Q = (2, 5)$. Our goal is to find all the position vectors that give the points of ℓ. In other words, we want to find all the vectors $\overrightarrow{OP}$ where $O = (0, 0)$ and $P \in \ell$. We begin by noting that the directed segment $\overrightarrow{PQ}$ specifies the vector

$$\vec{v} = \begin{pmatrix} 2 \\ 5 \end{pmatrix} - \begin{pmatrix} 0 \\ 2 \end{pmatrix} = \begin{pmatrix} 2-0 \\ 5-2 \end{pmatrix} = \begin{pmatrix} 2 \\ 3 \end{pmatrix}.$$

Algebraically, this vector $\vec{v}$ was found by subtracting the coordinates of P from those of Q. All the points on the line ℓ arise as the end points of multiples of the directed segment $\overrightarrow{PQ}$ (with starting point P). Each of these directed segments can be found by translating the possible vectors $t\vec{v}$ to P, where t is a real number. This is accomplished by adding the vectors $\overrightarrow{OP}$ and $t\vec{v}$. This is illustrated in Fig. 2.10. Since

$$\overrightarrow{OP} = \begin{pmatrix} 0 \\ 2 \end{pmatrix} \quad \text{and} \quad t\vec{v} = \begin{pmatrix} 2t \\ 3t \end{pmatrix},$$

we find that

$$\overrightarrow{OP} + t\vec{v} = \begin{pmatrix} 0 + 2t \\ 2 + 3t \end{pmatrix}$$

where $t \in \mathbf{R}$. For example, when $t = -1$, we obtain the point $S = (-2, -1)$ on the line ℓ as shown in Fig. 2.10. We have found that the line ℓ can be

[1]The written expression $\{(-2t, 3t) \mid t \in \mathbf{R}\}$ should be read as "the set of all pairs $(-2t, 3t)$ such that t is a real number."

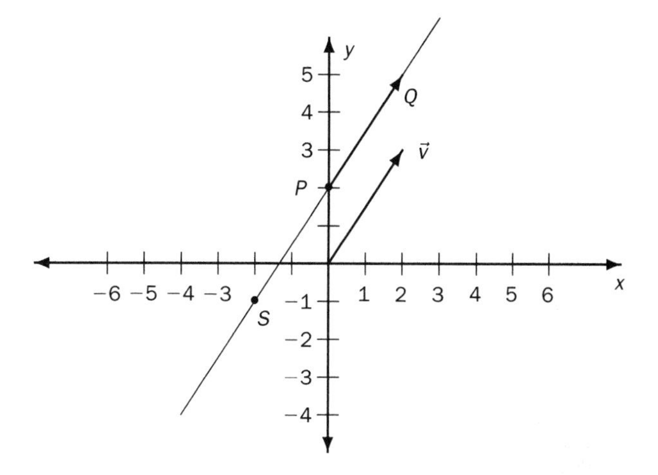

Fig. 2.10. Parametric representation of $\ell = \overleftrightarrow{PQ}$

described in Cartesian coordinates by $\ell = \{(2t, 2 + 3t) \mid t \in \mathbf{R}\}$. This is called a *parametric representation* of ℓ. Often one writes such a parametric representation in the form $x = 2t$, $y = 2 + 3t$, where t is understood as a *parameter* ranging over all real numbers.

We can generalize the discussion in the previous example as follows. If

$$\vec{v} = \begin{pmatrix} c \\ d \end{pmatrix}$$

is a nonzero vector and $P = (a, b)$ is a point in $\mathbf{R}^2$, then there is a unique line ℓ parallel to $\vec{v}$ passing through P. This is the line whose points are given by the position vector

$$\begin{pmatrix} a \\ b \end{pmatrix} + t \begin{pmatrix} c \\ d \end{pmatrix} = \begin{pmatrix} a + tc \\ b + td \end{pmatrix}.$$

In other words, $\ell = \{(a + tc, b + td) \mid t \in \mathbf{R}\}$, which is the parametric description of ℓ. Observe that whenever a line ℓ is described in parametric form one immediately knows both a point on ℓ and a direction vector for ℓ.

The Cartesian Descriptions of Lines

We next indicate how parametric representations of lines relate to the usual Cartesian descriptions found, say, in high school mathematics. Suppose that a line ℓ is specified by

$$\ell = \{(a + tc, b + td) \mid t \in \mathbf{R}\}.$$

This means that the points on ℓ can be expressed as (x, y), where $x = a + tc$ and $y = b + td$ for some real number t. We emphasize that the same t value is used to determine both the coordinates x and y.

Suppose for the moment that both $c \neq 0$ and $d \neq 0$ in the above parameterization of ℓ. Then solving each equation for t gives

$$\begin{aligned} x &= a + tc, \\ x - a &= tc, \\ t &= (x-a)/c. \end{aligned} \qquad \text{and} \qquad \begin{aligned} y &= b + td, \\ y - b &= td, \\ t &= (y-b)/d. \end{aligned}$$

This shows that for every point (x, y) on ℓ,

$$\frac{y-b}{d} = \frac{x-a}{c},$$

or

$$(y-b) = \frac{d}{c}(x-a).$$

This last expression is the point-slope equation for the line ℓ. Evidently, the point (a, b) satisfies this equation (both sides of the equality become 0), and the slope of the line is given by the ratio $m = \frac{d}{c}$. Using the point-slope equation for a line, one can readily obtain the slope-intercept equation by a small bit of algebraic manipulation. One finds that

$$y = \frac{d}{c}(x-a) + b = \frac{d}{c}x + \left(b - \frac{da}{c}\right).$$

Here again $m = \frac{d}{c}$ is the slope of the line. When $x = 0$ we find that $y = b - \frac{da}{c}$ which shows that $(0, b - \frac{da}{c})$ lies on ℓ. The y-coordinate of this point, $b - \frac{da}{c}$, is called the *y-intercept* of the line ℓ. (It is where the line ℓ intersects the y-axis.) Whenever a line is described by a slope-intercept equation $y = mx + k$, m is the slope of the line and k is the y-intercept.

In the special case where $d = 0$ in our parametric representation for ℓ, we find that $y = b$ for all values of t. In this case ℓ is the horizontal line $y = b$, which has slope 0 and y-intercept b. Note that the slope formula $m = \frac{d}{c}$ also applies in this case. In the special case where $c = 0$ in our parametric representation for ℓ, we find that $x = a$ for all values of t. In this case ℓ is the vertical line $x = a$, which has infinite slope ∞ and does not have a y-intercept. The slope formula $m = \frac{d}{c}$ applies in this case, provided you are willing to interpret division by zero as giving ∞.

Example. Find parametric, point-slope, and slope-intercept equations for each of the following lines:

(a) the line through $(0, 0)$ and $(3, -1)$,

(b) the line through $(2, 1)$ and $(-1, 1)$,

(c) the line through $(3, -1)$ and in the direction of the vector

$$\vec{v} = \begin{pmatrix} 4 \\ 7 \end{pmatrix},$$

(d) the line given by the equation $x = -6$.

Solution. (a) Since this line passes through the origin $(0, 0)$, we know that every point of the line has a position vector that is a multiple of

$$\begin{pmatrix} 3 \\ -1 \end{pmatrix}.$$

This shows that a parametric representation is given by $x = 3t$ and $y = (-1)t$. (We emphasize that there can be different parametric descriptions of the same line. For example, the line under consideration is also parameterized by $x = -6u$ and $y = 2u$, where u is the parameter.) Solving for the parameter t, we find that $t = \frac{1}{3}x = -y$. This shows that the slope of the line is $-\frac{1}{3}$ and therefore the point-slope equation is $(y - 0) = -\frac{1}{3}(x - 0)$. Applying algebra, the slope-intercept equation is $y = -\frac{1}{3}x + 0$.

(b) Since this line does not pass through the origin, we must find a direction vector by subtracting the coordinates of the two known points. This gives the direction vector

$$\begin{pmatrix} -1 \\ 1 \end{pmatrix} - \begin{pmatrix} 2 \\ 1 \end{pmatrix} = \begin{pmatrix} -1-2 \\ 1-1 \end{pmatrix} = \begin{pmatrix} -3 \\ 0 \end{pmatrix}.$$

Since the line passes through $(-1, 1)$, this shows that a parametric representation is given by $x = -1 - 3t$ and $y = 1 + 0t$. In other words, this is the horizontal line $y = 1$. The point-slope equation is $(y - 1) = 0(x - 0)$, and the slope-intercept equation is $y = 1$.

(c) The direction vector for the line is $\vec{v}$, and the line passes through $(3, -1)$. This shows that a parametric representation is given by $x = 3 + 4t$ and $y = -1+7t$. Solving for the parameter t, we find that $t = \frac{1}{4}(x-3) = \frac{1}{7}(y+1)$. This shows that the point-slope equation is $(y+1) = \frac{7}{4}(x-3)$ and the slope-intercept equation is $y = \frac{7}{4}x - \frac{25}{4}$.

(d) The line given by the equation $x = 6$ consists of the points in $\mathbf{R}^2$ with arbitrary y-coordinate and constant x-coordinate 6. Therefore, a parametric representation for this line is given by $y = t$ and $x = 6$. Since this line has infinite slope (it is vertical), it does not have a point-slope or slope-intercept equation.

Intersections of Lines

If descriptions of two nonparallel lines in $\mathbf{R}^2$ are known, then the point of intersection can be found by finding common solutions to the equations describing the two lines. This is illustrated in our final example of this section.

Example. Find the point of intersection of the given pairs of lines:

(a) the line through the points $(2, 0)$ and $(1, 0)$ and the line given by the equation $y = 3x - 1$,

(b) the lines parameterized by $\{(t, t) \mid t \in \mathbf{R}\}$ and $\{(t - 1, -t - 3) \mid t \in \mathbf{R}\}$.

Solution. (a) In this case one sees that the line through the points $(2, 0)$ and $(1, 0)$ is the line whose equation is $y = 0$. The point that solves both equations $y = 0$ and $y = 3x - 1$ must have y-coordinate 0 and x-coordinate solving $0 = 3x - 1$. This is the point $(\frac{1}{3}, 0)$.

(b) To solve this problem, we change the variable in the second parameterization to u. So now the second line is parameterized by $\{(u - 1, -u - 3) \mid u \in \mathbf{R}\}$. We did this so we can look for the point that this line has in common with the first line parameterized by $\{(t, t) \mid t \in \mathbf{R}\}$. This point is given by the possible t and u for which $(t, t) = (u - 1, -u - 3)$. We find that $t = u - 1$ and $t = -u - 3$. Consequently, $u - 1 = -u - 3$. We find that $2u = -2$, that is, $u = -1$. Thus $t = -2$, and the point in common is $(-2, -2)$.

Problems

1. Find a parametric, a point-slope, and the slope-intercept equations for each of the following lines:
   (a) the line through $(3, 2)$ and $(1, -1)$,
   (b) the line through $(2, 1)$ and $(0, 0)$,
   (c) the line through $(0, 0)$ and in the direction of $\begin{pmatrix} 1 \\ 3 \end{pmatrix}$,
   (d) the line given by the equation $y + x = -6$.
2. Find a parametric, a point-slope, and the slope-intercept equations for each of the following lines:
   (a) the line given by the equation $3(x - 1) = 2(y + 1)$,
   (b) the line through the points $(1, 2)$ and $(3, 4)$,
   (c) the line parallel to the vector through $(1, 2)$ and $(3, 4)$ passing through $(1, 1)$.
3. Find the point of intersection of the given pairs of lines:
   (a) the line through the points $(1, 1)$ and $(0, 0)$ and the line given by the equation $2x - y = 1$,
   (b) the lines parameterized by $\{(t - 1, t) \mid t \in \mathbf{R}\}$ and $\{(2t - 1, -t - 3) \mid t \in \mathbf{R}\}$,
   (c) the line through $(1, 1)$ parallel to the vector $\begin{pmatrix} 1 \\ 2 \end{pmatrix}$ and the line given by the equation $y = x$.
4. Give a parameterization of all the points on the following line segments (note that your parameter will have to range over an interval instead of all of $\mathbf{R}$):
   (a) the segment $\overline{OP}$, where $O = (0, 0)$ and $P = (2, 1)$,

(b) the segment $\overline{RS}$, where $R = (3, 1)$ and $S = (5, 2)$,

(c) the segment starting at $(-1, 2)$, which has the same direction and length as the vector $\begin{pmatrix} 2 \\ -1 \end{pmatrix}$. What is the end point of this segment?

5. Recall that lines are perpendicular if their slopes are negative reciprocals. Find an equation and a parametric representation of the line through $(2, 3)$ and perpendicular to $\ell = \{(6 + t, 2 - 2t) \mid t \in \mathbf{R}\}$.

6. Express, as a function of time, the position of the boat crossing the river described in Prob. 4(a) of Sec. 2.1. Relate your answer to a parametric representation of the segment that is the boat's path.

Group Project: Families of Lines

In this problem you will find descriptions for interesting collections of lines.

(a) When are two vectors $\vec{u} = \begin{pmatrix} a \\ b \end{pmatrix}$ and $\vec{v} = \begin{pmatrix} c \\ d \end{pmatrix}$ perpendicular? Write your answer as a single equation in a, b, c, and d.

(b) Recall from trigonometry that the set of points $(\cos(t), \sin(t))$ for all real numbers t forms a circle of radius 1. Call this circle C. Using your answer to (a), write down parametric equations that represent all possible lines tangent to C. (Your answer will have two parameters. Why?)

(c) The collection of points (x, y) in the xy-plane that satisfy $y^2 = x^3 + x^2$ is a curve called a *nodal cubic*. Give a parametric description of this curve in terms of a single variable t. Sketch the curve.

(d) If you have had calculus, write down parametric equations that represent all possible lines tangent to the nodal cubic drawn in (c).

2.3 Linear Geometry in Space

In this section we study vectors in space by generalizing the ideas from the plane. We assume that the reader is familiar with the basic geometry of three-dimensional space as well as Cartesian coordinates. This includes the notions of points, lines, planes, parallelism, and perpendicularity in space. We will use the following facts:

(a) There is a unique line through any two distinct points.

(b) There is a unique plane through any three distinct, noncollinear points.

(c) Two different lines in space meet either at a point or not at all.

(d) Two different planes in space meet either in a line or not at all.

(e) Given any plane π and a point P not on π, there is a unique plane λ parallel to π containing P.

As in the plane, we shall define a vector in $\mathbf{R}^3$ to be a directed line segment $\overrightarrow{OP}$, where $O = (0,0,0)$ is the origin of $\mathbf{R}^3$ and P is a point in $\mathbf{R}^3$. If P is the point with coordinates (p, q, r), then we use the column matrix

$$\vec{v} = \begin{pmatrix} p \\ q \\ r \end{pmatrix}$$

to denote the vector $\overrightarrow{OP}$. We also call $\vec{v}$ the *position vector* of the point P. As in the case of the plane, this column notation is useful in helping us distinguish between vectors and the coordinates of points.

Parameterizing Lines in Space

Suppose next that ℓ is a line in space. How do we describe its coordinates algebraically? In the previous section we explored two methods for lines in the plane. One method was to use vectors and parameterize the line, and the other, more familiar, method was to view the line as the graph of a function $y = mx + k$. In space, the second method doesn't work, because if one graphs a linear function $f(X, Y) = aX + bY + c$ in $\mathbf{R}^3$, the resulting graph is a plane, not a line. So it is best to use vectors when describing lines in space. As in the last section, we will use vector addition and scalar multiplication, which is defined for their matrix representations.

The process of parameterizing a line in $\mathbf{R}^3$ is essentially the same as in $\mathbf{R}^2$. Suppose that the line ℓ contains the point $P = (p, q, r)$ and is parallel to the vector

$$\vec{v} = \begin{pmatrix} a \\ b \\ c \end{pmatrix}.$$

Then every position vector $\vec{w}$ of points on ℓ can be expressed as a sum

$$\vec{w} = t\begin{pmatrix} a \\ b \\ c \end{pmatrix} + \begin{pmatrix} p \\ q \\ r \end{pmatrix} = \begin{pmatrix} ta + p \\ tb + q \\ tc + r \end{pmatrix} \quad \text{where } t \in \mathbf{R}.$$

In other words, the line ℓ can be parameterized as the set of all points

$$\{(ta + p, tb + q, tc + r) \mid t \in \mathbf{R}\}$$

inside $\mathbf{R}^3$.

Example. Consider $P = (2, 3, -7)$ and $Q = (-3, 4, 0)$ in $\mathbf{R}^3$. Let ℓ be the line through P and Q. A parametric representation for ℓ can be found as follows.

By subtracting the coordinates of Q from P, we obtain the vector

$$\vec{v} = \begin{pmatrix} 2-(-3) \\ 3-4 \\ -7-0 \end{pmatrix} = \begin{pmatrix} 5 \\ -1 \\ -7 \end{pmatrix}.$$

The line ℓ is parallel to the vector $\vec{v}$. Since ℓ passes through P, we see that the position vectors of all the points on ℓ are given by

$$\vec{w} = t\begin{pmatrix} 5 \\ -1 \\ -7 \end{pmatrix} + \begin{pmatrix} 2 \\ 3 \\ -7 \end{pmatrix} = \begin{pmatrix} 5t+2 \\ -t+3 \\ -7t-7 \end{pmatrix} \quad \text{where } t \in \mathbf{R}.$$

So ℓ is parameterized as

$$\{(5t+2, -t+3, -7t-7) \mid t \in \mathbf{R}\}.$$

Planes Through the Origin

It is also possible to parameterize planes in space using vectors. One knows that any plane in $\mathbf{R}^3$ is uniquely determined by three noncollinear points. Suppose first that one of these three points is the origin $O = (0, 0, 0)$ and the other two points are $P = (a, b, c)$ and $Q = (d, e, f)$. Let the plane that passes through the points O, P, and Q be called $\mathcal{P}$. We claim that the point $R = (a+d, b+e, c+f)$ also lies on the plane $\mathcal{P}$. To see this, first note that the vectors

$$\vec{v} = \begin{pmatrix} a \\ b \\ c \end{pmatrix} \quad \text{and} \quad \vec{w} = \begin{pmatrix} d \\ e \\ f \end{pmatrix}$$

both lie in the plane $\mathcal{P}$. The directed segment $\overrightarrow{QR}$ is the translation of the vector $\vec{v}$ to the starting point Q. Since Q lies in $\mathcal{P}$, this translated vector must also lie inside $\mathcal{P}$, and consequently we see that $R \in \mathcal{P}$. This is illustrated in Fig. 2.11.

In order to complete our parametric description of the plane $\mathcal{P}$, we next note that the line ℓ_1 through O and P and the line ℓ_2 through O and Q are contained in the plane $\mathcal{P}$. The line ℓ_1 is the set of all points $\{(ta, tb, tc) \mid t \in \mathbf{R}\}$, and the line ℓ_2 is the set of all points $\{(ud, ue, uf) \mid u \in \mathbf{R}\}$. As we just saw, if we add the coordinates of a point on ℓ_1 to the coordinates of a point on ℓ_2, we obtain a point on $\mathcal{P}$. In fact, all points on $\mathcal{P}$ arise in this way. Putting this together, we have shown that

$$\mathcal{P} = \{(ta+ud, tb+ue, tc+uf) \mid t, u \in \mathbf{R}\}.$$

We have shown how to describe parametrically any plane that contains the origin in terms of any two nonzero points that lie on the plane. Note that two variables (in this case t and u) are needed to parameterize a plane.

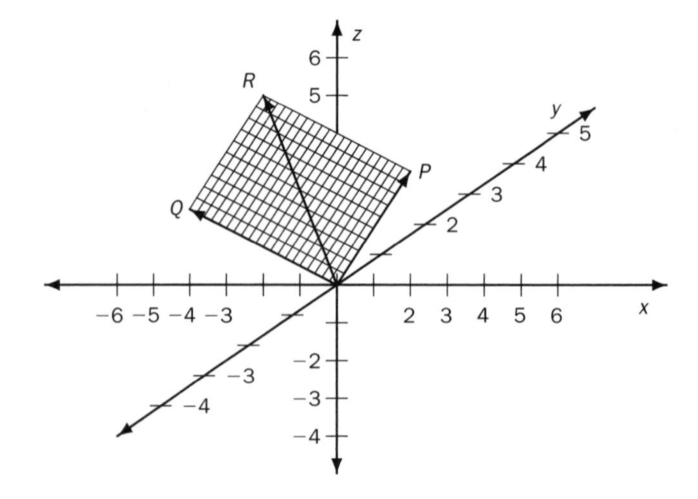

Fig. 2.11. Plane $\mathcal{P}$ through origin, *P* and *Q*

Example. Suppose that $\mathcal{P}$ is the plane through the origin O and the two points $P = (\frac{5}{2}, -1, 3)$ and $Q = (-4, 0, 2)$. Then the set of all points on the plane $\mathcal{P}$ can be described parametrically by

$$\begin{aligned}\mathcal{P} &= \left\{\left(\frac{5}{2}u, -u, 3u\right) + (-4t, 0t, 2t) \mid t, u \in \mathbf{R}\right\} \\ &= \left\{\left(\frac{5}{2}u - 4t, -u, 3u + 2t\right) \mid t, u \in \mathbf{R}\right\}.\end{aligned}$$

The region of this plane inside the parallelogram with edges $\overrightarrow{OP}$ and $\overrightarrow{OQ}$ is shaded in Fig. 2.11.

Parameterizing Planes in Space

Of course, we also want to be able to describe planes that may not contain the origin. Recall what we did in the case of lines. If we knew how to represent parametrically the line ℓ_0 through the origin that was parallel to the line ℓ of interest, then the parametric representation of ℓ could be obtained by adding a position vector from ℓ to the representation of ℓ_0. This same process works for planes. Suppose we know how to represent parametrically the plane $\mathcal{P}_0$ through the origin that is parallel to the plane $\mathcal{P}$ of interest. Then the parametric representation of $\mathcal{P}$ can be obtained by translating the representation of $\mathcal{P}_0$ by any vector on $\mathcal{P}$. This is illustrated in the next example.

Example. Consider the plane $\mathcal{P}$ that passes through the three points $P = (2, 0, 1)$, $Q = (0, 3, -1)$, and $R = (3, -1, 1)$. We first need to find a parametric representation for the plane $\mathcal{P}_0$ that is parallel to $\mathcal{P}$ but passes through the origin. For this we have to find the coordinates of two points on $\mathcal{P}_0$. For exactly the same reasons as in the case of lines (Sec. 2.2), this can be accomplished by subtracting the coordinates of P from the points Q and R. Hence $A = (0-2, 3-0, -1-1) = (-2, 3, -2)$ and $B = (3-2, -1-0, 1-1) = (1, -1, 0)$ are both points of $\mathcal{P}_0$. Using the coordinates of A and B, we find that $\mathcal{P}_0$ has the parametric description

$$\mathcal{P}_0 = \{(-2t + u, 3t - u, -2t) \mid t, u \in \mathbf{R}\}.$$

Finally, we know that the points on $\mathcal{P}$ can be found by translating the points on $\mathcal{P}_0$ by Q. This shows that

$$\begin{aligned}\mathcal{P} &= \{(0, 3, -1) + (-2t + u, 3t - u, -2t) \mid t, u \in \mathbf{R}\}\\ &= \{(-2t + u, 3 + 3t - u, -1 - 2t) \mid t, u \in \mathbf{R}\}.\end{aligned}$$

The Cartesian Description of Planes

Our last task in this section is to relate what we have learned about parametric representations of planes and lines to the more familiar representations you may have seen earlier. Suppose that $a, b, c,$ and d are real numbers. Then the graph of the linear function $z = ax + by + c$ inside $\mathbf{R}^3$ is a plane. More generally, the set of all solutions to an equation of the form $ax + by + cz = d$ in xyz-space is a plane. For example, the plane $\mathcal{P}$ whose points are the solutions of $2x - y + z = 4$ is indicated in Fig. 2.12. The three axes intersect this plane at $(2, 0, 0)$, $(0, -4, 0)$, and $(0, 0, 4)$, which form the corners of the triangle on $\mathcal{P}$ that is shaded in the figure.

Suppose we wish to find the parametric representation of this plane $\mathcal{P}$. For this we must find two points on the plane $\mathcal{P}_0$ that is parallel to $\mathcal{P}$

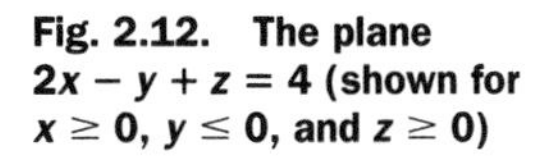

Fig. 2.12. The plane $2x - y + z = 4$ (shown for $x \geq 0$, $y \leq 0$, and $z \geq 0$)

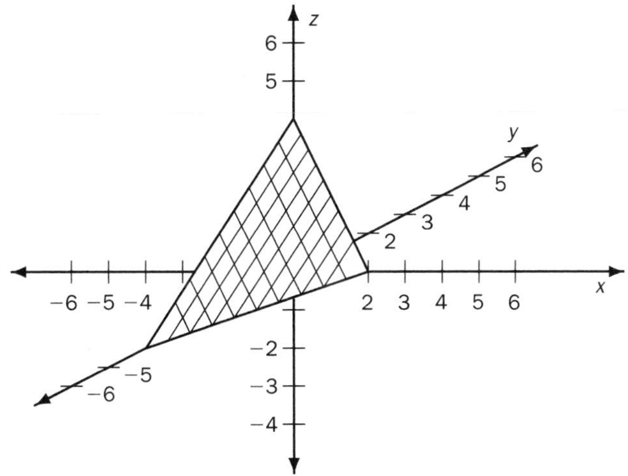

and passes through the origin. The plane $\mathcal{P}_0$ is the set of solutions to the equation $2x-y+z=0$. (To see this, note that the equations $2x-y+z=4$ and $2x-y+z=0$ do not have any common solutions, so the planes they describe cannot intersect. Since $(0,0,0)$ is a solution to $2x-y+z=0$, this equation must describe $\mathcal{P}_0$.) In order to find two points of $\mathcal{P}_0$, we must find two different nonzero solutions to the equation $2x-y+z=0$. We rewrite this equation as $-z=2x-y$. Setting $x=1$ and $y=0$ gives $z=-2$ and therefore the point $(1,0,-2)$ of $\mathcal{P}_0$. Setting $x=0$ and $y=1$ gives $z=1$ and therefore the point $(0,1,1)$ of $\mathcal{P}_0$. It follows that the plane $\mathcal{P}_0$ can be parametrically described as

$$\begin{aligned}\mathcal{P}_0 &= \{t(1,0,-2)+u(0,1,1) \mid t,u\in\mathbf{R}\}\\ &= \{(t,u,-2t+u) \mid t,u\in\mathbf{R}\}.\end{aligned}$$

Finally, we know that the points on $\mathcal{P}$ can be found by translating the points on $\mathcal{P}_0$ by any point on $\mathcal{P}$. We note, for example, that $(0,0,4)$ is a solution to $2x-y+z=4$, the equation that gave $\mathcal{P}$. This shows that

$$\begin{aligned}\mathcal{P} &= \{(0,0,4)+(t,u,-2t+u) \mid t,u\in\mathbf{R}\}\\ &= \{(t,u,4-2t+u) \mid t,u\in\mathbf{R}\}.\end{aligned}$$

More Examples

Many problems involving lines and planes in $\mathbf{R}^3$ can be answered by the parametric representations developed in this section.

Example 1. The two planes described by the equations $3X-3Y+2Z=6$ and $3X+6Y-2Z=18$ intersect in a line ℓ (see Fig. 2.13). Describe this line ℓ parametrically.

Solution. There are many ways to approach this. Here is one. The planes described by the equations $3X-3Y+2Z=0$ and $3X+6Y-2Z=0$ intersect in a line ℓ_0 through the origin which is parallel to the line in question. (These planes are parallel to the planes in question. The vector $\vec{v}$ in Fig. 2.13 lies on ℓ_0.) Next we find a nonzero point on this intersection. Such a point must solve $3X-3Y+2Z=0$ and $3X+6Y-2Z=0$. Adding these two equations gives $6X+3Y=0$. Choosing the solution $X=2$ and $Y=-4$ to this equation and substituting these values into the original equations gives $Z=-9$. This shows that $(2,-4,-9)$ is a point on the line ℓ_0. In particular, the line ℓ_0 is described parametrically by

$$\ell_0 = \{(2t,-4t,-9t) \mid t\in\mathbf{R}\}.$$

To conclude we must find a point on the original line ℓ. Adding the original equations $3X-3Y+2Z=6$ and $3X+6Y-2Z=18$, we find that

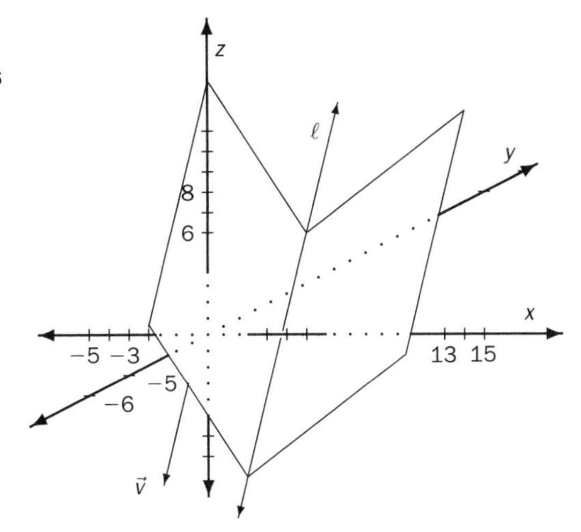

Fig. 2.13. The intersection ℓ of the planes $3X - 3Y + 2Z = 6$ and $3X + 6Y - 2Z = 18$

$6X + 3Y = 24$. If we set $X = 0$ (an arbitrary choice—any choice will give a point on ℓ), we find that $Y = \frac{24}{3} = 8$. As $-2Z = 6 - 3X + 3Y$, we find for this point that $Z = 15$, and consequently $(0, 8, 15)$ is a point of ℓ. We conclude that a parameterization of ℓ is given by

$$\begin{aligned}\ell &= \{(0, 8, 15) + (2t, -4t, -9t) \mid t \in \mathbf{R}\} \\ &= \{(2t, 8 - 4t, 15 - 9t) \mid t \in \mathbf{R}\}.\end{aligned}$$

Alternate Solution. This solution is suggested by the algebra in the second part of the preceding calculation. Adding the equations $3X - 3Y + 2Z = 6$ and $3X + 6Y - 2Z = 18$ gave us $6X + 3Y = 24$, which simplifies to $2X + Y = 8$, from which $Y = -2X + 8$ follows. We can then use $X = t$ as a parameter, from which we obtain $Y = -2t + 8$. Since $3X - 3Y + 2Z = 6$, we substitute our parameterized X and Y values to obtain $2Z = 6 - 3X + 3Y = 6 - 3t + 3(-2t + 8) = -9t + 30$. Dividing this latter equation by 2 shows that ℓ is parameterized by $\{(t, -2t + 8, -\frac{9}{2}t + 15) \mid t \in \mathbf{R}\}$. This is expression is essentially the same as our first, the only difference being that our parameter has been multiplied by $\frac{1}{2}$. This process will be studied in greater detail in the next chapter.

Example 2. The line ℓ_1, which is parameterized by $\{(1+t, 0, 3-2t) \mid t \in \mathbf{R}\}$, and the line ℓ_2, which is parameterized by $\{(1 + 2t, -t, 3 + 2t) \mid t \in \mathbf{R}\}$, both contain the point $(1, 0, 3)$. Find a parametric representation of the plane $\mathcal{P}$ that contains the lines ℓ_1 and ℓ_2.

Solution. Observe that the lines ℓ_1 and ℓ_2 are different. If we decompose the parametric representation of the line ℓ_1, we see that an arbitrary point on ℓ_1 can be expressed as $(1, 0, 3) + (t, 0, -2t)$. Similarly, an arbitrary point on ℓ_2 can be expressed as $(1, 0, 3) + (2t, -t, 2t)$. This means that if we consider the vectors

$$\vec{v}_1 = \begin{pmatrix} 1 \\ 0 \\ -2 \end{pmatrix} \quad \text{and} \quad \vec{v}_2 = \begin{pmatrix} 2 \\ -1 \\ 2 \end{pmatrix},$$

we see that ℓ_1 is parallel to $\vec{v}_1$ and that ℓ_2 is parallel to $\vec{v}_2$. Now this means that the plane $\mathcal{P}_0$, which is parallel to $\mathcal{P}$ and passes through the origin, consists of all the points whose position vectors are given by $t\vec{v}_1 + u\vec{v}_2$, where $t, u \in \mathbf{R}$. In other words,

$$\mathcal{P}_0 = \{(t + 2u, -u, -2t + 2u) \mid t, u \in \mathbf{R}\}.$$

If we translate this parametric description of $\mathcal{P}_0$ by the point $(1, 0, 3)$, we obtain the parametric description of $\mathcal{P}$,

$$\begin{aligned} \mathcal{P} &= \{(1, 0, 3) + (t + 2u, -u, -2t + 2u) \mid t, u \in \mathbf{R}\} \\ &= \{(1 + t + 2u, -u, 3 - 2t + 2u) \mid t, u \in \mathbf{R}\}. \end{aligned}$$

Example 3. Find two equations in the variables x, y, z that describe the line ℓ parameterized by $\{(\frac{5}{3}, t, \frac{1}{3} + t) \mid t \in \mathbf{R}\}$.

Solution. If such a system of equations is desired, the procedure of solving for t used in the last section for obtaining equations for lines in the plane can be used—only we must be careful to obtain *two* linear equations and not just one. This means that an arbitrary point (x, y, z) on ℓ solves $x = \frac{5}{3}$, $y = t$, and $z = \frac{1}{3} + t$ for some $t \in \mathbf{R}$. Solving for t gives $y = t = z - \frac{1}{3}$. This shows that the system of equations

$$\begin{aligned} x \qquad &= \tfrac{5}{3} \\ y - z &= -\tfrac{1}{3} \end{aligned}$$

has as solutions the line ℓ.

Problems

Note: Some of these problems are different than the examples given in the text. Students should try to use the type of geometric methods introduced in this section when looking for a method of solution.

1. Let $P = (1, 1, 3)$, $Q = (2, 4, 0)$, $R = (0, -1, 2)$, and $S = (1, 1, 1)$. Find a parametric description of the following lines in $\mathbf{R}^3$:
   (a) the line $\overleftrightarrow{PQ}$,
   (b) the line $\overleftrightarrow{RS}$,
   (c) the line parallel to $\overleftrightarrow{PQ}$ and containing S.

2. Consider the points $A = (1, 0, 3)$, $B = (2, 2, 0)$, $C = (3, 4, -3)$, and $D = (2, 4, 6)$ in $\mathbf{R}^3$.
   (a) Find a parametric description of the line $\overleftrightarrow{AB}$.
   (b) Using your answer to (a), does either C or D lie on $\overleftrightarrow{AB}$? (For this, check to see if they can be represented by your parametric equation.)
3. Find parametric descriptions of the following planes in $\mathbf{R}^3$:
   (a) the plane through $O = (0, 0, 0)$, $P = (1, 1, 1)$, and $Q = (2, 4, 1)$,
   (b) the plane through $A = (1, 2, 4)$, $B = (1, -1, 1)$, and $C = (0, 0, 1)$,
   (c) the plane described by the equation $x - y + z = 4$,
   (d) the plane through $O = (0, 0, 0)$, and the line described parametrically by $\{(t + 1, t + 2, t) \mid t \in \mathbf{R}\}$.
4. Find parametric descriptions of the following planes in $\mathbf{R}^3$:
   (a) the plane through $R = (1, 0, 0)$, $S = (0, 0, 1)$, and $T = (2, 0, 1)$,
   (b) the plane through $L = (1, 0, 4)$, $M = (-1, -1, 0)$, and $N = (1, 0, 1)$,
   (c) the plane described by the equation $y - z = 0$,
   (d) the plane through the two lines described parametrically by $\{(t, t, t) \mid t \in \mathbf{R}\}$ and $\{(t, 2t, 5t) \mid t \in \mathbf{R}\}$.
5. Find an equation in the variables x, y, and z that describes each plane in $\mathbf{R}^3$ below:
   (a) the plane described parametrically by $\{(t+u, t-u+1, t) \mid t, u \in \mathbf{R}\}$,
   (b) the plane through the points $O = (0, 0, 0)$, $P = (1, 3, 5)$, and $Q = (-1, 0, 2)$.
6. Find both a parametric description and an equation in the variables x, y, and z for each plane in $\mathbf{R}^3$ below:
   (a) the plane containing the point $P = (1, 3, 2)$ and the line ℓ parameterized by $\{(t, 2t - 1, 3) \mid t \in \mathbf{R}\}$,
   (b) the plane through the points $A = (2, 1, 0)$, $B = (1, 0, 1)$, and $C = (1, 1, 1)$.
7. Find parametric descriptions of the following lines in $\mathbf{R}^3$:
   (a) the line that is the intersection of the planes given by the equations $x + z = 0$ and $x + 2y - 3z = 0$,
   (b) the line that is the intersection of the planes given by the equations $x - y + z = 1$ and $x + z = 6$.
8. Find parametric descriptions of the following lines in $\mathbf{R}^3$:
   (a) the line that is the intersection of the planes described parametrically by $\{(t - u, 3t + u, u) \mid t, u \in \mathbf{R}\}$ and $\{(4r - s, 0, s + r) \mid r, s \in \mathbf{R}\}$,
   (b) the line that is the intersection of the planes described parametrically by $\{(t + u + 1, t, u + 1) \mid t, u \in \mathbf{R}\}$ and $\{(4r - s + 1, r, s + r + 1) \mid r, s \in \mathbf{R}\}$.

Group Project: Understanding Surfaces by Slicing with Planes

In this problem you will study the surface $T \subset \mathbf{R}^3$ described parametrically by

$$T = \{(\cos(s)(2+\cos(r)), \sin(s)(2+\cos(r)), \sin(r)) \mid r, s \in \mathbf{R}\}.$$

Before you begin, you should discuss what it means to parameterize a surface using two variables.

(a) Give a general parametric description of all planes parallel to the xy-plane.

(b) Using your answer to (a), if $\mathcal{Q}$ is a plane parallel to the xy-plane, what does $\mathcal{Q} \cap T$ look like?

(c) Your answer to (b) should tell you what T looks like. What is T?

(d) Give a general parametric description of all planes that pass through the z-axis.

(e) Using your answer to (d), if $\mathcal{P}$ is a plane containing the z-axis, what does $\mathcal{P} \cap T$ look like?

(f) Is your answer to (e) consistent with your answer to (c)?

2.4 An Introduction to Linear Perspective

Although we live in a three-dimensional world, most of the representations of our surroundings are two-dimensional. Paintings, photographs, and drawings all lie in a plane. However, this restriction to two dimensions is not a big problem. Photographs look quite real, and skilled artists can create the illusion of depth on their canvas. A reason for this is *perspective.* Discovered during the Renaissance, perspective is the theory of how three-dimensional objects correspond to points on a plane in our field of vision. In this section we will use the language of vectors to study some of the basic concepts of perspective.

The setup we shall use for understanding perspective is the following. Suppose that an artist is standing in a fixed position in front of a large window. On the other side of the window is an ocean view the artist desires to paint. In order to create a realistic painting, the artist decides to paint directly on the glass, placing every object exactly where it appears to be. Our problem is to help the artist determine where to locate certain objects on the glass. We shall shortly see that interesting things can happen.

A version of this process is illustrated in Fig. 2.14, which is a woodcut by Albrecht Dürer from his book *Instruction in Measurement.*

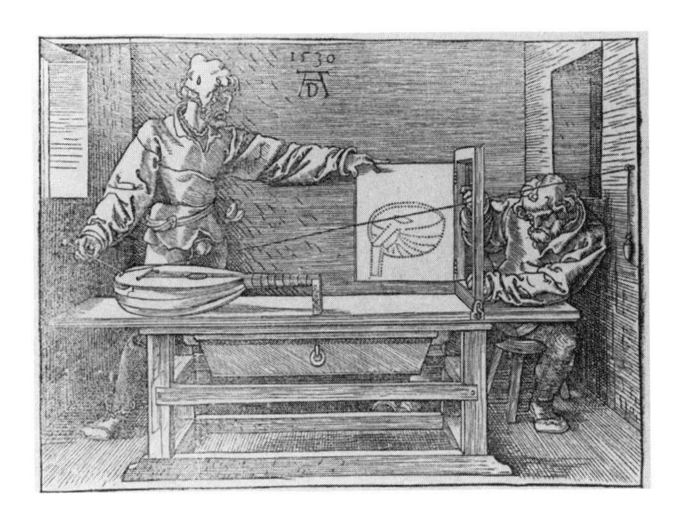

Fig. 2.14. Illustration from *Instruction in Measurement* by Albrecht Dürer (1525)

The Perspective Correspondence

In order to apply vector geometry, we need to set up our problem in xyz-space. We shall assume that the plane of the glass is the xz-plane $\mathcal{P}_{xz}$ and that the plane of the beach is the xy-plane $\mathcal{P}_{xy}$. The artist is approximately 5 feet tall and is standing 10 feet behind the glass. Therefore we shall assume that the artist's eye is located at the point $E = (0, -10, 5)$ in xyz-space. The scene to be painted lies on the other side of the glass, with all y-coordinates positive. Suppose the artist decides to paint an object that is positioned at $P = (a, b, c)$. To what point on the glass does this point correspond? In order to find out, we consider the segment $\overline{EP}$ and its intersection with the glass plane $\mathcal{P}_{xz}$. This is where the object should be painted on the glass in order for it to appear where the artist actually sees it. If Q is the point in the intersection $\overline{EP} \cap \mathcal{P}_{xz}$, we say that Q *corresponds perspectively* to P from E. This is illustrated in Fig. 2.15.

We next determine the coordinates of $\overline{EP} \cap \mathcal{P}_{xz}$ in terms of the coordinates of P. The line ℓ_P between the point $E = (0, -10, 5)$ (the "eye") and a point $P = (a, b, c)$ can be parameterized as follows. The direction vector is

$$\overrightarrow{EP} = \begin{pmatrix} a \\ b + 10 \\ c - 5 \end{pmatrix},$$

and consequently

$$\ell_P = \{(ta, -10 + t(b + 10), 5 + t(c - 5)) \mid t \in \mathbf{R}\}.$$

The intersection of ℓ_P with $\mathcal{P}_{xz}$ is the point on ℓ_P with y-coordinate zero, and this occurs where $t = \frac{10}{b+10}$. Substituting this t value into the parameterization of ℓ_P shows that the point $P = (a, b, c)$ corresponds perspectively

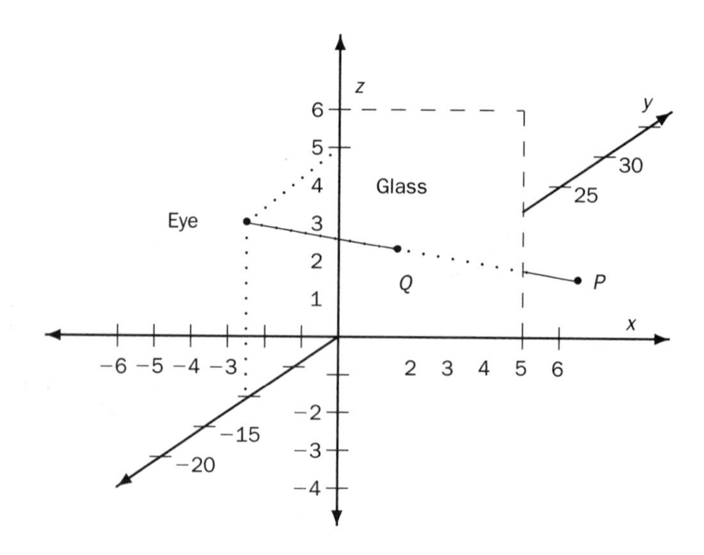

Fig. 2.15. Perspective correspondence

to the point

$$\left(\frac{10a}{b+10}, 0, 5 + \frac{10(c-5)}{b+10}\right) \in \mathcal{P}_{xz}.$$

Now we tabulate some points on the y-axis in $\mathbf{R}^3$ and the points in the plane $\mathcal{P}_{xz}$ to which they perspectively correspond:

Point on y-axis	Point on $\mathcal{P}_{xz}$
$(0, 0, 0)$	$(0, 0, 0)$
$(0, \frac{5}{2}, 0)$	$(0, 0, 1)$
$(0, 5, 0)$	$(0, 0, \frac{5}{3})$
$(0, 10, 0)$	$(0, 0, \frac{5}{2})$
$(0, 15, 0)$	$(0, 0, 3)$
$(0, 40, 0)$	$(0, 0, 4)$
$(0, 90, 0)$	$(0, 0, \frac{9}{2})$

Observe that as the points on the y-axis recede from the observer, the corresponding points on $\mathcal{P}_{xz}$ rise more and more slowly along the z-axis, approaching the value $z = 5$.

Collinear Points Correspond to Collinear Points

If we knew all the coordinates of the objects on the beach we wanted to paint, our formula would tell us exactly what to do. However, this is clearly impossible! Instead, we shall use our vector geometry in order to learn how different shapes transform in perspective. In the end this will save a great deal of time, and we will be able to paint more efficiently.

The first question asked by our artist is how to paint lines. We shall restrict our attention to the values $y \geq 0$, since the artist is painting objects on the opposite side of the window. As a first case we consider two parallel rays on the xy-plane (which is the ground level), say $\mathcal{R}_1$ given by $x = 0$ and $\mathcal{R}_2$ given by $x = 2$. These rays are parameterized as $(0, t, 0)$ and $(2, t, 0)$, respectively, where $t \geq 0$. We apply our calculations above. For the ray $\mathcal{R}_1$, we substitute $a = 0$, $b = t$, and $c = 0$, and find that it perspectively corresponds to set of points $(0, 0, 5 - \frac{50}{t+10})$ where $t \geq 0$. Similarly, the second ray $\mathcal{R}_2$ perspectively corresponds to set of points $(\frac{20}{t+10}, 0, 5 - \frac{50}{t+10})$ where $t \geq 0$.

In both of these formulas, the fraction $\frac{1}{t+10}$ occurs. Since we are considering only the values of t with $t \geq 0$, the fraction $\frac{1}{t+10}$ varies between $\frac{1}{10}$ and 0. Hence the fraction $\frac{50}{t+10}$ varies between 5 and 0 and the fraction $\frac{20}{t+10}$ varies between 2 and 0. If we set $s = \frac{10}{t+10}$, we find that the first line perspectively corresponds to the line segment $\{(0, 0, 5 - 5s) \mid 0 \leq s \leq 1\}$, and the second line perspectively corresponds to the line segment $\{(2s, 0, 5 - 5s) \mid 0 \leq s \leq 1\}$. We have found that both rays $\mathcal{R}_1$ and $\mathcal{R}_2$ correspond to line segments on the plane $\mathcal{P}_{xz}$. This illustrates a general fact about perspective:

Fact. *Any collection of collinear points perspectively corresponds to another collection of collinear points.*

Note that these two segments corresponding to $\mathcal{R}_1$ and $\mathcal{R}_2$ have the common end point $(0, 0, 5)$. This point is called the *infinite point* of the parallel rays $\mathcal{R}_1$ and $\mathcal{R}_2$. The reason for this is that if your eye looks down either ray as far as you can see (toward "infinity"), then you will be looking through the glass very close to the point $(0, 0, 5)$. If you look at a railroad track running straight away from you toward the horizon, the rails appear closer and closer together as they recede and in fact appear to join at the horizon. We just noted this phenomenon algebraically in this calculation for the parallel pair of rays $\mathcal{R}_1$ and $\mathcal{R}_2$.

The perspective correspondence of the rays $\mathcal{R}_1$ and $\mathcal{R}_2$ is illustrated in Fig. 2.16.

The Infinite Line

We next consider two more parallel rays on $\mathcal{P}_{xz}$. Suppose $\mathcal{R}_3$ is parameterized by $(t, 2t, 0)$ and $\mathcal{R}_4$ is parameterized by $(1 + t, 2t, 0)$, both where $t > 0$. By a calculation analogous to the preceding, we find that $\mathcal{R}_3$ corresponds perspectively to set of points $(\frac{10t}{2t+10}, 0, 5 - \frac{50}{2t+10})$ where $t \geq 0$. Since $\frac{10t}{2t+10} = 5 - \frac{50}{2t+10}$, we find that $\mathcal{R}_3$ corresponds to the line segment $\{(5u, 0, 5u) \mid 0 \leq u \leq 1\}$. Similarly, we find that $\mathcal{R}_4$ corresponds perspectively to set of points $(\frac{10(1+t)}{2t+10}, 0, 5 - \frac{50}{2t+10})$ where $t \geq 0$. Rewrit-

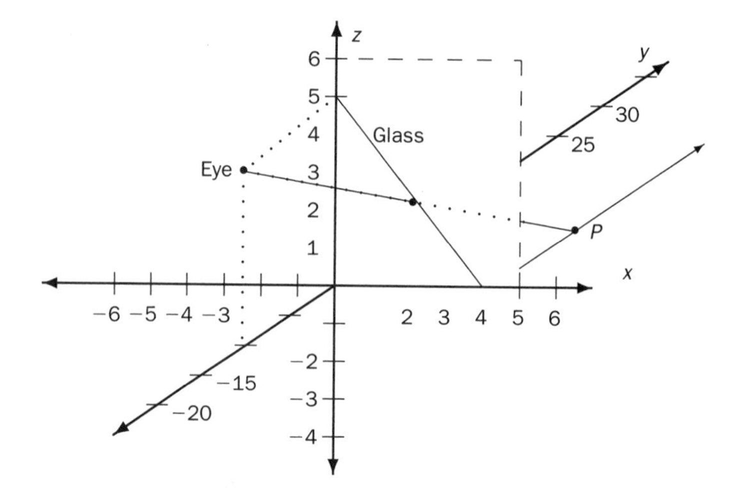

Fig. 2.16. Collinear points perspectively correspond to collinear points

ing $\frac{10(1+t)}{2t+10} = 5 - \frac{40}{2t+10}$ shows that $\mathcal{R}_4$ corresponds to the line segment $\{(5-4v, 0, 5-5v) \mid 0 \leq v \leq 1\}$.

We find that the parallel rays $\mathcal{R}_3$ and $\mathcal{R}_4$ correspond to segments that have the common point $(5,0,5) \in \mathcal{P}_{xz}$. This is their infinite point. Observe that the infinite point for the parallel rays $\mathcal{R}_1$ and $\mathcal{R}_2$ also had z-coordinate 5. In general, each collection of parallel rays in the xy-plane (with $y > 0$) correspond to segments in $\mathcal{P}_{xz}$ that meet at an infinite point with z-coordinate 5. This is because the line $z = 5$ on $\mathcal{P}_{xz}$ is where the horizon is viewed on the picture plane. For this reason it is called the *horizon* or *infinite line*. Observe that the height (z-coordinate) of the infinite line is the same height as the artist's eye.

More Perspective Correspondences

Students of perspective must learn how to draw many different three-dimensional shapes. We will not give any further calculations in this section but will instead describe how some specific objects correspond. You will need to use your imagination to supply some of the reasons.

First we ask you to imagine that the ground on the other side of our artist's glass is covered with square tiles. How will they be drawn in perspective on the glass? Of course, each tile will be drawn as a quadrilateral, since a square has four line segments as sides and line segments perspectively correspond to line segments. Since there are two directions of parallel sides in the squares, when drawing these square tiles on the glass the artist will have to draw two collections of lines that intersect at two different infinite points. The resulting representation of the tiles is shown in Fig. 2.17. Whenever the dominant features of a drawing are determined by two collections of parallel lines (as are the tiles in the figure), the drawing is said to be *two-point perspective*.

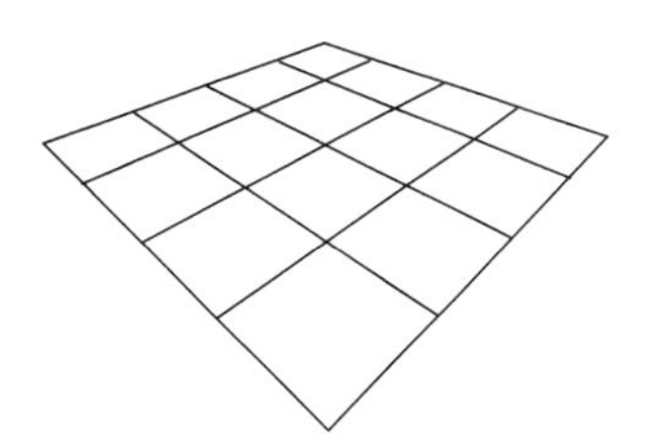

Fig. 2.17. Square tiles drawn in two-point perspective. The horizon line is the dotted line. The two infinite points on the horizon line are indicated.

We next ask, what happens to circles? Imagine a circle centered in the four tiles in Fig. 2.17. You can see by inspecting the representation of the tiles that circles need not perspectively correspond to circles. In fact, these circles turn out to correspond to ellipses!

Finally, we ask the reader to think about how a sphere must be drawn. How must the artist draw the moon? If you think about the different times you have viewed a full moon, you know that it appears in the sky as a circle, regardless of the angle of view. In fact, any sphere will always perspectively correspond to a circle when drawn on the plane.

Problems

1. In this problem you are not expected to carry out any calculations. Instead you should use your geometric imagination—a tool that will be extremely valuable as you continue in this course. What can happen to the following geometric shapes and relationships when they are viewed in linear perspective?
   (a) Angles. Are they drawn as they really are in perspective?
   (b) Triangles. How do they change in perspective?
   (c) Squares. How do they change in perspective?
   (d) Lengths. Can they increase or decrease?
2. Consider the perspective correspondence set up in this section.
   (a) Which line in $\mathcal{P}_{xz}$ corresponds to the line $\{(t, t, 0) \mid t \in \mathbf{R}\}$ in $\mathcal{P}_{xy}$?
   (b) Which line in $\mathcal{P}_{xz}$ corresponds to the line $\{(t, t + 3, 0) \mid t \in \mathbf{R}\}$ in $\mathcal{P}_{xy}$?
   (c) What is the infinite point on $\mathcal{P}_{xz}$ that arises when viewing all lines in $\mathcal{P}_{xy}$ parallel to the line $\{(t, t, 0) \mid t \in \mathbf{R}\}$ from E?
   (d) What is the infinite line on $\mathcal{P}_{xz}$ that arises when viewing $\mathcal{P}_{xy}$?

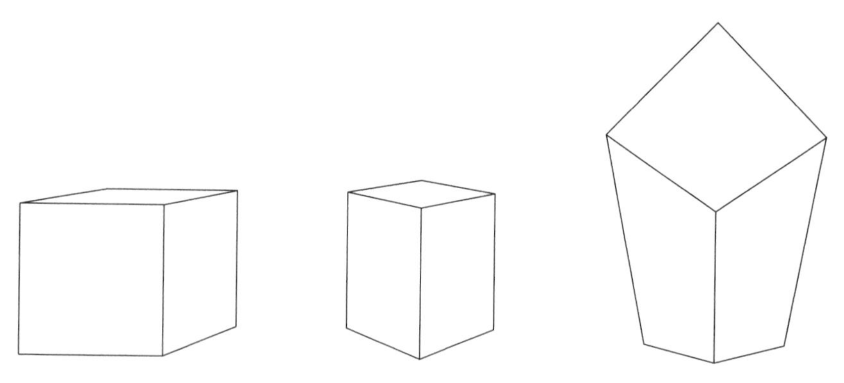

Fig. 2.18. A cube drawn in one-, two-, and three-point perspective

Fig. 2.19. Nativity (1504) by Albrecht Dürer

Group Project: Perspective Drawing

Figure 2.18 contains three drawings of a cube. The left-hand drawing is in one-point perspective, the middle drawing is in two-point perspective, and the right-hand drawing is in three-point perspective. Each drawing is made of nine line segments.

(a) By considering the parallel edges of the above representations of a cube, give a precise definition of what you think one-, two- and three-point perspective is.
(b) Which drawing of the cube looks the most realistic. Why?
(c) What impression of your position relative to the cube do these drawings give you? Why do you think this happens?
(d) Imagine a four-step staircase carved out of the cube. Draw this staircase in each of one-, two-, and three-point perspective.

Group Project: Find the Floor Plan

Figure 2.19 shows a famous engraving by Albrecht Dürer. The scene is staged in what presumably were real buildings for Dürer, and the picture is drawn according to the principles of linear perspective as described in this section.

Your task is to estimate distances and draw a scaled floor plan representing the courtyard and the placement of the buildings in the picture. Hint: the courtyard is basically rectangular, but it is much deeper than it is wide (even though the width, as drawn on the paper, is longer). Don't try to use the messy coordinate calculations found in this section. Instead, use the proportionality of similar triangles as described in Sec. 2.1, along with what you have learned about perspective. Base your estimates on realistic measurements of things you see (for example, people or roof heights). Be sure to label your scale on your floor plan and be able to defend your distance estimates.

CHAPTER 3

SYSTEMS OF LINEAR EQUATIONS

3.1 Systems of Linear Equations

We have already seen many linear equations in our study of linear functions, and on a number of occasions we had to find their solution sets. In this chapter we will develop systematic methods for doing so. Throughout this chapter we will consider systems of linear equations where the number of equations and the number of variables will *not necessarily be the same.* This means that they may not have a single solution, as is often the case when this subject is discussed in high school texts.

Linearity in Electrical Circuits

In Fig. 3.1 we have illustrated the simplest electric circuit, that of a battery connected to a single resistor. The resistor could represent, for example, a light bulb, an electronic game, or any appliance that uses battery power. Linear equations describe the relationship between the amount of current flowing throughout the circuit and the voltage applied by the battery. This relationship is known as *Ohm's law.* Ohm's law states that the current flow

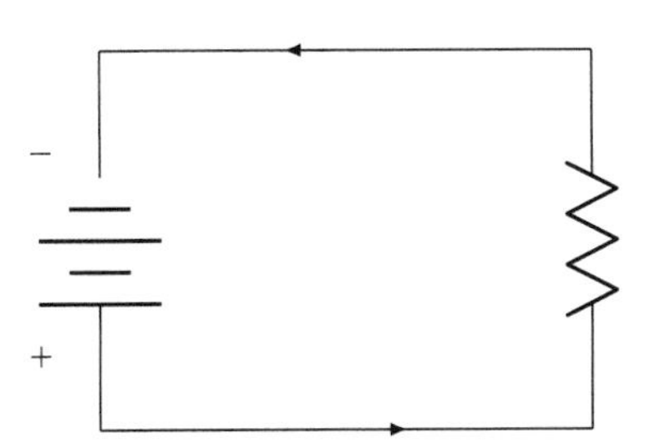

Fig. 3.1. A simple circuit with battery (left) and resistor (right)

through the resistor is proportional to the (battery) voltage applied. It is customary to denote the current by I and the voltage by V. With this notation, Ohm's law reads $V = rI$, where r is a constant, known as the *resistance* of the resistor. Ohm's law shows that given a fixed resistance, the voltage required is a linear function of the current desired.

The units measuring voltage, V, as you are probably aware, are *volts*. The units measuring the current, I, are called *amperes*, or *amps*, and the units of the resistance constant, r, are called *ohms*. So, for example, if $r = 100$ ohms (denoted $r = 100\Omega$), and if we desire 15 amperes of current to flow through our circuit, then Ohm's laws says $V = 100I$ and consequently we need to apply $100 \cdot 15 = 1500$ volts from our battery. Similarly, if we desire 1 ampere of current, then we need 100 volts. To put the meaning of these units in better perspective, we remark that power dissipated through the resistor is the product (volts × amps) = *watts*. So if our 100-ohm resistor was really a light bulb, then applying 100 volts would give 1 amp of current and produce the light given by a 100-watt bulb.

Conversely, if we know the voltage, Ohm's law enables us to determine current flow. For example, if we know that 6 volts are applied to a circuit with a resistance of 24 ohms, then the current I is determined by the linear equation $24I = 6$. In this case $I = .25$ amps. We will study more complex circuits shortly, and instead of a single linear equation we will have to consider systems of linear equations.

Linear Equations and Systems of Linear Equations

We recall that a *linear equation* in the variables $X_1, X_2, \ldots, X_n$ is an equation of the form

$$a_1X_1 + a_2X_2 + \cdots + a_nX_n = b,$$

where the constants $a_1, a_2, \ldots, a_n, b$ are real numbers. For example, the equations $3X - \sqrt{2}Y + Z = 4$ and $2X_1 - 3X_2 = -1$ are both linear equations, while the equation $X - Y + 3XY = 0$ is not linear since it has a product of variables.

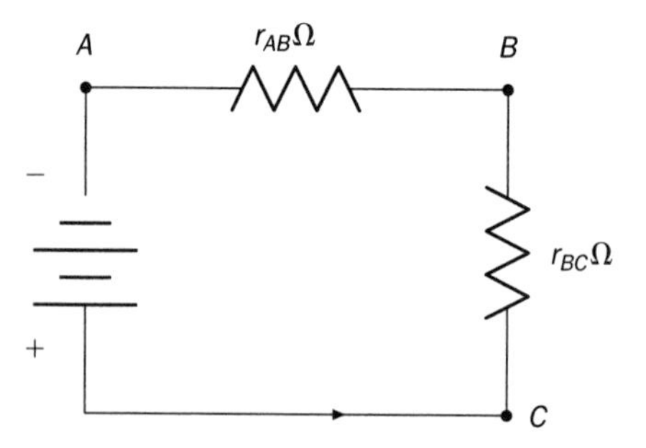

Fig. 3.2. A circuit with two resistors

A *system* of m linear equations in the n variables $X_1, X_2, \ldots, X_n$ is usually represented as

$$\begin{array}{ccccccccc} a_{11}X_1 & + & a_{12}X_2 & + \cdots + & a_{1n}X_n & = & b_1 \\ a_{21}X_1 & + & a_{22}X_2 & + \cdots + & a_{2n}X_n & = & b_2 \\ \vdots & & \vdots & \ddots & \vdots & & \vdots \\ a_{m1}X_1 & + & a_{m2}X_2 & + \cdots + & a_{mn}X_n & = & b_m. \end{array}$$

Here, a_{ij} and b_i are real numbers. In order to simplify terminology, we will often use the phrase "a system of equations" instead of "a system of linear equations."

Two Resistors in a Circuit

We saw how to use Ohm's law to compute the current through a simple circuit containing a battery and a single resistor. We next analyze the case where the circuit has two resistors instead of one.

In order to compute the current flow through this circuit, it is necessary to know what the total resistance is between the points A and C. Figure 3.2 indicates that the resistor between A and B has resistance r_{AB} ohms (denoted $r_{AB}\Omega$) and the resistor between B and C has resistance $r_{BC}\Omega$. Since the resistors are connected in series, the total resistance between A and C is known to be $(r_{AB} + r_{BC})\Omega$. Therefore, by Ohm's law we have that $V = (r_{AB} + r_{BC})I$ for this circuit.

Suppose we know that the voltage applied by our battery is V. Then, if we connect a voltmeter between the points labeled A and C, we would read V volts. It is important to understand what happens if we instead measure the voltage between A and B, or between B and C. It turns out that Ohm's law applies to these pieces of the circuit as well. If the current flowing in this circuit is I, then the voltage read between A and B is given as $V_{AB} = r_{AB}I$, and the voltage read between B and C is given as $V_{BC} = r_{BC}I$. We have now obtained a system of three equations in the four variables I, V, V_{AB}, and V_{BC}:

$$\begin{aligned} V &= r_{AB}I + r_{BC}I \\ V_{AB} &= r_{AB}I \\ V_{BC} &= \qquad\quad\; r_{BC}I. \end{aligned}$$

If we add the second and third equations, we find $V_{AB} + V_{BC} = r_{AB}I + r_{BC}I = (r_{AB} + r_{BC})I$. But the first equation says that $V = (r_{AB} + r_{BC})I$, so we obtain that $V_{AB} + V_{BC} = V$. This observation is known as *Kirchoff's voltage law*.

Solutions to Systems of Equations

We denote by $\mathbf{R}^n$ the set of all n-tuples of real numbers, that is $\mathbf{R}^n = \{(r_1, r_2, \ldots, r_n) \mid r_1, r_2, \ldots, r_n \in \mathbf{R}\}$. In the past two chapters we made an effort to use column vectors when considering elements of $\mathbf{R}^n$. When we consider solutions to systems of linear equations, it will be convenient to use n-tuples in rows as well. We will add and scalar multiply these n-tuples as

$$(r_1, r_2, \ldots, r_n) + (s_1, s_2, \ldots, s_n) = (r_1 + s_1, r_2 + s_2, \ldots, r_n + s_n)$$

and

$$t(r_1, r_2, \ldots, r_n) = (tr_1, tr_2, \ldots, tr_n).$$

In the last chapter we used $\mathbf{R}^2$ to give the coordinates of elements of the Euclidean plane, and $\mathbf{R}^3$ to give the coordinates of elements in three-dimensional space. Often people ask, "What does $\mathbf{R}^4$ look like?, and how about $\mathbf{R}^5$?" Very few people have any clear picture in mind of these higher dimensions (the author doesn't), but this is not crucial. One does not have to worry about any possible geometric interpretation of $\mathbf{R}^n$ for $n > 3$. What you need to remember is that $\mathbf{R}^n$ consists of n-tuples of real numbers and as such is a place where we can encode n-tuples of information. However, we do use our intuition from $\mathbf{R}^2$ and $\mathbf{R}^3$ as a guide when working with $\mathbf{R}^n$, and consequently some geometric language is carried over.

Our first definition is the following. The *set of solutions* to the linear equation

$$a_1X_1 + a_2X_2 + \cdots + a_nX_n = b$$

is the set

$$\{(r_1, r_2, \ldots, r_n) \in \mathbf{R}^n \mid a_1r_1 + a_2r_2 + \cdots + a_nr_n = b\}.$$

For example, consider the equation

$$X + 2Y = 1.$$

The pairs of real numbers $(1, 0)$, $(3, -1)$, and $(-3, 2)$ are each solutions to this equation. In fact, there are infinitely many possible solutions, because any solution must be a solution to the equation $X = 1 - 2Y$, and the collection of solutions to this equation can be parameterized by setting $Y = t$ and $X = 1 - 2t$. In other words, the solution set is $\{(1 - 2t, t) \mid t \in \mathbf{R}\}$. We recognize that we have just parameterized a line in $\mathbf{R}^2$. Of course, the fact that the set of solutions is a line should be familiar from the last chapter (as well as high school mathematics).

We next consider systems of linear equations. The set of solutions to the system of linear equations

$$\begin{array}{ccccccccc} a_{11}X_1 & + & a_{12}X_2 & + & \cdots & + & a_{1n}X_n & = & b_1 \\ a_{21}X_1 & + & a_{22}X_2 & + & \cdots & + & a_{2n}X_n & = & b_2 \\ \vdots & & \vdots & & \ddots & & \vdots & & \vdots \\ a_{m1}X_1 & + & a_{m2}X_2 & + & \cdots & + & a_{mn}X_n & = & b_m \end{array}$$

is $\{(r_1, r_2, \ldots, r_n) \in \mathbf{R}^n \mid (r_1, r_2, \ldots, r_n)$ is a solution to each of the m equations$\}$. In other words, the set of solutions to the system is the intersection of the solution sets of *each* of the m individual equations.

In Example 1 of Sec. 2.3 we studied the solutions to the system of equations

$$\begin{aligned} 3X - 3Y + 2Z &= 6 \\ 3X + 6Y - 2Z &= 18, \end{aligned}$$

and we found that its solution set in $\mathbf{R}^3$ is the line $\{(t, -2t + 8, -\frac{9}{2}t + 15) \mid t \in \mathbf{R}\}$. Solution sets to systems often look like this. Whenever there is more than one solution to a system of equations, the solutions can be expressed using some parameters, with each coordinate of a solution expressed as a linear function of the parameters.

A system of equations need not have a solution. For example,

$$\begin{aligned} X + 2Y &= 1 \\ 2X + 4Y &= 3 \end{aligned}$$

does not have any solutions. This is because any solution (r, s) to the first equation must satisfy $r + 2s = 1$. Therefore, multiplying by 2 shows $2r + 4s = 2$, so (r, s) cannot be a solution to the second equation.

If a system of equations has at least one solution, it is called *consistent.* If it has no solutions, it is called *inconsistent.*

An Electrical Network

Recall that Ohm's law can be used to determine the current in a simple electric circuit if we know the resistance and the voltage applied. Also recall that when a circuit has two resistors, the sum of the voltages between the portions of the circuit loop is the total voltage. We utilize these ideas and consider a slightly more complicated circuit with two batteries and three resistors in Fig. 3.3.

Our problem is to determine the current flow in the above circuit where the voltages and resistances are as indicated. In order to do this we assign currents to the three circuit sections: I_{BA} for the section between B and A,

Fig. 3.3. A circuit with two loops

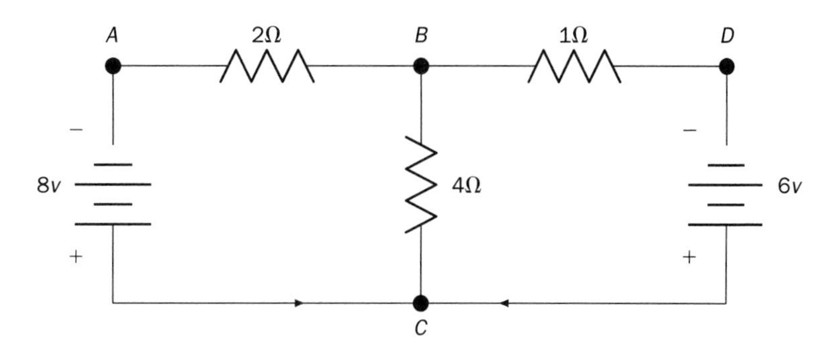

I_{BD} for the section between B and D, and I_{CB} for the section between C and B. Ohm's law tells us that the voltage between B and A is $2I_{BA}$, the voltage between C and B is $4I_{CB}$, and the voltage between D and B is $1I_{DB}$. The fact that we have 8 volts between C and A now shows by Kirchoff's voltage law that $8 = 4I_{CB} + 2I_{BA}$; similarly, since we have 6 volts between C and D, we find $6 = 4I_{CB} + I_{BD}$. Finally, we also know that the current I_{CB} between C and B must be the sum of the currents $I_{BA} + I_{BD}$. This is called *Kirchoff's current law.*

We have obtained three linear equations in three unknowns, which we write as

$$\begin{aligned} 4I_{CB} + 2I_{BA} \qquad\quad &= 8 \\ 4I_{CB} \qquad\quad + I_{BD} &= 6 \\ I_{CB} - \ I_{BA} - I_{BD} &= 0. \end{aligned}$$

Adding the second and third equations together shows that $5I_{CB} - I_{BA} = 6$. Adding twice this new equation to the first equation shows that $14I_{CB} = 20$. This shows that $I_{CB} = \frac{20}{14} = \frac{10}{7}$ amp. The second equation now shows that $2I_{BA} = 8 - \frac{40}{7} = \frac{16}{7}$, so $I_{BA} = \frac{8}{7}$ amp. Finally, the third equation shows $I_{BD} = 6 - \frac{40}{7} = \frac{2}{7}$ amp. We have determined the solution to our system of equations (it is unique) and have therefore found how current flows through each section of our electrical network.

The Geometry of Unique Solutions

We consider three equations in three unknowns:

$$\begin{aligned} X \qquad - Z &= 0 \\ X + Y - Z &= 0 \\ Z &= 1. \end{aligned}$$

Since we know that $Z = 1$ in any solution (by the third equation), the first equation shows that $X = Z = 1$ in any solution. The second equation gives $Y = Z - X = 1 - 1 = 0$. Hence the only solution to the system is the point $(1, 0, 1)$. This is reasonable, since each of the equations in this

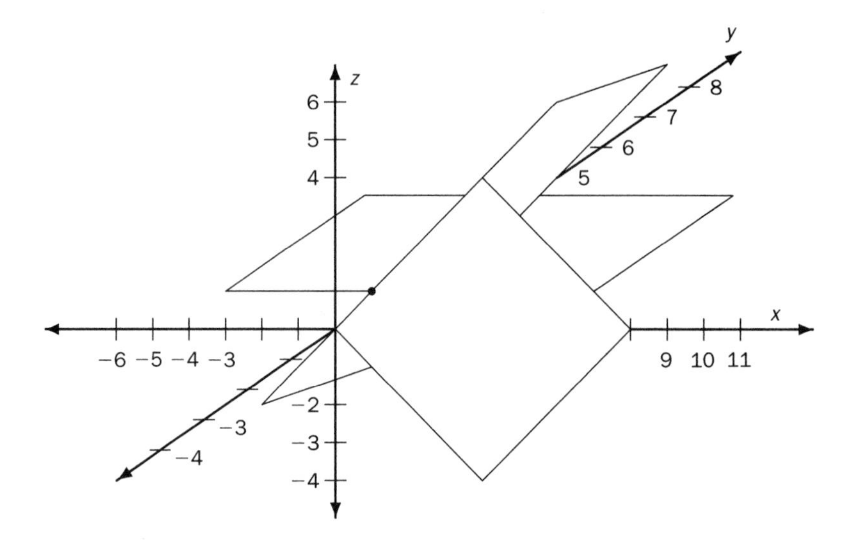

Fig. 3.4. Solutions to $X - Z = 0$, $X + Y - Z = 0$, and $Z = 1$.

system is the equation of a plane in $\mathbf{R}^3$. These three planes in $\mathbf{R}^3$ intersect pairwise in nonparallel lines, and consequently the intersection of all three is a single point. Portions of these planes and their point intersection $(1, 0, 1)$ are illustrated in Fig. 3.4.

Inconsistent Systems

Consider the system

$$\begin{aligned} X \quad\quad + Z &= 0 \\ Y + Z &= 0 \\ X - Y \quad\quad &= 1. \end{aligned}$$

It can quickly be checked that any two of these equations have common solutions. However, the entire system is inconsistent. To see this, add the second and third equations to obtain $X + Z = 1$. But the first equation says that $X + Z = 0$. So the system can never have a solution. Geometrically (in $\mathbf{R}^3$), each equation in the system defines a plane. Each pair of planes intersect in a line, but the three lines (given by the three pairs) are all parallel. These three parallel lines do not intersect, and consequently the three planes taken together do not have any points in common. This is why the entire system is inconsistent.

In the next section we shall study a systematic procedure for finding the set of solutions to arbitrary systems of linear equations. This will enable us to handle large numbers of equations in many unknowns. But you will need to keep in mind the key examples from $\mathbf{R}^2$ and $\mathbf{R}^3$ illustrated here in order to have a geometric picture of what the solutions mean.

Problems

1. Determine the solution set of each of the following systems of equations.

   (a) $\begin{aligned} X - Y &= 5 \\ 2X + 2Y &= 8 \end{aligned}$ (b) $\begin{aligned} X + 5Y &= 0 \\ X - 3Y &= 0 \end{aligned}$

   (c) $\begin{aligned} X - 2Y &= -2 \\ -3X + 6Y &= 6 \end{aligned}$ (d) $\begin{aligned} 2X + 6Y &= 1 \\ X + 3Y &= 3 \end{aligned}$

2. Determine the solution set of each of the following systems of equations.

   (a) $\begin{aligned} X - Y &= 1 \\ 2X - 2Y &= 3 \end{aligned}$ (b) $\begin{aligned} 2X + 5Y &= 0 \\ X - 4Y &= 0 \end{aligned}$

   (c) $\begin{aligned} 3X - 2Y &= -2 \\ -3X + 6Y &= 0 \end{aligned}$ (d) $\begin{aligned} 2X + 8Y &= -6 \\ -X - 4Y &= 3 \end{aligned}$

3. Determine the solution set of each of the following systems of equations.

   (a) $\begin{aligned} 6X - 2Y - 2Z &= 10 \\ 2X \qquad + Z &= 5 \end{aligned}$ (b) $\begin{aligned} X - 3Y + Z &= 0 \\ X + 5Y - 2Z &= 0 \\ 2X + 10Y - 2Z &= 0 \end{aligned}$

4. Determine the solution set of each of the following systems of equations in four variables.

   (a) $\begin{aligned} 3X_1 + X_2 - X_3 \qquad &= 0 \\ X_1 - 2X_2 \qquad - X_4 &= 0 \\ X_1 + 3X_2 \qquad &= 0 \end{aligned}$ (b) $\begin{aligned} 2X_1 - X_2 \qquad &= 0 \\ X_1 \qquad - 2X_3 + X_4 &= 1 \\ X_2 \qquad + 2X_3 - 3X_4 &= 2 \end{aligned}$

5. Determine the solution set of each of the following systems of equations in four variables.

   (a) $X + Y + Z + W = 7$ (b) $\begin{aligned} X - Z &= 3 \\ Y + W &= 2 \end{aligned}$

6. Determine the solution set of each of the following systems of equations. Explain, in geometric terms, what the solution set looks like.

   (a) $\begin{aligned} X + 3Y + Z &= 0 \\ X - Y + 3Z &= 0 \\ 2Y - Z &= 0 \end{aligned}$ (b) $\begin{aligned} X + 3Y + Z &= 1 \\ X - Y + 3Z &= 1 \\ 2Y - Z &= 1 \end{aligned}$

7. Consider the system of equations where u and v are real numbers,

$$\begin{aligned} 3X - Y &= u \\ 6X - 2Y &= v. \end{aligned}$$

   For what values of u and v does the system have a solution? For what values does it have a unique solution?

8. True or false, and why? (Use your geometric thinking.)

   (a) If two linear equations in two variables have a common solution, then this solution is unique.

(b) Three linear equations in three variables never have exactly two common solutions.

(c) Two linear equations in three variables never have exactly one common solution.

(d) Two linear equations in three variables always have at least one common solution.

9. What happens to the current flow in the electrical network shown in Fig. 3.3 if the polarity of the 8 volt battery is reversed?

10. Consider the electrical network shown in Fig. 3.3. Suppose that the resistors in that circuit have resistance values r, s, and t instead of the values 2, 1, and 4. Express the current flow from each battery in the circuit as a linear function of the variables r, s, and t.

Group Project

Consider the system of equations where $a, b, c, d, e, f \in \mathbf{R}$:

$$\begin{aligned} aX + bY + eZ &= 0 \\ cX + dY + fZ &= 0. \end{aligned}$$

(a) Does this system always have a solution? If so, how many?

(b) Show that if (r, s, t) and (u, v, w) are both solutions to the system, then $(r + u, s + v, t + w)$ is also a solution.

(c) Show that if (r, s, t) is a solution to the system and $k \in \mathbf{R}$, then (kr, ks, kt) is also a solution.

(d) What happens to statements (a), (b), and (c) if we instead consider the system of equations:

$$\begin{aligned} aX + bY + eZ &= 1 \\ cX + dY + fZ &= 1. \end{aligned}$$

(e) How can you generalize your ideas developed above?

Group Project

These questions involve the electrical network shown in Fig. 3.5.

(a) Find a system of five equations in five unknowns that enables you to determine the current flowing in each segment of the circuit.

(b) Shorten this system to three equations in three unknowns. Why can you do this?

(c) Find an equivalent network using only three resistors. By "equivalent" we mean that if the same voltages are used in both networks, then the current flowing is the same in each.

Fig. 3.5. Group project circuit

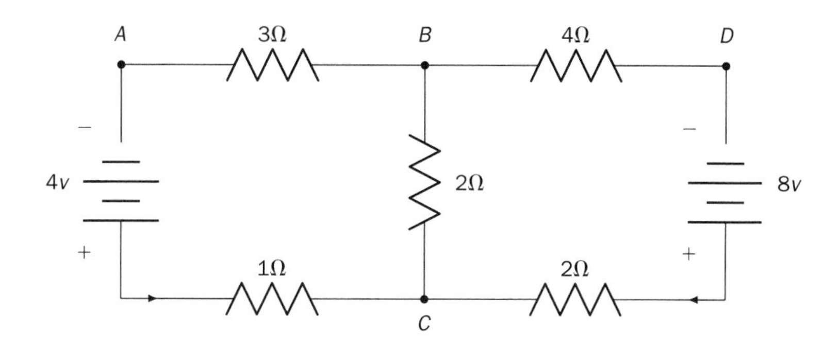

(d) Can you find an equivalent network using only two resistors? Why or why not?

3.2 Gaussian Elimination

We begin by isolating the basic steps used in solving systems of linear equations. These are called elementary operations.

Elementary Operations

Definition. An *elementary operation* applied to a system of linear equations is one of the following three operations:

(i) Multiply any single equation by a nonzero real number, leaving all the other equations the same.

(ii) Add a multiple of one equation to a second, leaving all but this second equation the same.

(iii) Interchange the position of any two equations on the list.

In Theorem 3 ahead, we point out that elementary operations do not change the set of solutions to a system of equations. One needs to know this in order to use elementary operations to solve systems of equations. This next example illustrates how to use elementary operations in solving systems of equations.

Example. Consider the system of equations

$$\begin{aligned} 3X - Y &= 4 \\ X + Y &= 1. \end{aligned}$$

Applying the first operation, we can multiply the second equation by 3 and obtain

$$\begin{aligned} 3X - Y &= 4 \\ 3X + 3Y &= 3. \end{aligned}$$

We next subtract the second equation from the first (this is allowed by the second elementary operation)

$$\begin{aligned} 3X - \quad Y &= 4 \\ -4Y &= 1. \end{aligned}$$

The second equation shows that $Y = -\frac{1}{4}$. Substituting this information into the first equation shows that $3X = 4 - \frac{1}{4} = \frac{15}{4}$. Hence $X = \frac{5}{4}$, and we have found the solution to the system.

These same operations can be applied to the rows of matrices. In this situation we will refer to them as *elementary row operations*. For example, if

$$A = \begin{pmatrix} 4 & 5 & 7 & 9 \\ 1 & 2 & 3 & 4 \\ 2 & 2 & 2 & 2 \end{pmatrix},$$

then multiplying the second row of A by 3 gives

$$B = \begin{pmatrix} 4 & 5 & 7 & 9 \\ 3 & 6 & 9 & 12 \\ 2 & 2 & 2 & 2 \end{pmatrix},$$

and subtracting 2 times the third row from the first row of B gives

$$C = \begin{pmatrix} 0 & 1 & 3 & 5 \\ 3 & 6 & 9 & 12 \\ 2 & 2 & 2 & 2 \end{pmatrix}.$$

This shows that the matrix C can be obtained from the matrix A by applying a sequence of two elementary row operations.

The Augmented Matrix of a System

Consider the system of equations

$$\begin{aligned} 3X - 4Y + \ Z &= 1 \\ 6X - 8Y + 4Z &= 12. \end{aligned}$$

In order to save the trouble involved in writing the variables in each line, we write this system in the matrix form

$$\begin{pmatrix} 3 & -4 & 1 \\ 6 & -8 & 4 \end{pmatrix} \begin{pmatrix} X \\ Y \\ Z \end{pmatrix} = \begin{pmatrix} 1 \\ 12 \end{pmatrix}.$$

The coefficients of the variables in a particular equation are represented by a row of the left-hand matrix. We can save more notation by replacing this equation by the *augmented matrix* of the system, in which we add a final column separated by a bar that gives the constants in each equation. The augmented matrix in this case is

$$\left(\begin{array}{ccc|c} 3 & -4 & 1 & 1 \\ 6 & -8 & 4 & 12 \end{array}\right).$$

We now apply row operations to our augmented matrix in order to solve the system of equations. In this setup, our elementary operations on the augmented matrix correspond exactly to the same operations on the system of equations. Subtracting 2 times the first equation (first row) from the second equation (second row) gives

$$\left(\begin{array}{ccc|c} 3 & -4 & 1 & 1 \\ 0 & 0 & 2 & 10 \end{array}\right).$$

Multiplying the second equation by $\frac{1}{2}$ gives

$$\left(\begin{array}{ccc|c} 3 & -4 & 1 & 1 \\ 0 & 0 & 1 & 5 \end{array}\right).$$

Subtracting the second equation from the first gives

$$\left(\begin{array}{ccc|c} 3 & -4 & 0 & -4 \\ 0 & 0 & 1 & 5 \end{array}\right),$$

and dividing the first equation by 3 gives

$$\left(\begin{array}{ccc|c} 1 & -\frac{4}{3} & 0 & -\frac{4}{3} \\ 0 & 0 & 1 & 5 \end{array}\right).$$

This augmented matrix represents the system

$$\begin{aligned} X - \tfrac{4}{3}Y \qquad &= -\tfrac{4}{3} \\ Z &= 5. \end{aligned}$$

At this point we have a system of equations whose solutions are precisely the same as the original system. In any solution of this system, the third variable Z must be 5, while the first and second variables are not uniquely determined. Representing the possible values of Y by the *parameter* t (ranging over all $\mathbf{R}$), we see that in any solution the value of X is determined by the Y value as $X = \frac{4}{3}t - \frac{4}{3}$. Hence, all the solutions to the original system have the form $(X, Y, Z) = (\frac{4}{3}t - \frac{4}{3}, t, 5)$, where t is a real parameter. Alternatively, this set of solutions can be expressed as $\{(-\frac{4}{3}, 0, 5) + t(\frac{4}{3}, 1, 0) \mid t \in \mathbf{R}\}$.

Equivalent Systems of Equations

Whenever elementary operations are applied, equivalent systems are obtained. We formalize this notion of equivalent systems.

Definition. Two systems of m equations in n variables are called *equivalent* if one system can be obtained from the other system by performing a sequence of elementary operations. Analogously, two matrices are called *row equivalent* if one can be obtained from the other by a sequence of elementary row operations.

Thus, for example, the three systems of equations listed in the example at the beginning of this section are all equivalent systems.

The concept of equivalence is useful because of the next result. It shows why the technique used in the previous section enabled us to solve systems of equations.

Theorem 3. *If two systems of equations are equivalent, then they have exactly the same set of solutions.*

Although we do not give a formal proof of this theorem, it should be quite believable. For example, if you multiply the equation $X + Y + Z = 1$ by 3 to obtain $3X + 3Y + 3Z = 3$, then the set of solutions does not change. After all, if $r, s, t \in \mathbf{R}$ and if $r + s + t = 1$, then multiplying this equation by 3 shows $3r + 3s + 3t = 3$. Hence (r, s, t) is a solution to each equation. Similar ideas show that the other elementary operations cannot change the set of solutions.

Note. We emphasize one point. Elementary operations do *not* allow you to multiply an equation by zero. If you do this, you may enlarge the solution set since it eliminates an equation. Each of our elementary operations is reversible. This means that after applying an elementary operation there is another elementary operation that will return you to the original system. Clearly, multiplication by zero is an irreversible operation, and therefore one does not obtain equivalent systems.

Gaussian Elimination

We next discuss the systematic process for solving systems of equations known as *Gaussian elimination.* In the subsection on augmented matrices we found that it was relatively easy to write down the set of solutions for the final system of equations considered there. This happened because in the second equation, the third variable Z was isolated by itself. One could then determine X in terms of Y using the first equation. In Gaussian elimination the strategy is to apply elementary operations in a systematic way to produce systems that are roughly triangular in shape. Then the solutions to the system can be understood by generalizing the ideas utilized in the subsection on augmented matrices.

As an example of this strategy, we consider the system

$$\begin{aligned} X + Y + Z + W &= 1 \\ 2X + 2Y - Z + W &= 2 \\ 3X + 3Y + Z + W &= 4. \end{aligned}$$

We begin by writing down the augmented matrix for this system and applying row operations to eliminate the variable X from the second and third equations:

$$\left(\begin{array}{cccc|c} 1 & 1 & 1 & 1 & 1 \\ 2 & 2 & -1 & 1 & 2 \\ 3 & 3 & 1 & 1 & 4 \end{array}\right) \mapsto \left(\begin{array}{cccc|c} 1 & 1 & 1 & 1 & 1 \\ 0 & 0 & -3 & -1 & 0 \\ 0 & 0 & -2 & -2 & 1 \end{array}\right).$$

At this point, the variable Y has also been eliminated from the second and third equations. This doesn't always happen. We now view Z as our leading variable in the second and third equations. Our strategy is apply operations so that its lead coefficient in the second equation is 1 and to eliminate it from the third equation:

$$\left(\begin{array}{cccc|c} 1 & 1 & 1 & 1 & 1 \\ 0 & 0 & -3 & -1 & 0 \\ 0 & 0 & -2 & -2 & 1 \end{array}\right) \mapsto \left(\begin{array}{cccc|c} 1 & 1 & 1 & 1 & 1 \\ 0 & 0 & 1 & \frac{1}{3} & 0 \\ 0 & 0 & -2 & -2 & 1 \end{array}\right)$$

$$\mapsto \left(\begin{array}{cccc|c} 1 & 1 & 1 & 1 & 1 \\ 0 & 0 & 1 & \frac{1}{3} & 0 \\ 0 & 0 & 0 & -\frac{4}{3} & 1 \end{array}\right).$$

We finally multiply the third equation by $-\frac{3}{4}$ so it has lead coefficient 1. This gives the augmented matrix

$$\left(\begin{array}{cccc|c} 1 & 1 & 1 & 1 & 1 \\ 0 & 0 & 1 & \frac{1}{3} & 0 \\ 0 & 0 & 0 & 1 & -\frac{3}{4} \end{array}\right),$$

whose solution set is easily determined. The last equation says that $W = -\frac{3}{4}$. Substituting this into the second equation shows $Z + (\frac{1}{3})(-\frac{3}{4}) = 0$, so $Z = \frac{1}{4}$. Finally, letting t be a parameter for the variable Y, the first equation says $X + t + \frac{1}{4} + (-\frac{3}{4}) = 1$, giving $X = -t + \frac{3}{2}$. Our solution set is $\{(-t + \frac{3}{2}, t, \frac{1}{4}, -\frac{3}{4}) \mid t \in \mathbf{R}\}$. When we describe the set of solutions in this fashion, Y is called a *free variable* and t is called its *parameter*, while X, Z, and W are called *determined variables.*

Echelon Form Systems

Now that we have seen an example of Gaussian elimination, we next put the key points of this strategy in words. The process is described in stages, with one stage for each row, starting at the top. The first stage begins by arranging the variables in some order and aligning them in columns.

Next, make sure that the first variable occurs in the top equation—if not, interchange equations (we use the phrase "variable occurs" to mean that the variable has a nonzero coefficient). Next, the first equation is multiplied by an appropriate real number to give the first variable coefficient 1. Now, using the second row operation, remove the first variable from all equations below it. At this point, the first variable of the system only occurs in the first equation.

In the second stage of Gaussian elimination, the process just described is repeated using the second equation as if it were the top equation, leaving the first equation alone. At the end of the second stage, the first occurring variable of the second equation does not occur in any equation below. Note that, as in the preceding example, *the first occurring variable in the second equation is not necessarily the second variable.* Also, one might have to interchange equations in order to bring the second occurring variable to the second equation. After this, the process is repeated for the third, fourth, . . . , nth equations until a "triangular" or "staircase" shape system of equations results. This staircase shape of coefficients is pictured in Fig. 3.6.

At the end of Gaussian elimination the resulting system of equations is in *echelon form.* In this definition it is crucial that an ordering of the variables be specified in advance.

Definition. A system of equations is in *echelon form* if

(EF 1) The first occurring variable of each equation has coefficient 1. (This variable, if any, is called the *leading variable* of that equation.)

(EF 2) The leading variable of any equation occurs to the right of the leading variable of any equation above it.

Solutions to Echelon Form Systems

The set of solutions to a system in echelon form is reasonably easy to understand. A system in echelon form will be consistent unless an equation of the form $0 = r$ occurs where r is nonzero. Suppose the system is consistent. Any variable that is not the leading variable of an equation will be called a *free variable.* The remaining variables (that is, the leading variables) are called *determined variables.* By assigning parameters to the free variables, and solving for the determined variables in terms of the parameters, one obtains a description of all the solutions to the system. This process is known as *back-substitution.*

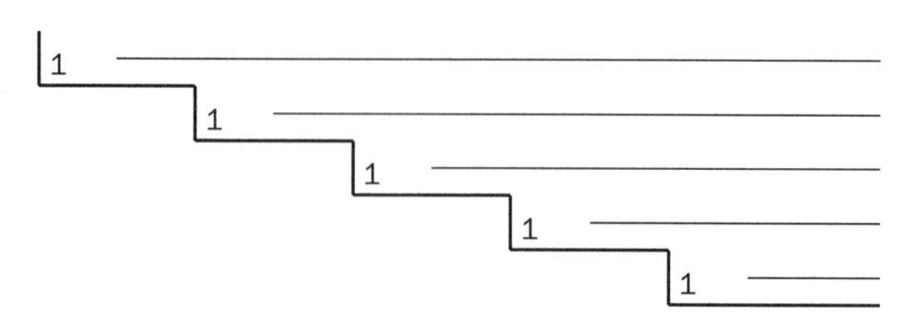

Fig. 3.6. The staircase shape of an echelon system. All entries below the staircase are zero.

Consider the following examples:

(i)
$$\begin{aligned} X - 2Y &= 2 \\ Y &= 1 \end{aligned}$$

(ii)
$$\begin{aligned} X + Y \qquad &= 1 \\ 0Z &= 1 \end{aligned}$$

(iii)
$$\begin{aligned} X + 2Y - Z + 2W &= 1 \\ Y - Z - W &= 0 \\ Z \qquad &= 1 \end{aligned}$$

(iv)
$$\begin{aligned} X_1 + X_2 \qquad\qquad\qquad &= 1 \\ X_3 - X_4 \qquad &= 2 \\ X_5 &= 1 \end{aligned}$$

We use back-substitution to find all solutions to these systems of equations in echelon form.

(i) The second equation says that $Y = 1$. Substituting this value into the first equation, we see $X - 2(1) = 2$. We find that $X = 4$. Hence $(4, 1)$ is the only solution to the system.

(ii) This system is inconsistent. In other words, the solution set is the empty set $\varnothing$.

(iii) The third equation says that $Z = 1$. The variable W is a free variable, and we choose the parameter $W = t$. Using what we know about Z, we find that $Y - (1) - t = 0$; that is, $Y = 1 + t$. Substituting this information into the first equation, we see that $X + 2(1 + t) - 1(1) + 2t = 1$; that is, $X = -4t$. It follows that the set of solutions is $\{(-4t, 1 + t, 1, t) \mid t \in \mathbf{R}\}$.

(iv) Necessarily, $X_5 = 1$ in any solution. Assigning the free variable X_4 the parameter t and substituting $X_5 = 1$, the second equation shows that $X_3 = t + 2$. Now, if we assign the free variable X_2 the parameter u, the first equation shows that $X_1 = -u + 1$. Consequently, the set of solutions to this system is $\{(-u + 1, u, t + 2, t, 1) \mid t, u \in \mathbf{R}\}$.

Row-Echelon Matrices

We shall say that a matrix is a *row-echelon matrix* if it is the matrix of coefficients of system of equations in echelon form. Specifically, this means the following.

Definition. A matrix is a *row-echelon matrix* if

(EM 1) The first nonzero entry of each row is 1. These are called *leading entries.*

(EM 2) The leading entry of any row occurs to the right of the leading entries of any row above it.

Thus, for example, the matrix obtained at the end of the subsection on Gaussian elimination is a row-echelon matrix. The process of Gaussian elimination shows that any matrix is row equivalent to some row-echelon matrix. Here are some more row-echelon matrices:

$$\begin{pmatrix} 1 & 0 & 3 & 0 \\ 0 & 1 & 3 & 5 \\ 0 & 0 & 0 & 1 \end{pmatrix}, \quad \begin{pmatrix} 1 & 0 \\ 0 & 1 \\ 0 & 0 \end{pmatrix}, \quad \begin{pmatrix} 0 & 1 & 1 & 3 \\ 0 & 0 & 1 & 1 \end{pmatrix}.$$

Two Remarks About Parameters

(1) In this text we use lowercase letters not used before as parameters in order to avoid confusing them with variables. For example, in Example (iii) just given we noted that W was a free variable and wrote $W = t$ to demonstrate that t is a parameter that can be substituted for W and ranges over all real numbers. Many people like to use the variable itself as the parameter. If this was done for the system just mentioned, its solution set would read $\{(-4W, 1 + W, 1, W) \mid W \in \mathbf{R}\}$. This notation is fine, and you may use it if you prefer. Just be careful only to use free variables as parameters in your solution sets.

(2) Note that prior to the definition of echelon systems we said that an ordering of the variables must be specified. This is crucial, for if the variables are reordered the echelon form can change. For example,

$$\begin{aligned} X + Y + Z &= 1 \\ Y + Z &= 1 \end{aligned}$$

is in echelon form, with X and Y determined variables and Z free. However, if we reorder so that Y comes first, then

$$\begin{aligned} Y + X + Z &= 1 \\ Y \qquad + Z &= 1 \end{aligned}$$

is not in echelon form. Further note that if we reorder as

$$\begin{aligned} X + Z + Y &= 1 \\ Z + Y &= 1, \end{aligned}$$

then the system is in echelon form but now X and Z are the determined variables with Y free.

Problems

1. Identify a sequence of row operations that transforms each of the following systems.

(a) The system

$$\begin{aligned} X + 2Y + Z &= 4 \\ X + Y - Z &= 1 \end{aligned}$$

transforms to

$$\begin{aligned} -X \qquad + 3Z &= 2 \\ X + Y - Z &= 1. \end{aligned}$$

(b) The system

$$\begin{aligned} X - Y &= 4 \\ X + Y &= 0 \end{aligned}$$

transforms to

$$\begin{aligned} 3X + Y &= 4 \\ 4X + 2Y &= 4. \end{aligned}$$

2. Use Gaussian elimination to find all solutions to the following systems:

(a) $$\begin{aligned} X \qquad + Z &= 0 \\ X + Y + Z &= 3 \end{aligned}$$

(b) $$\begin{aligned} X + Y + Z &= 1 \\ X - Y \qquad &= -3 \\ 2X - Y + Z &= 0 \end{aligned}$$

(c) $$\begin{aligned} X \qquad + Z &= 0 \\ 2X + Y + Z &= 0 \\ X - Y \qquad &= 0 \end{aligned}$$

(d) $$\begin{aligned} X + Y + Z &= 1 \\ X - Y \qquad &= 3 \\ 2X - Y + Z &= 1 \\ 2Y + Z &= 2 \end{aligned}$$

3. Use Gaussian elimination to find all solutions to the following systems:

(a) $$\begin{aligned} 3X + Y + 2Z &= 7 \\ 2X + 2Y + 2Z &= 6 \\ 2X + 4Y + 3Z &= 8 \end{aligned}$$

(b) $$\begin{aligned} -X - 2Y - Z &= 1 \\ 2X - 3Y + Z &= 0 \\ 5X - Y \qquad &= -3 \\ X - Y \qquad &= 1 \end{aligned}$$

(c) $$\begin{aligned} X \qquad - 2Z + 3W &= 13 \\ -Y + Z - 2W &= -8 \\ 4X - 2Y - 6Z + 8W &= 36 \end{aligned}$$

(d) $$\begin{aligned} X + Y - \tfrac{3}{2}Z + 2W &= 0 \\ X - 2Y + Z - W &= 0 \\ 2X - Y - \tfrac{1}{2}Z + W &= 0 \end{aligned}$$

4. Explain how you can determine if the following systems of equations equivalent. Are they?

$$\begin{aligned} -X + Y &= 0 \\ 3X + 2Y &= 0 \end{aligned} \quad \text{and} \quad \begin{aligned} X - 4Y &= 0 \\ 2X - 2Y &= 0. \end{aligned}$$

5. The following transformation of equations is *not* a sequence of elementary operations. Why? Consider the system

$$\begin{aligned} X - Y + 2Z &= 2 \\ 2X + Y + 4Z &= 0. \end{aligned}$$

We subtract half of the second equation from the first and at the same time subtract twice the first equation from the second:

$$\begin{aligned} -\tfrac{3}{2}Y &= 2 \\ +3Y &= -4. \end{aligned}$$

6. Use Gaussian elimination to find a row-echelon matrix row equivalent to each of the following matrices.

(a) $\begin{pmatrix} 1 & 4 \\ 2 & 3 \end{pmatrix}$ (b) $\begin{pmatrix} 1 & 2 & 3 & 4 \\ 5 & 6 & 7 & 8 \\ 9 & 10 & 11 & 12 \end{pmatrix}$

(c) $\begin{pmatrix} 1 & 4 & 7 \\ 0 & 1 & 2 \\ 2 & 0 & -2 \\ 2 & 1 & 0 \end{pmatrix}$ (d) $\begin{pmatrix} 3 & 7 & 3 & 7 \\ 6 & 15 & 6 & 15 \end{pmatrix}$

7. Give an example of two inconsistent systems in the variables X and Y that are not row equivalent.
   (a) Can you find a 3×3 row-echelon matrix that is not row equivalent to any other row-echelon matrix? Give an example or explain why such cannot exist.
   (b) Can you find a 3×4 row-echelon matrix that is not row equivalent to any other row-echelon matrix? Give an example or explain why such cannot exist.

Group Project

(a) The third elementary row operation can, in fact, be obtained from the other two. (For this reason, it is sometimes omitted as an elementary row operation.) This is illustrated below, where R and R' denote matrix rows:

$$\begin{pmatrix} R \\ R' \end{pmatrix} \mapsto \begin{pmatrix} R - R' \\ R' \end{pmatrix} \mapsto \begin{pmatrix} R - R' \\ R' + (R - R') \end{pmatrix} = \begin{pmatrix} R - R' \\ R \end{pmatrix}$$
$$\mapsto \begin{pmatrix} -R' \\ R \end{pmatrix} \mapsto \begin{pmatrix} R' \\ R \end{pmatrix}.$$

Identify the row operations illustrated and show how to interchange the first and third rows of the matrix

$$\begin{pmatrix} 1 & 3 & 2 & 5 \\ 2 & 3 & 2 & 1 \\ 5 & 1 & 1 & 2 \end{pmatrix}.$$

(b) As remarked in this section, each of the three row operations can be reversed. That is, whenever a single row operation is applied to a system of equations, another row operation can be applied to return the new system to the initial system. Explain how this is done.

3.3 Gauss-Jordan Elimination

The Further Gauss-Jordan Strategy

Consider the following sequence of row operations. It begins like Gaussian elimination but goes further. These addition steps are part of what is called *Gauss-Jordan elimination.*

$$\begin{pmatrix} 1 & 1 & 1 & 0 & | & 1 \\ 1 & 1 & 3 & 7 & | & 0 \end{pmatrix} \mapsto \begin{pmatrix} 1 & 1 & 1 & 0 & | & 1 \\ 0 & 0 & 2 & 7 & | & -1 \end{pmatrix}$$
$$\mapsto \begin{pmatrix} 1 & 1 & 1 & 0 & | & 1 \\ 0 & 0 & 1 & \frac{7}{2} & | & -\frac{1}{2} \end{pmatrix}$$
$$\mapsto \begin{pmatrix} 1 & 1 & 0 & -\frac{7}{2} & | & \frac{3}{2} \\ 0 & 0 & 1 & \frac{7}{2} & | & -\frac{1}{2} \end{pmatrix}$$

The final augmented matrix of this sequence represents the system of equations

$$\begin{aligned} X + Y \qquad - \tfrac{7}{2}W &= \tfrac{3}{2} \\ Z + \tfrac{7}{2}W &= -\tfrac{1}{2}. \end{aligned}$$

Applying back-substitution, it readily follows that the set of solutions to this system is

$$\left\{ \left(-u + \frac{7}{2}t + \frac{3}{2}, u, -\frac{7}{2}t - \frac{1}{2}, t \right) \mid s, t \in \mathbf{R} \right\}.$$

In Gauss-Jordan elimination, in addition to applying Gaussian elimination, one also eliminates all the entries *above* the leading 1 in any column.

The Reduced Row-Echelon Form

In Sec. 3.2 we described the Gaussian elimination process in a sequence of stages. The first stage of Gauss-Jordan elimination is the same as the first stage of Gaussian elimination. In the second stage, Gauss-Jordan elimination concludes by subtracting an appropriate multiple of the second equation from the first equation to eliminate (if necessary) the second leading variable from the first equation. In the conclusion of the third stage, appropriate multiples of the third equation are subtracted from the first and second equations to eliminate the third leading variable from these equations. This process is continued so that the leading variable of each equation of the system occurs only in the single equation for which it is the leading variable.

The system of equations resulting from Gauss-Jordan elimination is said to be in *reduced echelon* form. They are characterized by the three conditions in the following definition. Again, as in the case of the echelon form, we shall assume that an ordering of the variables is specified.

Definition. A system of equations is in *reduced echelon form* if:

(REF 1) The leading variable of each equation has coefficient 1.

(REF 2) The leading variable of any equation occurs to the right of the leading variable of any equation above it.

(REF 3) The leading variable of each equation occurs in no other equation.

The advantage of finding the reduced echelon form of a system of equations is that the solution set can be found without computation in back-substitution. Note, for example, in the system considered at the beginning of this section all we had to do was assign parameters to the free variables Y and W, and the two equations in the reduced echelon form enabled us immediately to write expressions for X and Z in terms of the parameters. Of course, the computation involved in the extra steps of Gauss-Jordan elimination are essentially the computations we would have had in the back substitution.

Another important property of the reduced echelon form of a system of equations is that in the consistent case it is uniquely determined (once an ordering of the variables has been fixed). This is discussed in Chap. 4.

The system

$$\begin{aligned} X \qquad\qquad &= 2 \\ Y \qquad &= 3 \\ Z &= 4 \end{aligned}$$

is in reduced echelon form. So is the system

$$X - 2Z + W = 4$$

which consists of a single equation. However, the system

$$-Z + 2X + W = 4,$$

is *not* in echelon form, since the lead coefficient is not 1. The system

$$\begin{array}{lllll} X & & -2Z & & = \ \ 1 \\ & Y & & +3W & = \ \ 0 \\ & & Z\ - & W & = -3, \end{array}$$

is in echelon form but not in reduced echelon form.

We shall say that a matrix is a *reduced row-echelon matrix* if it is the matrix of coefficients of system of equations in reduced echelon form. The process of Gauss-Jordan elimination shows that any matrix is row equivalent to some reduced row-echelon matrix.

We give below some reduced row-echelon matrices that are row equivalent to the echelon matrices given in Sec. 3.2:

$$\begin{pmatrix} 1 & 0 & 3 & 0 \\ 0 & 1 & 3 & 0 \\ 0 & 0 & 0 & 1 \end{pmatrix}, \quad \begin{pmatrix} 1 & 0 \\ 0 & 1 \\ 0 & 0 \end{pmatrix}, \quad \begin{pmatrix} 0 & 1 & 0 & 2 \\ 0 & 0 & 1 & 1 \end{pmatrix}.$$

Ill-Conditioned Systems of Equations

Many computer programs and calculators enable you to perform row operations on matrices. If possible you should try to learn how to use such a system. This will greatly reduce the risk of computational errors! We do have one caution, however. Keep in mind that computers will round off entries and at times this can lead to a problem. For example, the simple system of equations

$$\begin{aligned} 5.3433\,X + 4.1245\,Y &= 3.1416 \\ 5.3432\,X + 4.1244\,Y &= 3.1416 \end{aligned}$$

has as reduced echelon form

$$\begin{aligned} X \qquad &= \ \ 2.5776 \\ Y &= -2.5776. \end{aligned}$$

However, if we drop one significant figure from our coefficients, we obtain

$$\begin{aligned} 5.343\,X + 4.124\,Y &= 3.142 \\ 5.343\,X + 4.124\,Y &= 3.142 \end{aligned}$$

whose reduced echelon form is

$$\begin{aligned} X + 0.771\,Y &= 0.588 \\ 0 &= 0. \end{aligned}$$

These two echelon forms are quite different! This doesn't always happen, but when it does the original system is said to be *ill-conditioned.* Unfortunately, in many practical applications of linear algebra ill-conditioned systems of equations do arise. You will explore ill-conditioned systems more in the group project at the end of this section.

Solving Matrix Equations

Suppose that A and B are matrices. It is important to know how to solve the equation $AM = B$ for the matrix M. The methods of Gaussian and Gauss-Jordan elimination are quite effective in this problem. For example, consider the matrices

$$A = \begin{pmatrix} 1 & 1 & 0 \\ 0 & 2 & 1 \end{pmatrix} \quad \text{and} \quad B = \begin{pmatrix} 3 & 5 \\ 2 & 4 \end{pmatrix}.$$

If we want to solve the matrix equation $AM = B$, we note that M must be a 3×2 matrix. Suppose that the entries of M are labeled as

$$M = \begin{pmatrix} X & R \\ Y & S \\ Z & T \end{pmatrix}$$

and we multiply

$$AM = \begin{pmatrix} X + Y & R + S \\ 2Y + Z & 2S + T \end{pmatrix}.$$

Then the expression $AM = B$ gives the two systems of equations:

$$\begin{aligned} X + Y \quad &= 3 \\ 2Y + Z &= 2 \end{aligned} \quad \text{and} \quad \begin{aligned} R + S \quad &= 5 \\ 2S + T &= 4. \end{aligned}$$

We observe that for each of these systems of equations the coefficients of the variables are the same, namely they are given by the entries of the matrix A. This means we can solve both systems at the same time using a single Gauss-Jordan elimination:

$$\left(\begin{array}{ccc|cc} 1 & 1 & 0 & 3 & 5 \\ 0 & 2 & 1 & 2 & 4 \end{array}\right) \mapsto \left(\begin{array}{ccc|cc} 1 & 1 & 0 & 3 & 5 \\ 0 & 1 & \frac{1}{2} & 1 & 2 \end{array}\right)$$
$$\mapsto \left(\begin{array}{ccc|cc} 1 & 0 & -\frac{1}{2} & 2 & 3 \\ 0 & 1 & \frac{1}{2} & 1 & 2 \end{array}\right).$$

The augmented matrix used in this elimination represents both systems of equations *at the same time*. The first column to the right of the augmentation bar represents the constants arising from the system in X, Y, and Z while the second column represents the constants arising from the system in R, S, and T.

We are now able to find solutions for our matrix M. We solve for X, Y, and Z by back-substitution using the first column on the right of the bar. For example, if $Z = 0$, we would then obtain $Y = 1$ and $X = 2$. For the second system, if $T = 0$ we would have $S = 2$ and $R = 3$. This shows that one possible solution for M is

$$M = \begin{pmatrix} 2 & 3 \\ 1 & 2 \\ 0 & 0 \end{pmatrix}.$$

In fact, there are infinitely many solutions for M. If we take u and v as parameters for Z and T, respectively, we can back-substitute to determine X, Y, R, and S in terms of these parameters. We find that the general solution for M looks like

$$M = \begin{pmatrix} \frac{v}{2} + 2 & \frac{u}{2} + 3 \\ \frac{-v}{2} + 1 & \frac{-u}{2} + 2 \\ v & u \end{pmatrix}.$$

Inverting Square Matrices

The $n \times n$ matrix with ones on the diagonal and zeros elsewhere is called the $n \times n$ *identity matrix*. It is denoted as

$$I_n = \begin{pmatrix} 1 & 0 & 0 & \cdots & 0 \\ 0 & 1 & 0 & \cdots & 0 \\ 0 & 0 & 1 & \cdots & 0 \\ \vdots & \vdots & \vdots & \ddots & \vdots \\ 0 & 0 & 0 & \cdots & 1 \end{pmatrix}$$

In other words, $I_n = (a_{ij})$, where $a_{ij} = 1$ for $i = j$ and $a_{ij} = 0$ otherwise.

Suppose that A is an $n \times n$ matrix. Then both $AI_n = A$ and $I_nA = A$. Thus the identity matrix I_n plays the same role in matrix multiplication as the real number 1 does in the multiplication of real numbers. The inverse of a nonzero real number r is the number r^{-1} with the special property that $r \cdot r^{-1} = 1$. If A is an $n \times n$ matrix, an *inverse* for A is a matrix A^{-1} for which $AA^{-1} = I_n = A^{-1}A$. A square matrix need not always have an inverse, but when the inverse exists it can be found by the process of Gauss-Jordan elimination. We conclude this section by illustrating this.

Suppose that you desire to find the inverse of the 2×2 matrix

$$A = \begin{pmatrix} 2 & 7 \\ 1 & 4 \end{pmatrix}.$$

Finding this inverse is equivalent to solving the matrix equation

$$\begin{pmatrix} 2 & 7 \\ 1 & 4 \end{pmatrix} \begin{pmatrix} x & y \\ z & w \end{pmatrix} = \begin{pmatrix} 1 & 0 \\ 0 & 1 \end{pmatrix}$$

for x, y, z, and w. The matrix equation we are trying to solve is equivalent to solving the following two equations:

$$\begin{pmatrix} 2 & 7 \\ 1 & 4 \end{pmatrix} \begin{pmatrix} x \\ z \end{pmatrix} = \begin{pmatrix} 1 \\ 0 \end{pmatrix} \quad \text{and} \quad \begin{pmatrix} 2 & 7 \\ 1 & 4 \end{pmatrix} \begin{pmatrix} y \\ w \end{pmatrix} = \begin{pmatrix} 0 \\ 1 \end{pmatrix}.$$

These two systems lead to the following augmented matrix, to which we apply Gauss-Jordan elimination:

$$\left(\begin{array}{cc|cc} 2 & 7 & 1 & 0 \\ 1 & 4 & 0 & 1 \end{array}\right) \mapsto \left(\begin{array}{cc|cc} 2 & 7 & 1 & 0 \\ 0 & \frac{1}{2} & -\frac{1}{2} & 1 \end{array}\right) \mapsto$$

$$\left(\begin{array}{cc|cc} 2 & 7 & 1 & 0 \\ 0 & 1 & -1 & 2 \end{array}\right) \mapsto \left(\begin{array}{cc|cc} 2 & 0 & 8 & -14 \\ 0 & 1 & -1 & 2 \end{array}\right)$$
$$\mapsto \left(\begin{array}{cc|cc} 1 & 0 & 4 & -7 \\ 0 & 1 & -1 & 2 \end{array}\right).$$

The solution to each of the systems of equations we are solving can be found by looking at the corresponding columns of the right-hand side of the augmented matrix. This shows that $x = 4$, $y = -7$, $z = -1$, and $w = 2$. We have found that the inverse for A is

$$A^{-1} = \begin{pmatrix} 4 & -7 \\ -1 & 2 \end{pmatrix}$$

We can readily check that

$$AA^{-1} = \begin{pmatrix} 2 & 7 \\ 1 & 4 \end{pmatrix} \begin{pmatrix} 4 & -7 \\ -1 & 2 \end{pmatrix} = \begin{pmatrix} 1 & 0 \\ 0 & 1 \end{pmatrix}.$$

We also note that

$$A^{-1}A = \begin{pmatrix} 4 & -7 \\ -1 & 2 \end{pmatrix} \begin{pmatrix} 2 & 7 \\ 1 & 4 \end{pmatrix} = \begin{pmatrix} 1 & 0 \\ 0 & 1 \end{pmatrix},$$

which shows that multiplication by G on both sides of A gives the identity.

In fact, the process just given can be applied to an arbitrary 2×2 matrix to find its inverse. If $ad - bc \neq 0$, one obtains

$$\begin{pmatrix} a & b \\ c & d \end{pmatrix}^{-1} = \frac{1}{ad - bc} \begin{pmatrix} d & -b \\ -c & a \end{pmatrix}.$$

In case $ad - bc = 0$, then the row operations show that the matrix cannot have an inverse.

Example. Here is 3×3 example. To find the inverse of

$$T = \begin{pmatrix} 1 & 0 & 1 \\ 1 & 1 & -2 \\ 0 & 1 & 1 \end{pmatrix},$$

we row-reduce

$$\left(\begin{array}{ccc|ccc} 1 & 0 & 1 & 1 & 0 & 0 \\ 1 & 1 & -2 & 0 & 1 & 0 \\ 0 & 1 & 1 & 0 & 0 & 1 \end{array}\right) \mapsto \left(\begin{array}{ccc|ccc} 1 & 0 & 1 & 1 & 0 & 0 \\ 0 & 1 & -3 & -1 & 1 & 0 \\ 0 & 1 & 1 & 0 & 0 & 1 \end{array}\right)$$
$$\mapsto \left(\begin{array}{ccc|ccc} 1 & 0 & 1 & 1 & 0 & 0 \\ 0 & 1 & -3 & -1 & 1 & 0 \\ 0 & 0 & 4 & 1 & -1 & 1 \end{array}\right) \mapsto \left(\begin{array}{ccc|ccc} 1 & 0 & 1 & 1 & 0 & 0 \\ 0 & 1 & -3 & -1 & 1 & 0 \\ 0 & 0 & 1 & \frac{1}{4} & -\frac{1}{4} & \frac{1}{4} \end{array}\right)$$
$$\mapsto \left(\begin{array}{ccc|ccc} 1 & 0 & 0 & \frac{3}{4} & \frac{1}{4} & -\frac{1}{4} \\ 0 & 1 & 0 & -\frac{1}{4} & \frac{1}{4} & \frac{3}{4} \\ 0 & 0 & 1 & \frac{1}{4} & -\frac{1}{4} & \frac{1}{4} \end{array}\right).$$

We find that the inverse of T is

$$T^{-1} = \begin{pmatrix} \frac{3}{4} & \frac{1}{4} & -\frac{1}{4} \\ -\frac{1}{4} & \frac{1}{4} & \frac{3}{4} \\ \frac{1}{4} & -\frac{1}{4} & \frac{1}{4} \end{pmatrix}.$$

A direct calculation shows that $TT^{-1} = I_3 = T^{-1}T$.

Our procedure for computing matrix inverses shows that an $n \times n$ matrix has an inverse precisely when its reduced row-echelon form is I_n. (In the next section we will see that this means the matrix must have rank n.) We record this observation in the next corollary.

Corollary. *An $n \times n$ matrix A has an inverse if and only if its reduced row-echelon form is I_n.*

You can also invert matrices on a calculator or a computer. You should learn how to do this. The only tricky point is that the computer will give decimal expressions, which means that if you want an exact expression with fractional entries you will have to do some extra work. For example, if

$$R = \begin{pmatrix} 1 & 0 & 4 \\ -6 & 0 & 7 \\ 3 & -1 & 0 \end{pmatrix},$$

then a calculator computes that

$$R^{-1} = \begin{pmatrix} .2258 & -.1290 & 0 \\ .6774 & -.3870 & -1 \\ .1935 & .0323 & 0 \end{pmatrix}.$$

If you applied row operations to compute R^{-1}, you would have obtained denominators of 31 in your expression. For most practical computations, outside of the ill-conditioned situation, the calculator result will work fine.

Using Inverses to Solve Systems of Equations

The inverse of a square matrix is useful in solving systems of equations. For example suppose we wish to solve the systems

$$A\begin{pmatrix} X \\ Y \end{pmatrix} = \begin{pmatrix} 4 \\ 6 \end{pmatrix} \quad \text{and} \quad T\begin{pmatrix} Z_1 \\ Z_2 \\ Z_3 \end{pmatrix} = \begin{pmatrix} 3 \\ 1 \\ 2 \end{pmatrix},$$

where A and T are the matrices whose inverses were found previously. If we premultiply these equations by the inverses, we find that

$$\begin{pmatrix} X \\ Y \end{pmatrix} = A^{-1}A\begin{pmatrix} X \\ Y \end{pmatrix} = A^{-1}\begin{pmatrix} 4 \\ 6 \end{pmatrix} = \begin{pmatrix} 4 & -7 \\ -1 & 2 \end{pmatrix}\begin{pmatrix} 4 \\ 6 \end{pmatrix} = \begin{pmatrix} -26 \\ 8 \end{pmatrix}$$

and

$$\begin{pmatrix} Z_1 \\ Z_2 \\ Z_3 \end{pmatrix} = \begin{pmatrix} \frac{3}{4} & \frac{1}{4} & -\frac{1}{4} \\ -\frac{1}{4} & \frac{1}{4} & \frac{3}{4} \\ \frac{1}{4} & -\frac{1}{4} & \frac{1}{4} \end{pmatrix} \begin{pmatrix} 3 \\ 1 \\ 2 \end{pmatrix} = \begin{pmatrix} 2 \\ 1 \\ 1 \end{pmatrix}.$$

This shows that the solutions are given by $X = -26$, $Y = 8$, and $Z_1 = 2$, $Z_2 = 1$, $Z_3 = 1$.

We close with a few useful remarks.

Remark 1. Observe that when we computed the inverse of the square matrix A, we really only solved the equation $AG = I_2$ for the matrix G. We subsequently checked that also $GA = I_2$ holds, to see that G was the inverse of A. It turns out that whenever S is a *square* matrix, and whenever R is another square matrix for which $SR = I$ is the identity, then $RS = I$ is necessarily also the identity. For this reason it is unnecessary to check the second equation whenever an inverse is found by row operations as above. The reasons for this are indicated in the first group project in the upcoming problems.

Remark 2. Suppose that D is a *diagonal matrix*, that is, $D = (d_{ij})$ where $d_{ij} = 0$ unless $i = j$. Then you can easily see that D is invertible if and only if each "diagonal entry" d_{ii} is nonzero. In this case, the inverse of D is readily checked to be the matrix $D^{-1} = (d'_{ij})$, where $d'_{ij} = 0$ if $i \neq j$ and $d'_{ii} = d_{ii}^{-1}$ otherwise. The reader should check this by a quick calculation.

Problems

1. Use Gauss-Jordan elimination to find all solutions to the following systems:

   (a) $\begin{aligned} X + Y - Z - W &= 1 \\ X - W &= 2 \end{aligned}$ (b) $\begin{aligned} X + 3Y - 4Z &= 0 \\ 2X + 6Y - 8Z &= 1 \end{aligned}$

2. Find all possible echelon forms for the system given in Prob. 3(a) of Sec. 3.2. Find the reduced echelon form of the system as well.

3. Explain why no system of three equations in four variables has a unique solution. (Use the existence of an equivalent reduced echelon form system.)

4. Consider the system of equations

$$\begin{aligned} X - Y + Z &= -1 \\ X + 2Y - Z &= 0 \\ X - Y + 2Z &= r. \end{aligned}$$

   For which real numbers r is this system consistent? Find all solutions in this case.

5. Find conditions on the real numbers a, b, c which guarantee that the following system has a solution:

$$\begin{aligned} Y - Z &= a \\ X \quad - Z &= b \\ X + Y - 2Z &= c. \end{aligned}$$

Explain why you might suspect that a linear equation will arise between a, b, and c even before you do any calculations. What else could happen?

6. What conditions must $a, b, c \in \mathbf{R}$ satisfy in order that the following system be consistent?

$$\begin{aligned} 3X - 2Y + 4Z - W &= a \\ -2X + 3Y - 2Z - 2W &= b \\ 5X \quad + 8Z - 7W &= c \end{aligned}$$

7. Explain why two consistent systems of two equations in two unknowns are equivalent if they have the same set of solutions.

8. Find a system of equations whose set of solutions is given by $X = 1 - 2t + u$, $Y = t$, $Z = 2 - u$, and $W = u$, where t and u are real numbers. (Hint: Think about the reduced echelon form.)

9. Consider the system of equations over $\mathbf{R}$

$$\begin{aligned} a_{11}X + a_{12}Y + a_{13}Z &= b_1 \\ a_{21}X + a_{22}Y + a_{23}Z &= b_2 \\ a_{31}X + a_{32}Y + a_{33}Z &= b_3. \end{aligned}$$

Suppose this system has two different solutions. Show it has infinitely many different solutions.

10. Find all solutions to the equation $A\vec{X} = \vec{b}$, where

$$A = \begin{pmatrix} 2 & -1 & -2 & -5 \\ -1 & 2 & -2 & 4 \\ 1 & 4 & -10 & 2 \end{pmatrix} \quad \text{and} \quad \vec{b} = \begin{pmatrix} -10 \\ -1 \\ -23 \end{pmatrix}.$$

11. Find all solutions to the equation $A\vec{X} = \vec{b}$, where

$$A = \begin{pmatrix} 9 & 8 \\ 1 & 2 \\ 2 & 3 \end{pmatrix} \quad \text{and} \quad \vec{b} = \begin{pmatrix} 2 \\ 2 \\ 2 \end{pmatrix}$$

12. Show that for any invertible square matrix A, $(A^{-1})^{-1} = A$.

13. Find the inverse, if it exists, of the following square matrices.

(a) $\begin{pmatrix} 1 & 3 \\ -1 & -3 \end{pmatrix}$ (b) $\begin{pmatrix} 2 & 5 \\ -1 & -3 \end{pmatrix}$

(c) $\begin{pmatrix} 1 & 1 & 0 \\ 2 & 0 & 1 \\ 1 & 1 & -1 \end{pmatrix}$ (d) $\begin{pmatrix} 1 & 1 & 3 \\ 0 & 1 & 4 \\ 1 & 2 & -2 \end{pmatrix}$

14. Suppose that C is invertible and $A = CBC^{-1}$. Show that A is invertible if and only if B is invertible.

15. If a square matrix A is invertible, show that A^n is invertible for all $n > 1$. What is $(A^n)^{-1}$?

16. Suppose there is a nonzero column matrix C such that $AC = \vec{0}$. Show that A cannot be invertible.

17. Show that an upper triangular matrix $A = (a_{ij})$ is invertible whenever $a_{ii} \neq 0$ for all i.

18. An $n \times n$ matrix N is called *nilpotent* if for some $k > 0$, $N^k = 0$, the $n \times n$ matrix of all zeros.
    (a) Show that a nilpotent matrix cannot be invertible.
    (b) If N is nilpotent, show that $(I_n - N)$ is invertible by verifying that its inverse is $I_n + N + N^2 + \cdots + N^{k-1}$.

19. Suppose A and B are square commuting matrices (that is, $AB = BA$). If A and B are invertible, show that A^{-1} and B^{-1} commute.

Group Project: Right and Left Inverses for Square Matrices

Give a proof that if an $n \times n$ matrix S has a right inverse T (that is, $ST = I_n$ is the identity), then T is also a left inverse for S (that is, $TS = I_n$ also). This verifies the fact mentioned in the first remark at the end of this section. Use the the hints in (a) and (b) below as a guide:

(a) First show that if S has a right inverse T, then every system of equations $S\vec{X} = \vec{b}$ has a unique solution. (Here, $\vec{X}$ is a column of n variables and $\vec{b}$ is a column of n constants.) For this, think about the row reduction that enables you to find the right inverse T.

(b) Next, let $\vec{e}_k$ be the column vector with all entries 0 except for a 1 in the kth row. Since $ST = I_n$, $S(TS\vec{e}_k) = S\vec{e}_k$. Since the system $S\vec{X} = S\vec{e}_k$ has a unique solution, we must have $TS\vec{e}_k = \vec{e}_k$. Deduce that $TS = I_n$.

(c) Find an example that shows a nonsquare matrix may have a right inverse but not a left inverse. Find another example that shows a nonsquare matrix may have a left inverse but not a right inverse.

Group Project: Ill-Conditioned Systems of Equations

When interpreting scientific data one must always be careful and take into account the accuracy of the measurements. It is *also* crucial to understand how various computations using the data can affect this accuracy. Suppose in some experiment the following vectors of data have been collected, representing three repetitions of the experiment:

$$\vec{v}_1 = \begin{pmatrix} 3.12 \\ 2.14 \\ 7.94 \end{pmatrix}, \quad \vec{v}_2 = \begin{pmatrix} 3.11 \\ 2.15 \\ 7.98 \end{pmatrix}, \quad \vec{v}_3 = \begin{pmatrix} 3.13 \\ 2.14 \\ 7.96 \end{pmatrix}.$$

You are happy with these results because the numbers seem quite close. However, in order to apply your results, you need to solve the following equations:

$$\begin{pmatrix} 2.1 & 3.1 & 4.0 \\ 2.6 & 1.0 & .70 \\ 4.1 & 2.3 & .81 \end{pmatrix} \begin{pmatrix} X \\ Y \\ Z \end{pmatrix} = \vec{v}_i \quad \text{and} \quad \begin{pmatrix} 1.1 & .70 & 2.1 \\ 2.2 & .40 & .80 \\ 5.4 & 1.5 & 3.7 \end{pmatrix} \begin{pmatrix} X \\ Y \\ Z \end{pmatrix} = \vec{v}_i$$

and obtain a 20% agreement among each collection of solutions.

(a) Use a graphing calculator or a computer to solve these three equations for the three data vectors $\vec{v}_1$, $\vec{v}_2$, and $\vec{v}_3$. Did you obtain 20% agreement? What are the implications for interpreting your experimental data?

(b) Graph the following two pairs of systems of equations (on a computer or a calculator, if possible).

$$\begin{pmatrix} 1 & .01 \\ 1 & 1 \end{pmatrix} \begin{pmatrix} X \\ Y \end{pmatrix} = \begin{pmatrix} 1 \\ 1 \end{pmatrix} \quad \text{and} \quad \begin{pmatrix} 1 & .01 \\ 1 & 1 \end{pmatrix} \begin{pmatrix} X \\ Y \end{pmatrix} = \begin{pmatrix} 1 \\ 1.1 \end{pmatrix}$$

and

$$\begin{pmatrix} 1 & 1.01 \\ 1 & 1 \end{pmatrix} \begin{pmatrix} X \\ Y \end{pmatrix} = \begin{pmatrix} 1 \\ 1 \end{pmatrix} \quad \text{and} \quad \begin{pmatrix} 1 & 1.01 \\ 1 & 1 \end{pmatrix} \begin{pmatrix} X \\ Y \end{pmatrix} = \begin{pmatrix} 1 \\ 1.1 \end{pmatrix}.$$

The first coefficient matrix of the pair is *well-conditioned*, while the second is ill-conditioned. What is the geometric difference between these graphs? Why, in geometric terms, are your solutions so different for the second pair? What do these equations look like after applying Gaussian elimination? How does that fit in?

(c) Using what you have learned in (b), give an explanation of what happened in part (a).

3.4 Matrix Rank and Systems of Linear Equations

Reduced Row-Echelon Matrices

We saw in the last section that Gauss-Jordan elimination and reduced row-echelon matrices were useful in a number of situations. Since we will often speak of a matrix and the row-echelon matrices that result from Gaussian and Gauss-Jordan elimination in the same context we have the following terminology.

Definition. Suppose that A and B are matrices and B is row equivalent to A. If B is a row-echelon matrix, we say that B is a *row-echelon form* of A. If B is a reduced row-echelon matrix, we say that B is *the reduced row-echelon form* of A.

For example, suppose

$$A = \begin{pmatrix} 1 & 3 & 8 & 0 \\ 0 & 1 & 2 & 1 \\ 0 & 1 & 2 & 4 \end{pmatrix}.$$

Applying elementary row operations, we obtain

$$A \mapsto \begin{pmatrix} 1 & 3 & 8 & 0 \\ 0 & 1 & 2 & 1 \\ 0 & 0 & 0 & 3 \end{pmatrix} \mapsto \begin{pmatrix} 1 & 3 & 8 & 0 \\ 0 & 1 & 2 & 1 \\ 0 & 0 & 0 & 1 \end{pmatrix}$$
$$\mapsto \begin{pmatrix} 1 & 3 & 8 & 0 \\ 0 & 1 & 2 & 0 \\ 0 & 0 & 0 & 1 \end{pmatrix} \mapsto \begin{pmatrix} 1 & 0 & 2 & 0 \\ 0 & 1 & 2 & 0 \\ 0 & 0 & 0 & 1 \end{pmatrix}.$$

This shows that both

$$\begin{pmatrix} 1 & 3 & 8 & 0 \\ 0 & 1 & 2 & 0 \\ 0 & 0 & 0 & 1 \end{pmatrix} \quad \text{and} \quad \begin{pmatrix} 1 & 0 & 2 & 0 \\ 0 & 1 & 2 & 0 \\ 0 & 0 & 0 & 1 \end{pmatrix}$$

are row-echelon forms of A. The latter matrix is the reduced row-echelon form of A.

Matrix Rank

It is important to note that in the previous definition we spoke of *the* reduced row-echelon form of a matrix. This language implies that there cannot be two different reduced row-echelon matrices row equivalent to any given matrix. This fact is crucial to the notion of rank. The *rank* of a matrix is defined to be the number of nonzero rows in its reduced row-echelon form. For example, the rank of the matrix A in the preceding example is 2. If a matrix could have different reduced row-echelon forms, the definition of rank would become ambiguous. Fortunately, this doesn't happen, as is summarized next.

Theorem 4. *If A is any matrix and R_1 and R_2 are both reduced row-echelon matrices that are row equivalent to A, then $R_1 = R_2$. In other words, the reduced row-echelon form of any matrix is unique.*

We will not give a detailed proof of this theorem. The main idea behind the proof is reasonably simple. Suppose that A is an $m \times n$ matrix, and that $\vec{X}$ is a column of n variables. If the matrices R_1 and R_2 are row equivalent to A, then we know that the three systems of equations $A\vec{X} = \vec{0}$, $R_1\vec{X} = \vec{0}$, and $R_2\vec{X} = \vec{0}$ each have the same set of solutions. By studying the nature of the reduced row-echelon form carefully, one can show that this will force $R_1 = R_2$.

For example, consider the two reduced echelon matrices

$$R_1 = \begin{pmatrix} 1 & -1 & 0 & 7 \\ 0 & 0 & 1 & 3 \end{pmatrix} \quad \text{and} \quad R_1 = \begin{pmatrix} 1 & -1 & 0 & 6 \\ 0 & 0 & 1 & 3 \end{pmatrix}.$$

The matrices R_1 and R_2 are close but not the same. We consider the systems of equations $R_1\vec{X} = \vec{0}$ and $R_2\vec{X} = \vec{0}$. Since the bottom rows of R_1 and R_2 are identical, the equations they represent have identical solutions. Looking at the first equation, we see that setting $X_4 = 1$ and $X_2 = 0$ necessarily means $X_1 = -7$ in the first system and $X_1 = -6$ in the second system. Since the variables X_2 and X_4 can be taken as parameters for either system, we find that the two systems of equations $R_1\vec{X} = \vec{0}$ and $R_2\vec{X} = \vec{0}$ have different solution sets. Hence R_1 and R_2 cannot be row equivalent, illustrating the theorem.

Here is the definition of the rank of a matrix. As noted previously, the definition makes sense because we can talk about *the* reduced row-echelon form of a matrix A.

Definition. The rank of any matrix A is the number of nonzero rows in the reduced row-echelon form of A. We denote the rank of A by $\text{rk}(A)$.

Observe that whenever A is an $m \times n$ matrix, $\text{rk}(A)$ is at most the smaller of m or n, for the number of nonzero rows of any echelon form of A cannot exceed the number of rows of A. Therefore, $\text{rk}(A) \leq m$. But each leading 1 in a row-echelon matrix occurs in a distinct column. Thus the number of such rows cannot exceed the number of columns of A. This shows $\text{rk}(A) \leq n$. Note that it is possible for $\text{rk}(A)$ to be strictly less than both n and m. Also, be careful and remember that $\text{rk}(A)$ is *not* the number of nonzero rows of A.

We defined the rank of a matrix to be the number of nonzero rows in its reduced row-echelon form. Note, however, that the number of nonzero rows in any row-echelon form of a matrix is the same. Therefore, in order to determine the rank of a matrix A, all you need to do is determine some row-echelon form for A. The rank of A will be number of nonzero rows in this row-echelon matrix.

Solutions to Systems and Associated Homogeneous Systems

Suppose that A is an $m \times n$ matrix and $\vec{X}$ is an $n \times 1$ column of variables. The system of equations $A\vec{X} = \vec{0}$ is called a *homogeneous system*. If B is an $m \times 1$ column matrix, we consider the system of equations $AX = B$. We call the system $A\vec{X} = \vec{0}$ the *associated homogeneous* system of equations of the system $AX = B$.

For example, the system of equations

$$\begin{pmatrix} 1 & 2 & 3 \\ 3 & 1 & 1 \end{pmatrix} \begin{pmatrix} X \\ Y \\ Z \end{pmatrix} = \begin{pmatrix} 0 \\ 1 \end{pmatrix}$$

is not homogeneous, while the system

$$\begin{pmatrix} 1 & 2 & 3 \\ 3 & 1 & 1 \end{pmatrix} \begin{pmatrix} X \\ Y \\ Z \end{pmatrix} = \begin{pmatrix} 0 \\ 0 \end{pmatrix}$$

is homogeneous. This second system is the associated homogeneous system of the first.

We can solve both of these systems by applying Gauss-Jordan elimination to their augmented matrices. For the first system we have

$$\left(\begin{array}{ccc|c} 1 & 2 & 3 & 0 \\ 3 & 1 & 1 & 1 \end{array}\right) \mapsto \left(\begin{array}{ccc|c} 1 & 2 & 3 & 0 \\ 0 & -5 & -8 & 1 \end{array}\right)$$

$$\mapsto \left(\begin{array}{ccc|c} 1 & 2 & 3 & 0 \\ 0 & 1 & \frac{8}{5} & -\frac{1}{5} \end{array}\right) \mapsto \left(\begin{array}{ccc|c} 1 & 0 & -\frac{1}{5} & \frac{2}{5} \\ 0 & 1 & \frac{8}{5} & -\frac{1}{5} \end{array}\right),$$

and we find that its solution set is $\{(\frac{2}{5} + \frac{1}{5}t, -\frac{1}{5} - \frac{8}{5}t, t) \mid t \in \mathbf{R}\}$. Similarly, for the second system we have

$$\left(\begin{array}{ccc|c} 1 & 2 & 3 & 0 \\ 3 & 1 & 1 & 0 \end{array}\right) \mapsto \left(\begin{array}{ccc|c} 1 & 2 & 3 & 0 \\ 0 & -5 & -8 & 0 \end{array}\right)$$

$$\mapsto \left(\begin{array}{ccc|c} 1 & 2 & 3 & 0 \\ 0 & 1 & \frac{8}{5} & 0 \end{array}\right) \mapsto \left(\begin{array}{ccc|c} 1 & 0 & -\frac{1}{5} & 0 \\ 0 & 1 & \frac{8}{5} & 0 \end{array}\right),$$

and we find that its solution set is $\{(\frac{1}{5}t, -\frac{8}{5}t, t) \mid t \in \mathbf{R}\}$.

We observe that the exact same row operations were used each time, the only difference being that in the second sequence the right-hand column consisted solely of zeros. Consequently, the solutions have similar expressions. Note that the homogeneous solutions were all multiples of the triple $(\frac{1}{5}, -\frac{8}{5}, 1)$ and the nonhomogeneous solutions could be obtained by adding the triple $(\frac{2}{5}, -\frac{1}{5}, 0)$ to these. This phenomenon is true more generally and is made explicit in the next two theorems.

Theorem 5. *Let $\mathcal{H}$ denote the set of solutions to a homogeneous system $A\vec{X} = \vec{0}$. Then $\mathcal{H}$ is nonempty. Further, $\mathcal{H}$ is closed under addition and scalar multiplication. This means that if $\vec{v}_1, \vec{v}_2 \in \mathcal{H}$ and r is a real number, then both $\vec{v}_1 + \vec{v}_2$, $r\vec{v}_1 \in \mathcal{H}$.*

Proof. Observe that a homogeneous system of linear equations always has a solution, namely $X_1 = 0, X_2 = 0, \ldots, X_n = 0$. This is called the *trivial* solution, and any other solution is called a *nontrivial* solution. In particular, we find that $\mathcal{H}$ is nonempty.

If $\vec{v}_1$ and $\vec{v}_2$ are solutions to $A\vec{X} = \vec{0}$, then by definition $A\vec{v}_1 = \vec{0}$ and $A\vec{v}_2 = \vec{0}$. But the distributive law of matrix multiplication shows that $A(\vec{v}_1 + \vec{v}_2) = A\vec{v}_1 + A\vec{v}_2 = \vec{0} + \vec{0} = \vec{0}$. Thus $(\vec{v}_1 + \vec{v}_2) \in \mathcal{H}$. Since scalar multiplication commutes with matrices, we have $A(r\vec{v}_1) = r(A\vec{v}_1) = r\vec{0} = \vec{0}$. This proves the theorem. □

Theorem 6. *Let $\mathcal{I}$ denote the set of solutions to the system of equations $A\vec{X} = B$ and $\mathcal{H}$ denote the set of solutions to the associated homogeneous system of equations $A\vec{X} = \vec{0}$. Suppose that $\mathcal{I}$ is nonempty, and let $\vec{u} \in \mathcal{I}$. Then every element $\vec{w} \in \mathcal{I}$ is a translation of the form $\vec{w} = \vec{u} + \vec{v}$ for some fixed vector $\vec{v} \in \mathcal{H}$.*

Proof. By assumption we know that $A\vec{u} = B$ and $A\vec{w} = B$. Setting $\vec{v} = \vec{w} - \vec{u}$, we find that $A\vec{v} = A(\vec{w} - \vec{u}) = A\vec{w} - A\vec{u} = B - B = \vec{0}$. Thus, $\vec{v} \in \mathcal{H}$ and this shows that $\vec{w} = \vec{u} + \vec{v}$ has the desired form. Conversely, suppose that $\vec{w} = \vec{u} + \vec{v}$, where $\vec{v} \in \mathcal{H}$. Then $A\vec{w} = A(\vec{u} + \vec{v}) = A\vec{u} + A\vec{v} = B + \vec{0} = B$, and $\vec{w} \in \mathcal{I}$ follows. Thus, $\mathcal{I}$ is precisely the set of n-tuples $\vec{u} + \vec{v}$ where $\vec{v} \in \mathcal{H}$. □

Observe that Theorem 6 requires that $\mathcal{I}$ be nonempty. Without this the theorem would be false. For example, consider the system of equations

$$\begin{aligned} 3X + 3Y &= 1 \\ 2X + 2Y &= 3. \end{aligned}$$

Clearly, this system has no solutions, while its associated homogeneous system has infinitely many solutions (all of the form $(t, -t)$ where t is a real number). Be careful about this.

The Relationship Between Rank and Solutions to Systems of Equations

The existence solutions to a system of linear equations can be determined by comparing the rank of the coefficient matrix, the rank of the augmented matrix, and the number of variables. This is explained next.

Theorem 7. *Suppose the system of equations $A\vec{X} = B$ has n variables. We denote by $A' = (A \mid B)$ the augmented matrix of the system. Then we have the following:*

(i) *The system $A\vec{X} = B$ has a solution if and only if* $\text{rk}(A) = \text{rk}(A')$.

(ii) *The system $A\vec{X} = B$ has a unique solution if and only if* $\text{rk}(A) = n = \text{rk}(A')$.

This theorem can be understood if one thinks about the shape of the reduced row-echelon form system in each case. For part (i), suppose that the reduced row-echelon form of $A' = (A \mid B)$ is $R' = (R \mid S)$. If $\text{rk}(A) < \text{rk}(A')$, then R' has a row of the form $(0 \quad 0 \quad \cdots \quad 0 \quad 0 \quad | \quad 1)$. This row corresponds to the equation $0X_1 + 0X_2 + \cdots + 0X_n = 1$, which does not have any solutions. Since $\text{rk}(A) \leq \text{rk}(A')$, we see that whenever $A\vec{X} = B$ has a solution, then $\text{rk}(A') = rk(A)$. Conversely, if $\text{rk}(A) = \text{rk}(A')$, then $A\vec{X} = B$ has a solution by back-substitution, which shows why (i) is true.

In order to understand part (ii), note that the system will have a unique solution precisely when its reduced echelon system has no free variables and no inconsistent equations. This can happen only in case $\text{rk}(A) = \text{rk}(A') = n$ = the number of variables.

Observe that if A is an $m \times n$ coefficient matrix of a consistent system of equations, then there are $\text{rk}(A)$ determined variables (comming from leading ones) when we solve the system using back substitution. This also shows there are $n - \text{rk}(A)$ free variables. In fact, the number $n - \text{rk}(A)$ is the smallest number of parameters that can describe the solution set, but to see that it is the "smallest" we will need some results developed in Chap. 5. We summarize this in the next corollary.

Corollary. *Suppose that A is an $m \times n$ matrix. The system $A\vec{X} = \vec{0}$ always has a solution and the collection of all solutions can be described using $n - \text{rk}(A)$ parameters. In particular, if the number of variables of a homogeneous system exceeds the number of equations, the system has a nontrivial solution. If $A\vec{X} = B$ has a solution, then the collection of all solutions can be described using $n - \text{rk}(A)$ parameters.*

Problems

1. Find the reduced row-echelon form and rank of the following matrices.

(a) $\begin{pmatrix} 2 & 1 & 3 \\ 0 & 2 & 3 \\ 2 & 3 & 6 \end{pmatrix}$ (b) $\begin{pmatrix} 1 & 0 & 0 \\ 2 & 4 & 6 \\ 5 & 8 & 12 \\ 0 & 8 & 12 \end{pmatrix}$

(c) $\begin{pmatrix} 1 & 1 & 1 & 1 \\ 3 & 3 & 3 & 3 \\ 4 & 4 & 4 & 2 \end{pmatrix}$ (d) $\begin{pmatrix} 1 & 0 & 1 \\ 0 & 1 & 0 \\ 1 & 1 & 1 \end{pmatrix}$

2. Consider the system of equations

$$\begin{pmatrix} 1 & 0 & 1 \\ 0 & 2 & -1 \\ 1 & -2 & 2 \end{pmatrix} \begin{pmatrix} X \\ Y \\ Z \end{pmatrix} = \begin{pmatrix} 0 \\ 1 \\ 0 \end{pmatrix}.$$

Find the solutions to the associated homogeneous system of equations, and relate this solution set to the collection of solutions to the original system. What does Theorem 6 say in this context?

3. Find all possible row-echelon and reduced row-echelon matrices equivalent to

$$\begin{pmatrix} 1 & 6 & 1 \\ 1 & 4 & 2 \\ 1 & 8 & 0 \end{pmatrix}.$$

4. For which real numbers a, b, and c are the following matrices row equivalent?

$$\begin{pmatrix} 2 & 0 & 1 \\ 1 & 3 & 1 \\ 5 & -3 & 2 \end{pmatrix} \quad \text{and} \quad \begin{pmatrix} 1 & a & b \\ a & 1 & c \\ b & c & a \end{pmatrix}$$

5. Consider the system of equations

$$\begin{pmatrix} 1 & 2 & 3 & 0 \\ 0 & 1 & 4 & 1 \end{pmatrix} \begin{pmatrix} X \\ Y \\ Z \\ W \end{pmatrix} = \begin{pmatrix} 6 \\ 6 \end{pmatrix}.$$

Explain what information can be obtained for this system by applying each of Theorems 5, 6, and 7. In your answer you should also give the solution set for this system as well as its associated homogeneous system.

6. (a) Suppose that both systems of equations $AX = \vec{b}$ and $AX = \vec{b'}$ have infinitely many solutions. Show that the system $AX = (\vec{b} + \vec{b'})$ has infinitely many solutions.

   (b) If $\text{rk}(A) = 1$, show that all nonzero rows of A are multiples of one another.

7. For each of the following systems of equations, first solve the associated homogeneous system and then indicate how the general solution is related to the homogeneous solution. Explain how the corollary in

this section tells you the number of parameters needed to express the solution set.

(a) $\begin{pmatrix} 1 & 0 \\ 1 & 2 \\ 1 & 3 \end{pmatrix} \begin{pmatrix} X \\ Y \end{pmatrix} = \begin{pmatrix} 1 \\ 1 \\ 1 \end{pmatrix}$

(b) $\begin{pmatrix} 2 & 2 & 0 & 0 \\ 0 & 2 & 3 & 1 \\ 1 & 1 & 2 & 1 \end{pmatrix} \begin{pmatrix} X \\ Y \\ Z \\ W \end{pmatrix} = \begin{pmatrix} 2 \\ 2 \\ 1 \end{pmatrix}$

(c) $\begin{pmatrix} 1 & 0 & 1 \\ 1 & 3 & 2 \\ 0 & 6 & 6 \end{pmatrix} \begin{pmatrix} X \\ Y \\ Z \end{pmatrix} = \begin{pmatrix} 1 \\ 1 \\ 6 \end{pmatrix}$

(d) $\begin{pmatrix} 1 & 1 & 7 \end{pmatrix} \begin{pmatrix} X \\ Y \\ Z \end{pmatrix} = (1)$

8. Find all numbers r, s, and t for which

$$\begin{pmatrix} 1 & 2 & 1 \\ 1 & 4 & 2 \\ 4 & 6 & 3 \end{pmatrix} \begin{pmatrix} X \\ Y \\ Z \end{pmatrix} = \begin{pmatrix} r \\ s \\ t \end{pmatrix}$$

has a solution.

9. Give an example of an inconsistent system of four equations in four variables whose associated homogeneous system has a solution set that requires two parameters to describe.

10. Suppose that M is a 3×3 matrix and the homogeneous system of equations $M\vec{X} = \vec{0}$ has a unique solution. Let B be a 3×1 matrix. Why do you know that the system $M\vec{X} = B$ has a solution?

11. Suppose that M is a 5×7 rank-5 matrix and let B be any 5×2 matrix. Explain why the matrix equation $MX = B$ can always be solved for X.

12. True or false? If true, give a reason, and if false give a counterexample.

    (a) If A is an $n \times m$ matrix where $n > m$, then the system of equations $A\vec{X} = \vec{0}$ has infinitely many solutions.

    (b) If A is an $n \times m$ matrix where $n > m$, then for any $\vec{b} \in \mathbf{R}^m$ the system of equations $A\vec{X} = \vec{b}$ has infinitely many solutions.

    (c) If A is a rank-n, $n \times m$ matrix, then the system of equations $A\vec{X} = \vec{0}$ has a unique solution.

    (d) If A is an $n \times n$ matrix, then the system of equations $A\vec{X} = \vec{b}$ has a unique solution.

    (e) For any matrix A, the system of equations $A\vec{X} = \vec{0}$ has a solution if and only if the system $A\vec{X} = \vec{b}$ has a solution for all $\vec{b}$.

Group Project

(a) Show that $(a_1, b_1), (a_2, b_2), \ldots, (a_n, b_n)$ are collinear in $\mathbf{R}^2$ if and only if

$$\operatorname{rk}\begin{pmatrix} 1 & a_1 & b_1 \\ 1 & a_2 & b_2 \\ \vdots & \vdots & \vdots \\ 1 & a_n & b_n \end{pmatrix} \leq 2.$$

(b) What is the appropriate generalization of part (a) to points (a_1, b_1, c_1), $(a_2, b_2, c_2), \ldots, (a_n, b_n, c_n) \in \mathbf{R}^3$?

3.5 The Simplex Algorithm

In this section we show how the technique of Gaussian elimination can be used to solve linear optimization problems with constraints. The extension of Gaussian elimination used for these problems is called the *simplex algorithm.* The simplex algorithm was developed so that computers could be used to solve optimization problems with a large number of variables. Since our goal is to get an idea of how and why this important algorithm works, we will only consider systems with two or three variables (such as those considered in Sec. 1.4).

The Feasible Region Using Equalities and Positivity

For the moment we consider two variables X and Y. The version of the simplex algorithm we study in this section is designed to solve linear optimization problems whose feasible region is described by the positivity conditions $X \geq 0$ and $Y \geq 0$, together with a collection of r linear inequalities $a_1X + b_1Y \leq d_1$, $a_2X + b_2Y \leq d_2, \ldots, a_rX + b_rY \leq d_r$. When we apply the simplex algorithm, we always start with feasible regions described in this way. Other optimization problems with different feasible regions can be solved by transforming their conditions to this form.

For example, recall the constraints on the assembly-line problem considered in Sec. 1.4:

$$T \geq 25, \; C \geq 25, \; T + C \leq 100, \; 4T + 6C \leq 500.$$

Since the variables are constrained to be greater than 25, these conditions do not have the required form. However, if we introduce new variables X and Y defined by $X = T - 25$ and $Y = C - 25$, then the conditions $T \geq 25$, $C \geq 25$ show that $X \geq 0$ and $Y \geq 0$. Transforming our other constraints gives that $T + C = X + 25 + Y + 25 \leq 100$, so $X + Y \leq 50$, and $4T + 6C = 4(X + 25) + 6(Y + 25) \leq 500$ so $4X + 6Y \leq 250$. Altogether, our original constraints, in terms of X and Y, read

$$X \geq 0, \; Y \geq 0, \; X + Y \leq 50, \; 4X + 6Y \leq 250.$$

We now work with this feasible region for X and Y, noting that we can recapture our original T and C in any problem solved for X and Y using the equations $T = X + 25$ and $C = Y + 25$.

Since we are interested in applying Gaussian elimination to study our optimization problem, we next need to replace our inequalities with equalities in order to obtain a system of equations. We can do this as long as we require all variables to represent positive numbers only. This is accomplished by introducing a so-called slack variable for each inequality. Observe, for example, that the inequality $X + Y \leq 50$ is equivalent to the equality $X + Y + S_1 = 50$ together with the positivity condition $S_1 \geq 0$. We call S_1 the *slack variable*. Using this idea for each equation, we find that our constraints

$$X \geq 0, \; Y \geq 0, \; X + Y \leq 50, \; 4X + 6Y \leq 250$$

can be replaced by the constraints

$$X \geq 0, \; Y \geq 0, \; S_1 \geq 0, \; S_2 \geq 0, \; X + Y + S_1 = 50, \; 4X + 6Y + S_2 = 250,$$

since restricting our attention to the variables X and Y only gives the same feasible region as before.

The Simplex Tableau

Suppose next that we have a linear function $F(X, Y) = aX + bY + d$ whose value is to be maximized subject to linear constraints. Introducing a new variable V and considering the equation $-aX - bY - d + V = 0$, we see that a solution to this equation, say $X = p$, $Y = q$, $V = s$ corresponds to the output value $F(p, q) = s$ of our linear function. The slack variables, together with V, reduce our problem to studying the solutions to a system of linear equations with only positivity conditions.

We illustrate this reduction by considering the linear function $P(C, T) = .6C + .8T - 25$ studied in the assembly-line problem. Recall our goal in Sec. 1.4 was to maximize $P(C, T)$, since it was a profit function. Using our change of variables given by $X = T - 25$ and $Y = C - 25$, our function $P(C, T)$ becomes $P(X, Y) = .6(X + 25) + .8(Y + 25) - 25 = .6X + .8Y + 10$. Introducing the *value variable*[1] V, we can replace our function $P(X, Y)$ by the equation $-.6X - .8Y + V = 10$. Taking this equation together with our

[1]The terminology of value variable is not common usage. Also, the appearance of V in the tableau is not standard but is included here because the author feels it helps clarify the ideas.

previously obtained constraint equations gives the system of equations

$$X + Y + S_1 = 50, \; 4X + 6Y + S_2 = 250, \; -.6X - .8Y + V = 10.$$

Any solution to this system of equations that satisfies the positivity condition $X \geq 0, Y \geq 0, S_1 \geq 0, S_2 \geq 0$ will give a value for V that is the output of our function $P(X, Y)$, where X and Y satisfy our original constraints. So our problem now is to find a positive solution to our new system of linear equations with the value of V as large as possible.

In order to find the choice of X and Y that maximizes the value of V, we consider the augmented matrix of our system, which is known as the *simplex tableau.* In order to keep track of all the information in the tableau, we list the variables in the top row and put the equation corresponding to the function to maximize on the bottom row. Our tableau in this case is the following table.

X	Y	S_1	S_2	V	
1	1	1	0	0	50
4	6	0	1	0	250
$-.6$	$-.8$	0	0	1	10

The Strategy for Maximizing V

We next look closely at the tableau just written. Any sequence of row operations on this tableau will not change the set of solutions to the system of equations it describes. How can we pick out a solution with V maximal by applying row operations to the tableau?

The key is to think about the bottom row and the equation it represents. Suppose it looked like $(0 \;\; 0 \;\; 3 \;\; 2 \;\; 1 \mid 25)$. This would correspond to the equation $3S_1 + 2S_2 + V = 25$. Then, since both $S_1 \geq 0$ and $S_2 \geq 0$, *and since the coefficients of S_1 and S_2 are both nonnegative,* we would find that the maximal V is obtained when $S_1 = 0 = S_2$, in which case $V = 25$. We could find the values of X and Y giving this V using back-substitution in the other rows of the tableau, and our optimization problem would be solved. So *our strategy is to obtain nonnegative coefficients on the bottom row of the tableau*.

Following this strategy, we now apply row operations to our initial tableau, trying to make all coefficients in the bottom row nonnegative. In each step we will choose a column containing a negative coefficient in the bottom row for elimination, and choose a row above whose multiples will be used in the reduction. We call the chosen row the *pivot row*, and we call the element in the column above the chosen negative coefficient the *pivot.*

Starting with

X	Y	S_1	S_2	V	
1	1	1	0	0	50
4	6	0	1	0	250
−.6	−.8	0	0	1	10

we use the first row as the pivot row with pivot element 1 to eliminate the coefficients of X in the second and third rows.

X	Y	S_1	S_2	V	
1	1	1	0	0	50
0	2	−4	1	0	50
0	−.2	.6	0	1	40

We next use the second row as a pivot row with pivot element 2 to eliminate the coefficients of Y in the first and third rows.

X	Y	S_1	S_2	V	
1	0	3	−.5	0	25
0	2	−4	1	0	50
0	0	.2	.1	1	45

This third tableau has a bottom row with only positive coefficients. We see, setting $S_1 = 0 = S_2$ and using back-substitution, that $Y = \frac{50}{2} = 25$ and $X = 25$ give the maximal value of $V = 45$. This is the maximum value of $.6X + .8Y$ subject to our given constraints. If we return to our original variables C and T, we find that $T = X + 25 = 50$ and $C = Y + 25 = 50$. This is the answer obtained for the problem in Sec. 1.4 using the principle of corners.

The Simplex Algorithm

Our strategy of obtaining a bottom row without negative coefficients is almost everything we need to understand the simplex algorithm. There is, however, one difficulty we must be careful about. Suppose our back-substitution gave us negative values for our variables? Then we would be in trouble since our constraints require them to be positive. We were lucky in the preceding example that this didn't happen. In order to avoid this difficulty, we must choose our row operations so that all entries above the bottom row in the right column are nonnegative. Then when we back-substitute, we will obtain nonnegative values for our variables. For this, we must choose our pivot row so that *the ratio of the right-hand entry to the pivot is nonnegative and minimal among all possible choices.* Then, as we note in the following example, the nonnegativity of the right-hand column will be maintained during our row operations.

For example, suppose we desired to maximize $6X + 8Y$ subject to the constraints

$$X \geq 0, \ Y \geq 0, \ X + Y \leq 10, \ X + 2Y \leq 16, \ 2X + Y \leq 18.$$

We set up the initial tableau.

$$\left[\begin{array}{cccccc|c} X & Y & S_1 & S_2 & S_3 & V & \\ \hline 1 & 1 & 1 & 0 & 0 & 0 & 10 \\ 1 & 2 & 0 & 1 & 0 & 0 & 16 \\ 2 & 1 & 0 & 0 & 1 & 0 & 18 \\ \hline -6 & -8 & 0 & 0 & 0 & 1 & 0 \end{array}\right]$$

We decide first to eliminate the -6 in the bottom row. Looking at the three ratios of last-to-first column entries, $10 : 1$, $16 : 1$, and $18 : 2$, we see that the third is minimal; hence the third row is our pivot row. Row operations give

$$\left[\begin{array}{cccccc|c} X & Y & S_1 & S_2 & S_3 & V & \\ \hline 0 & \frac{1}{2} & 1 & 0 & -\frac{1}{2} & 0 & 1 \\ 0 & \frac{3}{2} & 0 & 1 & -\frac{1}{2} & 0 & 7 \\ 2 & 1 & 0 & 0 & 1 & 0 & 18 \\ \hline 0 & -5 & 0 & 0 & 3 & 1 & 54 \end{array}\right]$$

We next eliminate the -5 in the bottom row using our first row as the pivot row (check that the ratio $1 : \frac{1}{2} = 2 : 1$ is smaller than the ratio $7 : \frac{3}{2} = 14 : 3$.) We find

$$\left[\begin{array}{cccccc|c} X & Y & S_1 & S_2 & S_3 & V & \\ \hline 0 & \frac{1}{2} & 1 & 0 & -\frac{1}{2} & 0 & 1 \\ 0 & 0 & -3 & 1 & 1 & 0 & 4 \\ 2 & 0 & -2 & 0 & 2 & 0 & 16 \\ \hline 0 & 0 & 10 & 0 & -2 & 1 & 64 \end{array}\right]$$

We might have hoped to be done at this point, but we still have a negative coefficient in the bottom row to eliminate. The second row is our pivot this time, since it has minimal nonnegative ratio. (We can't use the first row since a negative pivot would bring new negative entries into the bottom row.) We find

$$\left[\begin{array}{cccccc|c} X & Y & S_1 & S_2 & S_3 & V & \\ \hline 0 & \frac{1}{2} & -\frac{1}{2} & \frac{1}{2} & 0 & 0 & 3 \\ 0 & 0 & -3 & 1 & 1 & 0 & 4 \\ 2 & 0 & 4 & -2 & 0 & 0 & 8 \\ \hline 0 & 0 & 4 & 2 & 0 & 1 & 72 \end{array}\right]$$

We conclude that the maximum value of $6X + 8Y$ on our region is 72 and occurs (by back-substitution) when $X = 4$ and $Y = 6$.

This example illustrates the simplex algorithm. We summarize its key ingredients next.

The Simplex Algorithm. Suppose a linear optimization problem has constraints as formulated in the beginning of this section. Assume that a tableau for this problem has a negative coefficient in its bottom row. The simplex algorithm produces another tableau according to the following steps:

(a) Choose a negative entry in the bottom row and call its column the pivot column.

(b) The pivot row is chosen to be a row whose ratio of the last column entry to the pivot column entry is positive and smallest among all such.

(c) Using as pivot element the entry in the pivot column and pivot row, row operations are applied to reduce all other entries in the pivot column to 0.

Steps (1), (2), and (3) are repeated until the bottom row has no negative entries. The maximum value can be read from the lower right-hand corner of the final tableau. The values of the variables giving the maximum can be found by back-substitution.

The simplex algorithm can be used to solve linear optimization problems in as many variables as needed. Some three-variable examples are given in the problems next. They can be solved using the same method as in the two-variable case. The only difference is that the tableau is larger.

Problems

1. (a) Use the simplex algorithm to find the maximum value of the linear function $f(X, Y) = 2X+5Y+1$ subject to the constraints that $X \geq 0$, $Y \geq 0$, $X + Y \leq 6$, and $2X + Y \leq 8$.

   (b) Use the simplex algorithm to find the maximum value of the linear function $g(X, Y) = 5X + 2Y$ subject to the constraints that $X \geq 0$, $Y \geq 0$, $X + Y \leq 6$, $4X + Y \leq 12$, and $2X + Y \leq 8$.

2. (a) Use the simplex algorithm to find the maximum values of $Z = 3X + 4Y$ subject to the constraints that $X \geq 0$, $Y \geq 0$, $X \leq 3$, $Y \leq 3$, and $6X + 8Y \leq 30$.

   (b) The maximum value in (a) occurs at more than one vertex of the feasible region. How does the simplex algorithm account for this?

3. Use the simplex algorithm to find the maximum value of the linear function $h(X, Y) = X - 2Y + 1$ subject to the constraints that $X \geq 0$, $Y \geq 0$, $3X + Y \leq 8$, and $X + 3Y \leq 8$. Why did the simplex algorithm work so quickly?

4. (a) Use the simplex algorithm to find the maximum value of $R(X, Y, Z) = 2X + Y + 4Z$ subject to the constraints that $X \geq 0$, $Y \geq 0$, $Z \geq 0$, and $X + Y + Z \leq 4$.

   (b) Use the simplex algorithm to find the maximum value of $R(X, Y, Z) = 2X + Y + 4Z$ subject to the constraints that $X \geq 0$, $Y \geq 0$, $Z \geq 0$, $X + Y + Z \leq 4$, and $2X + 2Y + Z \leq 6$.

5. Use the simplex algorithm to find the maximum value of $T(X, Y, Z) = 6X - 9Y + 3Z$ subject to the constraints that $X \geq 0$, $Y \geq 0$, $Z \geq 0$, $X + Y + 2Z \leq 10$, $X - Z \leq 6$, and $2Y + 3Z \leq 12$.

6. (a) Solve the optimization problem considered in the previous subsection by graphing the feasible region and using the principle of corners.

   (b) Now compare your feasible region to the four tableaus constructed in the previous subsection. If you back-substituted in the initial tableau to get $V = 0$, you would have required $X = 0$ and $Y = 0$, the lower left vertex of the feasible region. Similarly, back-substitution in the second tableau to obtain $V = 54$ gives $Y = 0$ and $X = 9$, the lower right vertex of the feasible region. To which vertices do the third and fourth tableaus correspond? Explain this correspondence more carefully than is described here. Why do you think these vertices arise?

7. Find the maximum value of $32X + 28Y + 35Z$ subject to $X \geq 0$, $Y \geq 0$, $Z \geq 0$, $X + 2Y + 7Z \leq 8$, $3X + 2Y + 4Z \leq 8$, and $2X + 3Y + 5Z \leq 7$ using the principle of corners. For this find all the vertices that you need to test using a calculator. Compare this process to the simplex algorithm.

Group Project: Minimum Optimization Problems and Duality

Consider the optimization problem of trying to find the *minimum* value of $F(X, Y) = rX + sY$ subject to the constraints

$$X \geq 0, \;\; Y \geq 0, \;\; aX + bY \geq c, \;\; dX + eY \geq f,$$

where a, b, c, d, e, and f are all positive real numbers. In this project you will find out how to transform this optimization problem into one of the types considered in this section.

(a) Consider the matrices

$$A = \begin{pmatrix} a & b \\ d & e \end{pmatrix}, \quad C = (\, c \quad f\,), \; D = (\, r \quad s\,).$$

Then we can say that $\vec{v}$ lies in the feasible region for our problem if $A\vec{v} \geq C$, and that the function we choose to minimize is $D\vec{v}$ among all feasible $\vec{v}$ with $\vec{v} \geq 0$. Explain why this makes sense.

(b) Glance ahead to Sec. 4.1 and read the definition of the transpose matrix. The dual problem to the problem just considered is the problem of maximizing $C^t\vec{w}$ subject to the constraint $A^t\vec{w} \leq D^t$ with $\vec{w} \geq 0$. Write out specifically what this dual problem is.

(c) Suppose that $\vec{v}$ and $\vec{w}$ are feasible vectors for the problems considered in (a) and (b) respectively. Show that $C^t\vec{w} \leq D\vec{v}$. To do this, explain why the following reasoning works: Since $A\vec{v} \geq C$, we find that $\vec{w}^t A\vec{v} \geq \vec{w}^t C$; and since $\vec{w}^t A \leq D$, we find that $\vec{w}^t A\vec{v} \leq D\vec{v}$. We conclude that $\vec{w}^t C \leq D\vec{v}$, which, using $\vec{w}^t C = C^t\vec{w}$, is what we want.

(d) Using the inequality $C^t\vec{w} \leq D\vec{v}$, show that the problem considered in (a) and its dual problem considered in (b) have the same solution.

(e) Use the ideas presented above to find the minimum value of $F(X, Y) = 4X + Y$ subject to the constraints

$$X \geq 0, \ \ Y \geq 0, \ \ X + 2Y \geq 6, \ \ 3X + 5Y \geq 7$$

by applying the simplex algorithm to the dual problem.

BASIC MATRIX ALGEBRA

4.1 The Matrix Product: A Closer Look

Matrices are useful in studying systems of linear equations precisely because matrix multiplication is defined so that the coefficient matrix times the column of variables gives the column of constants for the system. We begin this chapter by investigating matrix multiplication from a more general point of view.

Products of Three Matrices

Suppose that a fruit dealer has a fleet of four large and six medium-sized vans that are used for deliveries. The van drivers have found that the most efficient way to pack a large van is to lay 12 large crates along the bottom and then stack 20 smaller crates on top. For a medium-sized van it is best to stack 8 large crates on the bottom and place 14 small crates on top. This fruit dealer always packs 40 pounds each of bananas and apples in his large crates (bananas are easily smashed and need strong crates) and 20 pounds each of apples and oranges in his small crates.

Suppose our problem is to find out how many pounds of apples, bananas, and oranges this fruit dealer can deliver each day. If we are counting apples, we can count in two ways. As a first strategy we could figure out the number of pounds of apples in each large van and each medium van, multiply these numbers by the numbers of large and medium vans (in this case 4 and 6), and add the result. Alternatively, we could figure out the number of large crates and the number of small crates carried by the fleet of vans, multiply these numbers by the number of pounds of apples each type of crate carries, and add these numbers to find the result. Both methods are valid strategies for counting the total pounds of apples carried.

Either of these strategies can be applied to bananas and oranges as well as apples. Let us express both of these computations in terms of matrices. Consider the matrices

$$A = \begin{pmatrix} 40 & 20 \\ 40 & 0 \\ 0 & 20 \end{pmatrix}, \quad B = \begin{pmatrix} 12 & 8 \\ 20 & 14 \end{pmatrix}, \quad C = \begin{pmatrix} 4 \\ 6 \end{pmatrix}.$$

In matrix A the first column entries are the pounds of apples, bananas, and oranges in the large crates, and the second column gives those numbers for the small crates. In matrix B the first column entries are the number of large and small crates stacked in the large vans, and the second column gives those numbers for the medium vans. Finally, matrix C is the column matrix denoting the number of large and medium vans.

Suppose we are counting according to our first strategy. Then we will compute the number of pounds of each fruit carried in each type of van. This is given by the matrix product AB, whose columns give this information:

$$AB = \begin{pmatrix} 40 & 20 \\ 40 & 0 \\ 0 & 20 \end{pmatrix} \begin{pmatrix} 12 & 8 \\ 20 & 14 \end{pmatrix} = \begin{pmatrix} 880 & 600 \\ 480 & 320 \\ 400 & 280 \end{pmatrix}.$$

For example, we see that each large van carries 880 pounds of apples and that the small vans each carry 600 pounds of apples. When we multiply this product AB by the column matrix C, we find the result of our first strategy, namely

$$(AB)C = \begin{pmatrix} 880 & 600 \\ 480 & 320 \\ 400 & 280 \end{pmatrix} \begin{pmatrix} 4 \\ 6 \end{pmatrix} = \begin{pmatrix} 7120 \\ 3840 \\ 3280 \end{pmatrix}.$$

In particular, we find a total of 7120 pounds of apples delivered daily.

Suppose we carried out the second strategy. In order to figure out the number of large and small crates carried by the entire van fleet, we would multiply the matrices B and C, producing

$$BC = \begin{pmatrix} 12 & 8 \\ 20 & 14 \end{pmatrix} \begin{pmatrix} 4 \\ 6 \end{pmatrix} = \begin{pmatrix} 96 \\ 164 \end{pmatrix}.$$

This means the fleet can carry a total of 96 large crates and 164 small crates. Multiplying by our matrix A gives us the total number of pounds the fleet can carry:

$$A(BC) = \begin{pmatrix} 40 & 20 \\ 40 & 0 \\ 0 & 20 \end{pmatrix} \begin{pmatrix} 96 \\ 164 \end{pmatrix} = \begin{pmatrix} 7120 \\ 3840 \\ 3280 \end{pmatrix}.$$

Since we noted earlier that each of these strategies should yield the same answer, we should not be surprised that the resulting matrix products agree! We have just demonstrated in this example the general algebraic fact that $(AB)C = A(BC)$, that is, *matrix multiplication is associative*. The associative law is the most important algebraic law about matrix multiplication. For emphasis we state it as the next theorem.

The Associativity of Matrix Multiplication

Theorem 8. *If A is an $m \times n$ matrix, B is an $n \times p$ matrix, and C is a $p \times q$ matrix, then $(AB)C = A(BC)$.*

The previous discussion demonstrated why this theorem is true. Next we give a purely algebraic (or symbolic) proof of this result. But we emphasize that the reason this associative law works is because one can count collections of collections in several ways, just as the fruit was counted above.

For the proof we recall the $\sum$ summation notation. We write $\sum_{i=1}^{n} a_i$ as an abbreviation for the sum of the n numbers $a_1 + a_2 + a_3 + \cdots + a_{n-1} + a_n$. When you read the next proof, look carefully at which index is used with each summation. It is either j or k. Also note that between the third and fourth lines in the string of equalities we have the adjacent summation signs $\sum_{k=1}^{p}\sum_{j=1}^{n}$ and $\sum_{j=1}^{n}\sum_{k=1}^{p}$. The proof uses the fact that these summations each describe the sum of the same pn elements, except in different orders.

Proof. Using the associative and distributive laws of the real numbers, we compute the irth entry of $(AB)C$:

$$\begin{aligned} (AB)C(i,r) &= \sum_{k=1}^{p} (AB)(i,k)C(k,r) \\ &= \sum_{k=1}^{p} \left[\sum_{j=1}^{n} A(i,j)B(j,k) \right] C(k,r) \\ &= \sum_{k=1}^{p} \sum_{j=1}^{n} A(i,j)B(j,k)C(k,r) \\ &= \sum_{j=1}^{n} \sum_{k=1}^{p} A(i,j)B(j,k)C(k,r) \end{aligned}$$

$$= \sum_{j=1}^{n} A(i,j)\left[\sum_{k=1}^{p} B(j,k)C(k,r)\right]$$

$$= \sum_{j=1}^{n} A(i,j)(BC)(j,r)$$

$$= A(BC)(i,r).$$

This proves the theorem. □

We must emphasize one more time that matrix multiplication is *not* commutative, that is, in general one does not have $AB = BA$. However, the associative law is extremely useful and can be applied to long products of matrices. For example, the associative law shows that

$$[(AB)C]D = (AB)(CD) = A[B(CD)].$$

This means that it is meaningful to write this product without parentheses or brackets as $ABCD$. In other words, we need not be concerned about the order in which the multiplications are performed as long as the order in which the matrices are listed remains the same.

Different Views of the Matrix Product

Consider the following expansion for the matrix product:

$$AB = \begin{pmatrix} 2 & 3 & 4 \\ 1 & 2 & 3 \end{pmatrix}\begin{pmatrix} 7 \\ 8 \\ 9 \end{pmatrix} = 7\begin{pmatrix} 2 \\ 1 \end{pmatrix} + 8\begin{pmatrix} 3 \\ 2 \end{pmatrix} + 9\begin{pmatrix} 4 \\ 3 \end{pmatrix} = \begin{pmatrix} 74 \\ 50 \end{pmatrix}.$$

At first glance this formulation looks strange and perhaps even wrong, because it is not the usual way you multiply matrices. What interests us is the expression following the second equal sign. It is a sum of multiples of the columns of matrix A. The numbers by which the columns of A are multiplied are the entries of matrix B. A sum of multiples of columns is called a *linear combination* of those columns. This example shows that if B is a column matrix, then the column matrix AB is a linear combination of the columns of A.

Next recall that if B has more than one column, then the matrix product AB was defined to be the matrix whose columns are the products of A with the corresponding columns of B. Combining this with the observation just made about the product of a matrix A with a column gives the following fact.

Fact. *The columns of a product matrix AB are linear combinations of the columns of A.*

We can also turn this observation around. What can we say about the rows of a product matrix AB? It turns out that the rows of a product AB are linear combinations of the rows of B. Here is an example illustrating this:

$$\begin{pmatrix} 1 & 2 & 3 \end{pmatrix}\begin{pmatrix} 1 & 2 \\ 1 & 3 \\ 0 & 4 \end{pmatrix} = \begin{pmatrix} 1\cdot 1 + 2\cdot 1 + 3\cdot 0 & 1\cdot 2 + 2\cdot 3 + 3\cdot 4 \end{pmatrix}$$

$$= 1\begin{pmatrix} 1 & 2 \end{pmatrix} + 2\begin{pmatrix} 1 & 3 \end{pmatrix} + 3\begin{pmatrix} 0 & 4 \end{pmatrix} = \begin{pmatrix} 3 & 20 \end{pmatrix}.$$

We have the following fact.

Fact. *The rows of a product matrix AB are linear combinations of the rows of B.*

Triangular and Diagonal Matrices

Two important types of matrices are the triangular and diagonal matrices.

Definition. Suppose A is an $n \times n$ (square) matrix. A is called *upper triangular* if $A(i, j) = 0$ whenever $i > j$. A is called *lower triangular* if $A(i, j) = 0$ whenever $i < j$. A is called *diagonal* if $A(i, j) = 0$ whenever $i \neq j$, that is, whenever A is both upper and lower triangular.

As the name suggests, upper triangular matrices are those matrices whose nonzero entries are contained in the "triangle" above the diagonal. For example,

$$\begin{pmatrix} 1 & 3 & 2 \\ 0 & 2 & 1 \\ 0 & 0 & 3 \end{pmatrix} \quad \text{and} \quad \begin{pmatrix} 4 & 0 & 0 \\ 0 & 0 & 0 \\ 0 & 0 & 2 \end{pmatrix}$$

are both upper triangular. The second matrix is also lower triangular, and so it is diagonal. The following properties of triangular and diagonal matrices will be useful in subsequent sections.

Theorem 9. (i) *Suppose that A and B are both upper triangular $n \times n$ matrices. Then AB is an upper triangular matrix. If A and B are both lower triangular, then so is AB.*
(ii) *If A and B are both diagonal $n \times n$ matrices, then $AB = BA$.*

The reason the theorem works is easily seen when investigating examples. For instance, consider the product of lower triangular matrices:

$$\begin{pmatrix} 3 & 0 & 0 \\ 1 & 1 & 0 \\ 0 & 2 & 1 \end{pmatrix}\begin{pmatrix} 1 & 0 & 0 \\ 0 & 1 & 0 \\ 1 & 0 & 1 \end{pmatrix} = \begin{pmatrix} 3 & 0 & 0 \\ 1 & 1 & 0 \\ 1 & 2 & 1 \end{pmatrix}.$$

The product is lower triangular because in the calculation of any upper right entry we are always adding together three zeros.

Elementary Matrices and Row Operations

It is useful to analyze elementary row operations on matrices in terms of matrix multiplication. The key concept is that of an *elementary matrix.* Each of the three basic row operations can be interpreted as an appropriate matrix multiplication.

Definition. An $n \times n$ *elementary matrix* is a matrix obtained from the $n \times n$ identity matrix I_n by applying a single elementary row operation.

We see that there are three types of elementary matrices. The 3×3 elementary matrices are displayed below. The first operation gives

$$\begin{pmatrix} k & 0 & 0 \\ 0 & 1 & 0 \\ 0 & 0 & 1 \end{pmatrix} \quad \text{or} \quad \begin{pmatrix} 1 & 0 & 0 \\ 0 & k & 0 \\ 0 & 0 & 1 \end{pmatrix} \quad \text{or} \quad \begin{pmatrix} 1 & 0 & 0 \\ 0 & 1 & 0 \\ 0 & 0 & k \end{pmatrix},$$

where k is a nonzero real number. The second operation gives, for any nonzero real number k, the six matrices

$$\begin{pmatrix} 1 & k & 0 \\ 0 & 1 & 0 \\ 0 & 0 & 1 \end{pmatrix}, \begin{pmatrix} 1 & 0 & k \\ 0 & 1 & 0 \\ 0 & 0 & 1 \end{pmatrix}, \begin{pmatrix} 1 & 0 & 0 \\ 0 & 1 & k \\ 0 & 0 & 1 \end{pmatrix},$$
$$\begin{pmatrix} 1 & 0 & 0 \\ k & 1 & 0 \\ 0 & 0 & 1 \end{pmatrix}, \begin{pmatrix} 1 & 0 & 0 \\ 0 & 1 & 0 \\ k & 0 & 1 \end{pmatrix}, \begin{pmatrix} 1 & 0 & 0 \\ 0 & 1 & 0 \\ 0 & k & 1 \end{pmatrix}.$$

Finally, the third operation gives the three matrices

$$\begin{pmatrix} 0 & 1 & 0 \\ 1 & 0 & 0 \\ 0 & 0 & 1 \end{pmatrix}, \begin{pmatrix} 0 & 0 & 1 \\ 0 & 1 & 0 \\ 1 & 0 & 0 \end{pmatrix}, \begin{pmatrix} 1 & 0 & 0 \\ 0 & 0 & 1 \\ 0 & 1 & 0 \end{pmatrix}.$$

Elementary matrices arising from this third operation are called *elementary permutation matrices.*

Elementary matrices connect row operations and matrix multiplication as is noted next.

Theorem 10. *Suppose that E is an $n \times n$ elementary matrix and A is an $n \times m$ matrix. Then the matrix EA is the matrix obtained from A by applying to A the elementary row operation that was used to obtain E from I_n.*

This result is best understood by looking at a few examples. Note for instance that the matrix multiplication

$$\begin{pmatrix} 3 & 0 \\ 0 & 1 \end{pmatrix} \begin{pmatrix} a & b & c & d \\ e & f & g & h \end{pmatrix} = \begin{pmatrix} 3a & 3b & 3c & 3d \\ e & f & g & h \end{pmatrix}$$

simply amounts to multiplying the first row of the right-hand matrix of the product by 3. This is precisely how the elementary matrix on the left was obtained; that is,

$$\begin{pmatrix} 3 & 0 \\ 0 & 1 \end{pmatrix}$$

is obtained by multiplying the first row of the 2×2 identity by 3.

As a second example,

$$\begin{pmatrix} 1 & 2 \\ 0 & 1 \end{pmatrix}$$

is the elementary matrix obtained from the identity by adding two times the second row to the first. If we perform the multiplication

$$\begin{pmatrix} 1 & 2 \\ 0 & 1 \end{pmatrix}\begin{pmatrix} 2 & 3 & 4 \\ 1 & 2 & 0 \end{pmatrix} = \begin{pmatrix} 4 & 7 & 4 \\ 1 & 2 & 0 \end{pmatrix},$$

we obtain the same result as adding two times the second row of

$$\begin{pmatrix} 2 & 3 & 4 \\ 1 & 2 & 0 \end{pmatrix}$$

to the first row.

Matrix Multiplication and Row Equivalence

In the next theorem we show how sequences of elementary operations (which up to now have been our main computational tool) can be understood as a matrix multiplication. In particular, we will be able to exploit properties of matrix algebra when studying row equivalence.

For example, consider the sequence of elementary row operations

$$\begin{pmatrix} 1 & 3 & 2 \\ 1 & 5 & 4 \end{pmatrix} \mapsto \begin{pmatrix} 1 & 3 & 2 \\ 0 & 2 & 2 \end{pmatrix}$$
$$\mapsto \begin{pmatrix} 1 & 3 & 2 \\ 0 & 1 & 1 \end{pmatrix} \mapsto \begin{pmatrix} 1 & 0 & -1 \\ 0 & 1 & 1 \end{pmatrix}.$$

The final matrix in this sequence is a reduced row-echelon matrix. According to Theorem 10, these three elementary row operations correspond to multiplication by the elementary matrices

$$\begin{pmatrix} 1 & 0 \\ -1 & 1 \end{pmatrix}, \quad \begin{pmatrix} 1 & 0 \\ 0 & \frac{1}{2} \end{pmatrix}, \quad \text{and} \quad \begin{pmatrix} 1 & -3 \\ 0 & 1 \end{pmatrix}.$$

Consequently, the following product describes the sequence of row operations just performed:

$$\begin{pmatrix} 1 & -3 \\ 0 & 1 \end{pmatrix}\begin{pmatrix} 1 & 0 \\ 0 & \frac{1}{2} \end{pmatrix}\begin{pmatrix} 1 & 0 \\ -1 & 1 \end{pmatrix}\begin{pmatrix} 1 & 3 & 2 \\ 1 & 5 & 4 \end{pmatrix} = \begin{pmatrix} 1 & 0 & -1 \\ 0 & 1 & 1 \end{pmatrix}.$$

Multiplying our three elementary matrices together shows

$$\begin{pmatrix} 1 & 0 & -1 \\ 0 & 1 & 1 \end{pmatrix} = \begin{pmatrix} \frac{5}{2} & -\frac{3}{2} \\ -\frac{1}{2} & \frac{1}{2} \end{pmatrix} \begin{pmatrix} 1 & 3 & 2 \\ 1 & 5 & 4 \end{pmatrix}.$$

In particular we see that the combined result of our sequence of three row operations is the same as multiplication on the left by a 2×2 matrix.

Theorem 11 summarizes these ideas.

Theorem 11. *Assume that A and B are row equivalent matrices. Then there exist elementary matrices $E_1, E_2, \ldots, E_s$ such that $B = E_s E_{s-1} \cdots E_1 A$. Consequently, there is an invertible matrix S such that $B = SA$.*

Proof. By hypothesis there is a sequence of elementary row operations that transform A into B. Suppose that the sequence of matrices resulting from these elementary operations is $A = A_0, A_1, A_2, \ldots, A_{s-1}, A_s = B$. Applying Theorem 10 we can find elementary matrices $E_1, E_2, \ldots, E_s$ such that $A_1 = E_1 A_0, A_2 = E_2 A_1, \ldots,$ and $B = A_s = E_s A_{s-1}$. From this we see that $B = E_s A_{s-1} = E_s(E_{s-1} A_{s-2}) = \cdots = E_s(E_{s-1}(\cdots(E_1 A)\cdots))$, as required. Applying the associativity of matrix multiplication (Theorem 8), we can omit all the parentheses. We set $S = E_s E_{s-1} \cdots E_1$, and the result follows. □

The theorem has the following corollary, which we will need later.

Corollary. *Suppose that A is an invertible $n \times n$ matrix. Then there exist elementary matrices $E_1, E_2, \ldots, E_s$ such that $A = E_s E_{s-1} \cdots E_1$.*

Proof. Since A is invertible, the results in the subsection on inverting matrices in Sec. 3.3 show that I_n and A are row equivalent. The theorem now shows that $A = E_s E_{s-1} \cdots E_1 I_n = E_s E_{s-1} \cdots E_1$ as required. □

The Transpose

Sometimes it is important to interchange the rows and columns of a matrix. We conclude this section by finding out what happens when this is done. We start with a definition.

Definition. Suppose A is an $n \times m$ matrix. The *transpose* of A, denoted A^t, is the $m \times n$ matrix with $A^t(j, i) = A(i, j)$ for all j and i, where $1 \leq j \leq m$ and $1 \leq i \leq n$. In other words, A^t is the matrix obtained from A by interchanging the rows and columns. If $A^t = A$, then A is called *symmetric.*

For example, we have

$$\begin{pmatrix} 1 & 4 \\ 1 & 3 \\ 8 & 2 \end{pmatrix}^t = \begin{pmatrix} 1 & 1 & 8 \\ 4 & 3 & 2 \end{pmatrix} \quad \text{and} \quad \begin{pmatrix} 0 & 0 & 1 \\ 0 & 1 & 0 \\ 1 & 0 & 0 \end{pmatrix}^t = \begin{pmatrix} 0 & 0 & 1 \\ 0 & 1 & 0 \\ 1 & 0 & 0 \end{pmatrix}.$$

The second of these two matrices is symmetric.

It turns out that the transpose of a product is the product of the transposes in reverse order. This fact turns out to be quite useful later.

Theorem 12. *If A is an $m \times n$ matrix and B is a $n \times p$ matrix, then $(AB)^t = (B^t)(A^t)$.*

Proof. Observe that B^t is a $p \times n$ matrix and A^t is an $n \times m$ matrix so that the product $(B^t)(A^t)$ makes sense. Let $A = (a_{ij})$ and $B = (b_{jk})$. The definition of matrix multiplication shows that the ikth entry of AB is $\sum_{j=1}^n a_{ij}b_{jk}$. Applying the transpose, this becomes the kith entry of $(AB)^t$. Next note that b_{jk} is the kjth entry of B^t and a_{ij} is the jith entry of A^t. Again using the definition of matrix multiplication, we see that the kith entry of $(B^t)(A^t)$ is $\sum_{j=1}^n (b_{jk})(a_{ij})$. The theorem follows since these two sums are identical. □

One nice application of the theorem is in constructing symmetric matrices, because if A is any matrix then taking $B = A^t$ in the theorem gives $(AA^t)^t = (A^t)^t A^t$. But clearly, $(A^t)^t = A$, so we find that $(AA^t)^t = AA^t$. This shows that AA^t is always symmetric. As an example, if

$$A = \begin{pmatrix} 2 & 3 & 4 \\ 6 & 5 & 4 \end{pmatrix}, \quad \text{then} \quad AA^t = \begin{pmatrix} 2 & 3 & 4 \\ 6 & 5 & 4 \end{pmatrix} \begin{pmatrix} 2 & 6 \\ 3 & 5 \\ 4 & 4 \end{pmatrix} = \begin{pmatrix} 29 & 43 \\ 43 & 77 \end{pmatrix}.$$

Problems

1. Are the following elementary matrices?

   (a) $\begin{pmatrix} 1 & 0 & 0 \\ 0 & 1 & 0 \end{pmatrix}$ (b) $\begin{pmatrix} 1 & 1 \\ 0 & 1 \end{pmatrix}$ (c) $\begin{pmatrix} 2 & 1 \\ 0 & 1 \end{pmatrix}$

2. Identify the sequence of elementary operations used below and then show how to obtain them by multiplication by a product of elementary matrices.

$$\begin{pmatrix} 1 & 2 & 3 \\ 3 & 6 & 8 \end{pmatrix} \mapsto \begin{pmatrix} 1 & 2 & 3 \\ 1 & 2 & 2 \end{pmatrix} \mapsto \begin{pmatrix} 0 & 0 & 1 \\ 1 & 2 & 2 \end{pmatrix}$$

3. (a) Consider the matrix

$$P = \begin{pmatrix} 0 & 0 & 1 \\ 1 & 0 & 0 \\ 0 & 1 & 0 \end{pmatrix}.$$

   For any $3 \times m$ matrix A, describe PA in terms of A.

(b) Consider the matrix

$$Q = \begin{pmatrix} 1 & 1 & 1 \\ 0 & 1 & 1 \\ 0 & 0 & 1 \end{pmatrix}.$$

For any $3 \times m$ matrix A, describe QA in terms of A.

4. Use elementary operations to transform A to B, and use your row operations to find an invertible matrix M such that $A = MB$, where A and B are as specified.

   (a) $A = \begin{pmatrix} 1 & 2 \\ 2 & 1 \end{pmatrix}$ and $B = \begin{pmatrix} 0 & 1 \\ 1 & 0 \end{pmatrix}$

   (b) $A = \begin{pmatrix} 1 & 1 & 2 \\ 0 & 2 & 1 \end{pmatrix}$ and $B = \begin{pmatrix} 1 & 1 & 2 \\ 1 & 3 & 3 \end{pmatrix}$

   (c) $A = \begin{pmatrix} 1 & 1 & 2 \\ 0 & 2 & 1 \end{pmatrix}$ and $B = \begin{pmatrix} 2 & 4 & 5 \\ 1 & 3 & 3 \end{pmatrix}$

5. Suppose that a 2×2 matrix A commutes with all other 2×2 matrices (that is, $AB = BA$ for all 2×2 matrices B). Show that

$$A = \begin{pmatrix} a & 0 \\ 0 & a \end{pmatrix} \quad \text{for some} \quad a \in \mathbf{R}.$$

6. Find six different 2×2 matrices A for which

$$A^2 = \begin{pmatrix} 1 & 0 \\ 0 & 1 \end{pmatrix}.$$

7. (a) If A and B are invertible $n \times n$ matrices, show that the product AB is invertible and that $(AB)^{-1} = B^{-1}A^{-1}$.

   (b) Suppose that A is an invertible matrix. Show that A^t is invertible and that $(A^t)^{-1} = (A^{-1})^t$.

8. Find a 2×2 matrix A for which $AA^t \neq A^tA$.

9. Suppose that the $n \times n$ matrix $T = (t_{ij})$ is upper triangular. If $T^n = 0$, show that $t_{ii} = 0$ for all i. Is the converse true?

10. Find A^t if A is

    (a) $\begin{pmatrix} 3 & 1 & 1 & 4 \\ 3 & 3 & 2 & 1 \\ 2 & 4 & 4 & 9 \end{pmatrix}$ (b) $\begin{pmatrix} 3 & 4 & 4 \\ 4 & 2 & 1 \\ 4 & 1 & 9 \end{pmatrix}$ (c) $\begin{pmatrix} 1 & 2 \\ 4 & 4 \\ 3 & 7 \end{pmatrix}$.

11. (a) If A is symmetric, show A is square.

    (b) Show that for any square matrix A, the matrix $A + A^t$ is symmetric.

    (c) Assume A and B are symmetric. Show that AB is symmetric if and only if A and B commute. (Two matrices A and B are said to *commute* if $AB = BA$.)

12. A matrix K is called *skew-symmetric* if $K^t = -K$.

(a) For any square matrix A show that $A - A^t$ is skew-symmetric.

(b) For any skew-symmetric matrix K, if $(I - K)$ is invertible show that the matrix $B = (I + K)(I - K)^{-1}$ satisfies $B^t B = I = BB^t$.

13. Express the following matrices as products of elementary matrices.

(a) $\begin{pmatrix} 1 & 4 \\ 3 & 2 \end{pmatrix}$ (b) $\begin{pmatrix} 1 & 0 & 1 \\ 1 & 1 & 0 \\ 0 & 0 & 2 \end{pmatrix}$

14. When is the product of two elementary matrices another elementary matrix?

15. What conditions characterize when an upper triangular, $n \times n$ matrix has rank n?

Group Project: Elementary Column Operations

You may have wondered if column operations are as useful as row operations. Although they are not used as frequently, there are some important uses. Here is one such application.

(a) Interpret multiplication on the right by elementary matrices as column operations.

(b) Use part (a) together with the existence of a row equivalent, reduced echelon matrix to show that for any $m \times n$ rank m matrix A, there exist invertible matrices P and Q such that $PAQ = (I_m \mid 0)$ is a "partitioned" matrix with left-hand block I_m and right-hand block the $m \times (n - m)$ matrix of zeros.

Group Project: The Chain Rule and Matrix Multiplication

This project requires some familiarity with multivariable calculus. Suppose that $f : \mathbf{R}^m \to \mathbf{R}^n$ is a differentiable function. The *Jacobian* of f is the $n \times m$ matrix of functions f' whose entries are the partial derivatives of the n component functions of F with respect to the m variables. More precisely, if we express $f(x_1, \ldots, x_m) = \big(f_1(x_1, \ldots, x_m), \ldots, f_n(x_1, \ldots, x_m)\big)$, then $f' = (f_{ij})$, where

$$f'_{ij} = \frac{\partial f_i}{\partial x_j}.$$

Recall from one-variable calculus that if $f, g : \mathbf{R} \to \mathbf{R}$ are both functions with continuous derivatives, then the derivative of the composite function $f \circ g : \mathbf{R} \to \mathbf{R}$ is given by the chain rule as $(f \circ g)'(x) = f'(g(x))g'(x)$. This exact same formula can be used to compute Jacobian matrices of composites of functions of several variables. The only difference is that multiplication of matrices is used instead of multiplication of real numbers. As a group, decipher what this means. Give some specific examples too, where m and n are small. Note that the sizes of matrices will be appropriate for multiplication in the chain rule only when the dimensions of the domains and ranges of the functions involved are such that composition makes sense. One thing to be careful about—be sure you don't forget to substitute the function g into f' before multiplying by g' when applying the chain rule!

4.2 Fibonacci Numbers and Difference Equations

Counting More Paths

We saw earlier in Sec. 1.3 that matrix arithmetic could be quite useful in the combinatorial problem of counting paths in a graph. In this example we will find a different relationship between the problem of path counting and matrix algebra. Consider the graph in Fig. 4.1. Suppose that you needed to calculate the number of paths between the points A and B, where you are allowed to travel only in the directions indicated by the arrows, and you are not concerned about the lengths of the paths.

If we try to tally the total number of paths, we will quickly discover that the list grows too rapidly. Not to say that this method won't work (it will, try it!), but it is best in this type of problem to find an organized method for counting. Observe that except for the left-hand vertices on this graph, there are precisely two paths coming into each vertex. The vertices on the top row have as possible previous vertices the two vertices in the column immediately to the left, and the vertices on the bottom row have as possible previous vertices the vertex above and the vertex to the left. This suggests that we number the vertices of the graph as indicated in Fig. 4.2.

With the vertices of the graph labeled as indicated, we see that any path through vertex n must come immediately from either vertex $n-1$ or $n-2$. In particular, if we know how many paths there are from A to the vertices labeled $n-2$ and $n-1$, then adding together these numbers gives the total number of paths to vertex n. For example, there is precisely one path to vertex 1 and only one path to vertex 2. This shows that there are $1+1=2$ paths to vertex 3. (These two paths are $A \to 1 \to 3$ and $A \to 1 \to 2 \to 3$.) Continuing this line of reasoning, we next see that there are $1+2=3$ paths to vertex 4, and $2+3=5$ paths to vertex 5. If we denote by $P(n)$ the number of paths to vertex n, our procedure gives the following result.

n	1	2	3	4	5	6	7	8	9	10	11	12
$P(n)$	1	1	2	3	5	8	13	21	34	55	89	144

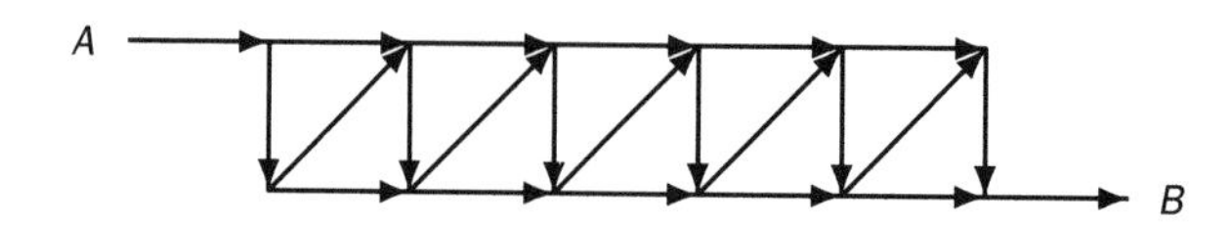

Fig. 4.1. How many paths form *A* to *B*?

Fig. 4.2. Ordering vertices for path counting

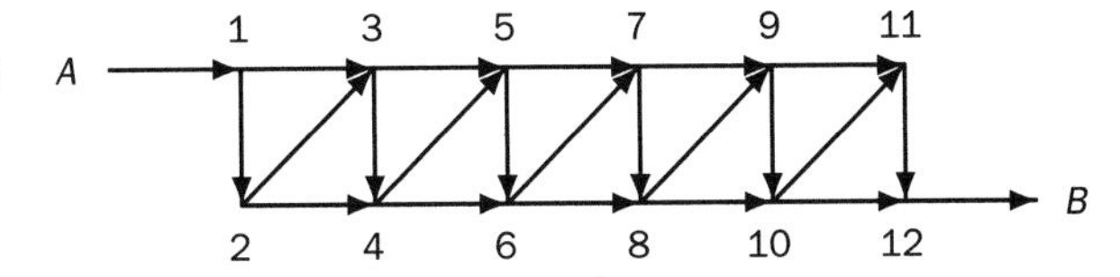

We conclude that there are 144 paths between A and B. It might have been possible to count them by making a list, but without an organized method the possibility for error is great.

The Fibonacci Numbers and Matrices

The sequence of numbers given by the function $P(n)$, $1, 1, 2, 3, 5, 8, \ldots$, is a famous sequence known as the Fibonacci sequence.[1] $P(n)$ is called the nth Fibonacci number. The facts that $P(1) = P(2) = 1$ and $P(n) = P(n-1) + P(n-2)$ for $n \geq 3$ tie the Fibonacci numbers to matrices in a fundamental way. Since each Fibonacci number is determined by its two predecessors, we consider the column vector of successive Fibonacci numbers

$$\begin{pmatrix} P(n+1) \\ P(n) \end{pmatrix}$$

and use matrix methods to analyze values of $P(n)$ for various n. The method for determining the next Fibonacci number shows

$$\begin{aligned} \begin{pmatrix} P(n+2) \\ P(n+1) \end{pmatrix} &= \begin{pmatrix} P(n+1) + P(n) \\ P(n+1) \end{pmatrix} \\ &= \begin{pmatrix} 1 & 1 \\ 1 & 0 \end{pmatrix} \begin{pmatrix} P(n+1) \\ P(n) \end{pmatrix}. \end{aligned}$$

From this we see that successive values of $P(n)$ can be obtained by repeated matrix multiplication. Using the fact that

$$\begin{pmatrix} 2 \\ 1 \end{pmatrix} = \begin{pmatrix} P(3) \\ P(2) \end{pmatrix} = \begin{pmatrix} 1 & 1 \\ 1 & 0 \end{pmatrix} \begin{pmatrix} 1 \\ 1 \end{pmatrix}$$

as our starting point, we obtain

$$\begin{aligned} \begin{pmatrix} 3 \\ 2 \end{pmatrix} = \begin{pmatrix} P(4) \\ P(3) \end{pmatrix} &= \begin{pmatrix} 1 & 1 \\ 1 & 0 \end{pmatrix} \begin{pmatrix} 2 \\ 1 \end{pmatrix} \\ &= \begin{pmatrix} 1 & 1 \\ 1 & 0 \end{pmatrix} \begin{pmatrix} 1 & 1 \\ 1 & 0 \end{pmatrix} \begin{pmatrix} 1 \\ 1 \end{pmatrix} = \begin{pmatrix} 1 & 1 \\ 1 & 0 \end{pmatrix}^2 \begin{pmatrix} 1 \\ 1 \end{pmatrix}. \end{aligned}$$

[1] Named in honor of Leonardo of Pisa (1175–1250), best known by his patronymic, Fibonacci, who first wrote down this sequence.

More generally, we find

$$\begin{pmatrix} P(n+2) \\ P(n+1) \end{pmatrix} = \begin{pmatrix} 1 & 1 \\ 1 & 0 \end{pmatrix}^n \begin{pmatrix} 1 \\ 1 \end{pmatrix}$$

for all natural numbers n.

The defining equation $P(n) = P(n-1) + P(n-2)$ is called a *difference equation*. (This terminology comes from the fact that they are closely related to differential equations.) Just as with the Fibonacci sequence, difference equations can be analyzed using repeated matrix methods. We shall see a bit later how important this type of matrix analysis is.

Difference Equations and Population Growth

A terrible flea-borne epidemic is beginning to attack the rabbit population on a small island with 12 million rabbits. Each month, one-third of the rabbits who have not had the disease contract it. Of these rabbits who contract the disease, one-fourth of them die a month later, but those that survive recover and become immune. It is further known that each month the population grows by one-twelfth, which takes into account both births and deaths due to causes other than the epidemic. We will assume that immunity to this disease is not passed on to rabbit children. Our problem is to determine what happens to the rabbit population after one year.

Observe that this problem can be solved on a month-by-month basis, where each month the numbers of uninfected, sick, and immune rabbits change. This suggests that we create the three functions $U(n)$, $S(n)$, and $I(n)$ to represent these segments of the rabbit population (in millions) after n months and then use a 3×3 matrix to relate the population changes in these categories each month. Our initial population is given by $(U(0), S(0), I(0)) = (12, 0, 0)$, which shows that before the epidemic there are 12 million uninfected rabbits. Each month, one-third of the uninfected rabbits become sick, which shows that

$$S(n+1) = \frac{1}{3}U(n).$$

Of those rabbits that were sick, one-fourth die and three-fourths join the immune rabbits. This shows that

$$I(n+1) = \frac{3}{4}S(n) + I(n).$$

Finally, the population increases by one-twelfth, which means that

$$\begin{aligned} U(n+1) &= \frac{2}{3}U(n) + \frac{1}{12}(U(n) + S(n) + I(n)) \\ &= \frac{3}{4}U(n) + \frac{1}{12}S(n) + \frac{1}{12}I(n). \end{aligned}$$

When these equations are assembled as a matrix equation we find

$$\begin{pmatrix} U(n+1) \\ S(n+1) \\ I(n+1) \end{pmatrix} = \begin{pmatrix} \frac{3}{4} & \frac{1}{12} & \frac{1}{12} \\ \frac{1}{3} & 0 & 0 \\ 0 & \frac{3}{4} & 1 \end{pmatrix} \begin{pmatrix} U(n) \\ S(n) \\ I(n) \end{pmatrix}.$$

The 3×3 matrix in this equation is called the *transition matrix* for this model. Repeatedly applying this equation shows that after n months the rabbit population vector is given by

$$\begin{pmatrix} U(n) \\ S(n) \\ I(n) \end{pmatrix} = \begin{pmatrix} \frac{3}{4} & \frac{1}{12} & \frac{1}{12} \\ \frac{1}{3} & 0 & 0 \\ 0 & \frac{3}{4} & 1 \end{pmatrix}^n \begin{pmatrix} 12 \\ 0 \\ 0 \end{pmatrix}.$$

Using a calculator with matrix multiplication capability to find the matrix power when $n = 6$, one obtains these results after 6 months:

$$\begin{pmatrix} U(6) \\ S(6) \\ I(6) \end{pmatrix} \approx \begin{pmatrix} 0.351 & 0.234 & 0.301 \\ 0.125 & 0.070 & 0.090 \\ 0.810 & 0.888 & 1.16 \end{pmatrix} \begin{pmatrix} 12 \\ 0 \\ 0 \end{pmatrix} = \begin{pmatrix} 4.22 \\ 1.50 \\ 9.72 \end{pmatrix}.$$

This population model predicts that after six months 9.72 million rabbits will be immune to the disease, 4.22 million rabbits will have not contracted the disease, and 1.50 million rabbits will be sick. Note that the rabbit population does successfully grow in spite of the deaths caused by the disease. Further calculation shows that the distribution of rabbits among the three categories at two-month intervals is the following.

$$\begin{array}{ccccccc} 0 \text{ mo.} & 2 \text{ mo.} & 4 \text{ mo.} & 6 \text{ mo.} & 8 \text{ mo.} & 10 \text{ mo.} & 12 \text{ mo.} \\ \begin{pmatrix} 12 \\ 0 \\ 0 \end{pmatrix} & \begin{pmatrix} 7.08 \\ 3.00 \\ 3.00 \end{pmatrix} & \begin{pmatrix} 4.99 \\ 1.94 \\ 7.02 \end{pmatrix} & \begin{pmatrix} 4.22 \\ 1.50 \\ 9.72 \end{pmatrix} & \begin{pmatrix} 4.09 \\ 1.36 \\ 11.9 \end{pmatrix} & \begin{pmatrix} 4.32 \\ 1.39 \\ 13.9 \end{pmatrix} & \begin{pmatrix} 4.76 \\ 1.51 \\ 16.1 \end{pmatrix} \end{array}$$

Observe that as time passes the number of immune rabbits increases more rapidly than any other group. In later sections we will consider other models where iterative matrix multiplication is used to describe changes over time.

Problems

1. Learn to use a calculator or a computer to compute matrix powers. Find A^2, A^5, A^{10}, A^{20}, and A^{30}, where A is the matrix below. In each case, (a)–(f), write down any special observations.

(a) $\begin{pmatrix} 2 & 4 & 0 \\ 5 & 0 & 2 \\ 1 & 2 & 1 \end{pmatrix}$ (b) $\begin{pmatrix} 0.3 & 0.6 \\ 0.7 & 0.4 \end{pmatrix}$ (c) $\begin{pmatrix} 1 & 0 & 1 & 0 \\ 0 & 1 & 0 & 1 \\ 1 & 0 & 1 & 0 \\ 0 & 1 & 0 & 1 \end{pmatrix}$

(d) $\begin{pmatrix} 0.0 & 0.1 & 0.3 \\ 0.1 & 2 & 0.1 \\ 0.2 & 0.6 & 0.2 \end{pmatrix}$ (e) $\begin{pmatrix} 0.1 & 0.3 & 0.1 \\ 0.4 & 0.2 & 0.6 \\ 0.5 & 0.5 & 0.3 \end{pmatrix}$ (f) $\begin{pmatrix} 1 & 2 & 3 \\ 4 & 5 & 6 \\ 7 & 8 & 9 \end{pmatrix}$

2. Using a calculator with matrix capability, compute

$$\begin{pmatrix} 1 & 1 \\ 1 & 0 \end{pmatrix}^N$$

for $N = 16$, 32, and 50. Use your matrix powers to find the Fibonacci numbers $P(17)$, $P(18)$, $P(33)$, $P(34)$, $P(51)$, and $P(52)$. Use a calculator to compute the ratios $P(17)/P(18)$, $P(33)/P(34)$, and $P(51)/P(52)$. Compute $\frac{1+\sqrt{5}}{2}$ on your calculator. What do you notice? This observation will be explained using eigenvalues in Chap. 7.

3. You deposit \$500 in a superbank that pays 1% interest each month. The bank compounds this interest monthly. How much money do you have in the bank after one month, two months, one year, 100 years? Even if you have seen a formula for this elsewhere, give an explanation of your answer that uses difference equations.

4. Bee families are structured a bit differently than those of other animals. A male bee develops from an unfertilized egg; in other words a male bee has only a mother, not a father. A female bee (the queen), on the other hand, develops from a fertilized egg and therefore has both a mother and a father. So in a bee's family tree, female parentage always branches in two, while the male lineage does not. Suppose you are studying the genetics of bees and you need to know how many ancestors a queen bee has. How would you count the number of ancestors the queen has going back 15 generations? How can difference equations and matrices help?

Group Project: How Many Trucks Where?

"Move Yurself Trucks" has rental distribution centers in San Francisco, Los Angeles, and Redding, California. They own 1000 trucks, for use by California families. Each month, half of the trucks rented in San Francisco are returned there, one-eighth are returned in Redding, and three-eighth are returned in Los Angeles. Also, each month three-fourths of the trucks rented in Redding are returned in San Francisco, while one-eighth are returned in Los Angeles, and five-eighths of the trucks in Los Angeles are returned in Los Angeles, while one-fourth are returned in San Francisco. How should Move Yurself distribute its trucks to minimize the number of trucks the company has to transport from one region to the next each month?

In writing up this project, explain the process that led you to your answer. Include any ideas you had that may have later been abandoned.

Group Project: More Paths

In each question below be sure to explain how you used difference equations and matrix algebra to find your solution.

(a) Find a graph whose number of paths from one end to the other is governed by the difference equation $P(n+2) = 2P(n+1) + P(n)$ for some ordering of the vertices.

(b) Find a difference equation that helps you find the number of paths from left to right in the following graph. Assume that paths join only at arrowheads.

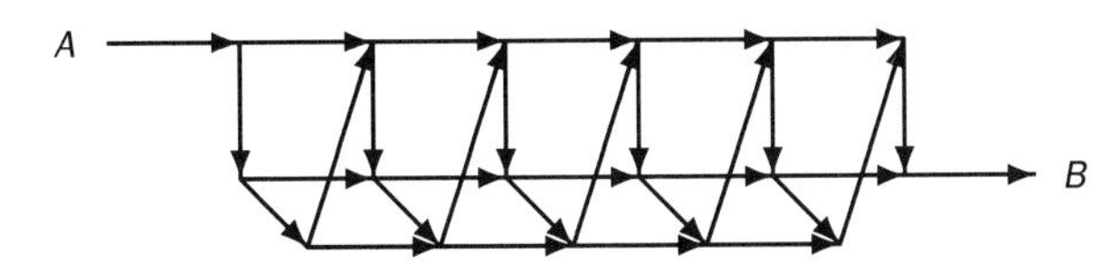

(c) Find a graph whose number of paths from one end to the other is governed by the difference equation $P(n+3) = P(n+2) + 2P(n+1) + P(n)$.

4.3 The Determinant

In this section we study the determinant of square matrices. The determinant of a square matrix A, denoted $\det(A)$, is an important number that gives some special information. We begin our look at the determinant by investigating the 2×2 and 3×3 cases using elementary operations.

The 2×2 Determinant

Some students are acquainted with the definition of the 2×2 and 3×3 determinants. Usually these small determinants are given by a formula. The $n \times n$ determinant generalizes these familiar cases. Unfortunately, one cannot understand the $n \times n$ determinant merely by manipulating the formula that defines it (as is often done in the 2×2 and 3×3 cases).

The 2×2 determinant arises when considering the invertibility question for general 2×2 matrices. For example, suppose we wish to determine if the matrix

$$A = \begin{pmatrix} a & b \\ c & d \end{pmatrix}$$

is invertible. For the sake of argument we suppose that $a \neq 0$, and we apply a row operation to A:

$$\begin{pmatrix} a & b \\ c & d \end{pmatrix} \mapsto \begin{pmatrix} a & b \\ 0 & d - \frac{bc}{a} \end{pmatrix}.$$

According to the process outlined in Sec. 3.3, the original matrix will be invertible whenever both diagonal entries of the second triangular matrix are nonzero, that is, whenever $a \neq 0$ and $d - \frac{bc}{a} \neq 0$. Since $a \neq 0$ by assumption, multiplying $d - \frac{bc}{a} \neq 0$ by a shows the matrix A is invertible precisely if $ad - bc \neq 0$. We call the expression $ad - bc$ the determinant, $\det(A)$, of matrix A. We have shown in this case that A is invertible if and only if $\det(A) \neq 0$. This calculation illustrates the origin of the 2×2 determinant.

The 3×3 Determinant

Having found the 2×2 determinant using one row operation, we turn to the 3×3 case. We carry out the following sequence of row operations:

$$B = \begin{pmatrix} a & b & c \\ d & e & f \\ g & h & i \end{pmatrix} \mapsto \begin{pmatrix} a & b & c \\ 0 & e - \frac{bd}{a} & f - \frac{cd}{a} \\ 0 & h - \frac{bg}{a} & i - \frac{cg}{a} \end{pmatrix}$$

$$\mapsto \begin{pmatrix} a & b & c \\ 0 & e - \frac{bd}{a} & f - \frac{cd}{a} \\ 0 & 0 & i - \frac{cg}{a} - (f - \frac{cd}{a})(h - \frac{bg}{a})/(e - \frac{bd}{a}) \end{pmatrix}.$$

First note that for this sequence of row operations to make sense, we need both $a \neq 0$ and $e - \frac{bd}{a} \neq 0$. With these assumptions, this calculation shows that matrix B is invertible precisely when the product of the diagonal entries of the last matrix is nonzero. We call this product the determinant of B, which after the following messy calculation is

$$\begin{aligned} \det(B) &= a\left(e - \frac{bd}{a}\right)\left[i - \frac{cg}{a} - \left(f - \frac{cd}{a}\right)\left(h - \frac{bg}{a}\right) \Big/ \left(e - \frac{bd}{a}\right)\right] \\ &= a\left(e - \frac{bd}{a}\right)\left(i - \frac{cg}{a}\right) - a\left(f - \frac{cd}{a}\right)\left(h - \frac{bg}{a}\right) \\ &= (ae - bd)i - \left(e - \frac{bd}{a}\right)cg - (af - cd)h + \left(f - \frac{cd}{a}\right)bg \\ &= (ae - bd)i - (af - cd)h + (bf - ce)g. \end{aligned}$$

This gives the formula for the 3×3 determinant. *If you didn't like this computation, don't worry!* You won't be asked to carry out this type of calculation any further. Instead we shall devote the rest of this section to a better method of understanding and computing the determinant. But do observe that we have shown in this case that B is invertible if and only if $\det(B) \neq 0$.

The Cofactor Expansion Formula for the Determinant

We next give an inductive description of the determinant known as the *cofactor expansion*.[2] The idea is to compute the determinant of an $n \times n$ matrix using determinants of smaller $(n-1) \times (n-1)$ submatrices. If we look again at the 3×3, case we see

$$\det \begin{pmatrix} a & b & c \\ d & e & f \\ g & h & i \end{pmatrix} = (ae - bd)i - (af - cd)h + (bf - ce)g$$

$$= i \cdot \det \begin{pmatrix} a & b \\ d & e \end{pmatrix} - h \cdot \det \begin{pmatrix} a & c \\ d & f \end{pmatrix} + g \cdot \det \begin{pmatrix} b & c \\ e & f \end{pmatrix},$$

which illustrates how the determinant of a 3×3 matrix is related to the determinants of three 2×2 submatrices.

This next definition is needed to generalize the idea from the 3×3 case and give the cofactor expansion of the determinant.

Definition. Let $A = (a_{ij})$ be an $n \times n$ matrix. We define $A(i \mid j)$ to be the $(n-1) \times (n-1)$ matrix obtained from A by deleting the ith row and jth column. $A(i \mid j)$ is called the ijth *maximal submatrix* of A.

For example, if

$$A = \begin{pmatrix} 3 & 1 & 2 \\ 1 & 1 & 7 \\ 2 & 7 & 1 \end{pmatrix},$$

the definition gives

$$A(1 \mid 1) = \begin{pmatrix} 1 & 7 \\ 7 & 1 \end{pmatrix}, \quad A(2 \mid 1) = \begin{pmatrix} 1 & 2 \\ 7 & 1 \end{pmatrix}, \quad \text{and} \quad A(3 \mid 1) = \begin{pmatrix} 1 & 2 \\ 1 & 7 \end{pmatrix}.$$

The cofactor expansions for the determinant give $\det(A)$ in terms of the determinants of the maximal submatrices of A taken along a specified row or column. Specifically, we shall introduce functions denoted $\mathcal{DET}_{n,i}$ and $\mathcal{DET}^n_j$, each of which can be shown to be the same function on $n \times n$ matrices. The function $\mathcal{DET}_{n,i}$ is the cofactor expansion along the ith row, and the function $\mathcal{DET}^n_j$ is the cofactor expansion along the jth column. We remark that this notation for the cofactor expansion is *not* standard and is introduced merely to emphasize the two different types of cofactor expansion. We will shortly abandon this notation and continue to use "det" to mean *the* determinant.

[2] Also known as the *Laplace expansion*.

Definition. If $A = (a)$, we define $\mathcal{DET}_{1,1}(A) = \mathcal{DET}_1^1(A) = a = \det(A)$. If $A = (a_{ij})$ is an $n \times n$ matrix with $n > 1$, for i with $1 \leq i \leq n$ we define

$$\mathcal{DET}_{n,i}(A) = \sum_{j=1}^{n} (-1)^{i+j} a_{ij} \det(A(i \mid j)),$$

which is called the ith *row cofactor expansion* of *det*. Similarly, for j with $1 \leq j \leq n$ we define

$$\mathcal{DET}_j^n(A) = \sum_{i=1}^{n} (-1)^{i+j} a_{ij} \det(A(i \mid j)),$$

which is called the jth *column expansion* of *det*.

The expression $\mathcal{DET}_{3,3}(B)$ is what we obtained earlier when we expressed the 3×3 determinant in terms of maximal 2×2 submatrices. In order to illustrate how the different cofactor expansions all compute the determinant, we give some examples.

We considered the matrix

$$A = \begin{pmatrix} 3 & 1 & 2 \\ 1 & 1 & 7 \\ 2 & 7 & 1 \end{pmatrix}$$

and found its maximal submatrices resulting from deleting the first column and the three possible rows. Computing the determinants of these maximal submatrices gives $\det(A(1 \mid 1)) = -48$, $\det(A(2 \mid 1)) = -13$, and $\det(A(3 \mid 1)) = 5$. The cofactor expansion along the first column gives

$$\mathcal{DET}_1^3(A) = 3(-48) + -1(-13) + 2(5) = -121.$$

We caution the reader that the most common error in applying cofactor expansions is failing to keep track of the $+$ and $-$ signs appropriately. The sign is $+1$ if $i + j$ (the sum of the row and column numbers) is even, and it is -1 if $i + j$ is odd.

As a second example we use the cofactor expansion to compute the determinant of a 4×4 matrix. Suppose

$$C = \begin{pmatrix} 4 & 2 & 1 & 0 \\ 0 & 3 & 2 & 0 \\ 0 & 1 & 4 & 0 \\ 2 & 1 & 0 & 1 \end{pmatrix}.$$

We will compute $\det(C)$ using a second row cofactor expansion, with further cofactor expansions of the 3×3 submatrices that arise. To save space, no summands equaling 0 will be listed:

$$\mathcal{DET}_{4,2}(C) = (-1)^{2+2} \cdot 3 \cdot \det \begin{pmatrix} 4 & 1 & 0 \\ 0 & 4 & 0 \\ 2 & 0 & 1 \end{pmatrix} + (-1)^{2+3} \cdot 2 \cdot \det \begin{pmatrix} 4 & 2 & 0 \\ 0 & 1 & 0 \\ 2 & 1 & 0 \end{pmatrix}$$

$$= 3 \left[(-1)^{2+2} \cdot 4 \cdot \det \begin{pmatrix} 4 & 0 \\ 2 & 1 \end{pmatrix} \right] - 2 \left[(-1)^{2+2} \cdot 1 \cdot \det \begin{pmatrix} 4 & 0 \\ 2 & 1 \end{pmatrix} \right]$$

$$= 3 \cdot 4 \cdot 4 - 2 \cdot 1 \cdot 4 = 40.$$

We next state the theorem that has been promised—that all the cofactor expansions agree and are called the determinant.

Theorem 13. *For any $n \times n$ matrix A all of the cofactor expansions agree. That is, for any i and j with $1 \le i, j \le n$,*

$$\mathcal{DET}_{n,i}(A) = \mathcal{DET}_j^n(A) = \det(A).$$

We will not detail the proof of this theorem here as we wish to emphasize applications. The idea behind the proof is to show that cofactor expansions have the same properties as do det with regard to row operations. Hence, any row reduction of a matrix to triangular form will show that all cofactor expansions $\mathcal{DET}_{n,i}$ or $\mathcal{DET}_j^n$ agree.

The theorem just given has some very useful corollaries. Note that a row cofactor expansion for the determinant of a matrix A is the same as the corresponding column cofactor expansion of the determinant of the transpose matrix A^t. Hence we obtain the following corollary.

Corollary. *For any square matrix A,* $\det(A) = \det(A^t)$.

We also note that if $A = (a_{ij})$ is an $n \times n$ upper triangular matrix, the cofactor expansion along the bottom row shows that $\det(A)$ is a_{nn} times the determinant of the $(n-1) \times (n-1)$ upper triangular matrix, which is obtained by deleting the bottom row and right column. Inductively this shows that the determinant of an upper triangular matrix is the product of its diagonal entries. Since a similar argument applies to lower triangular matrices, we have the next corollary.

Corollary. *If A is an triangular matrix, then* $\det(A)$ *is the product of A's diagonal entries.*

Calculating Determinants Using Gaussian Elimination

Unless there are many zeros in the matrix, cofactor expansions can be a very tiresome method for computing determinants. We next show how a determinant can be computed using row operations. The objective of this method is to row-reduce the matrix to a triangular matrix, keeping track of how the determinant changes. Since the determinant of a triangular matrix is the product of its diagonal entries, the computation will be complete.

Theorem 14. *Suppose that A is a square matrix. Then*

(i) *If B is obtained from A by multiplying a single row of A by a nonzero real number k, then* $\det(B) = k \cdot \det(A)$.

(ii) *If B is obtained from A by adding a multiple of one row to another, then* $\det(B) = \det(A)$.

(iii) *If B is obtained from A by interchanging two rows, then* $\det(B) = -\det(A)$.

(iv) *If A is triangular (upper or lower) then* $\det(A)$ *is the product of its diagonal entries.*

(v) *If A has a row or column of zeros, then* $\det(A) = 0$.

The result can be proved by induction using the cofactor expansion, and you will have a chance to explore this in the group project at the end of the section. The next examples illustrate how to use the theorem to evaluate determinants. You will get to practice this technique in the homework problems. Observe in these examples we do not compute any cofactor expansions; all we need to do is apply Gaussian elimination.

Examples. (a) Find the determinant of

$$M = \begin{pmatrix} 1 & 0 & 1 \\ 2 & 0 & 4 \\ 1 & 3 & 1 \end{pmatrix}.$$

We could apply the 3×3 formula developed above, but instead we shall use row operations and apply the theorem. We begin by adding multiples of the first row to the second and third rows. According to property (ii) of Theorem 14 this doesn't change the determinant. We find

$$\det \begin{pmatrix} 1 & 0 & 1 \\ 2 & 0 & 4 \\ 1 & 3 & 1 \end{pmatrix} = \det \begin{pmatrix} 1 & 0 & 1 \\ 0 & 0 & 2 \\ 0 & 3 & 0 \end{pmatrix}.$$

We next interchange the second and third rows to find by part (iii) of the theorem that

$$\det \begin{pmatrix} 1 & 0 & 1 \\ 0 & 0 & 2 \\ 0 & 3 & 0 \end{pmatrix} = -\det \begin{pmatrix} 1 & 0 & 1 \\ 0 & 3 & 0 \\ 0 & 0 & 2 \end{pmatrix}.$$

Finally applying part (iv) of the theorem, we obtain

$$\det\begin{pmatrix} 1 & 0 & 1 \\ 0 & 3 & 0 \\ 0 & 0 & 2 \end{pmatrix} = 6.$$

which shows that $\det(M) = -6$.

(b) Consider

$$A = \begin{pmatrix} 2 & 1 & 4 & 7 \\ 3 & 0 & 1 & 1 \\ 7 & 2 & 9 & 15 \\ 0 & 1 & 1 & 1 \end{pmatrix}.$$

We apply row operations. Eliminating the nonzero entries from the first column gives

$$B = \begin{pmatrix} 2 & 1 & 4 & 7 \\ 0 & -\frac{3}{2} & -5 & -\frac{19}{2} \\ 0 & -\frac{3}{2} & -5 & -\frac{19}{2} \\ 0 & 1 & 1 & 1 \end{pmatrix},$$

and subtracting the second row from the third row of B gives

$$C = \begin{pmatrix} 2 & 1 & 4 & 7 \\ 0 & -\frac{3}{2} & -5 & -\frac{19}{2} \\ 0 & 0 & 0 & 0 \\ 0 & 1 & 1 & 1 \end{pmatrix}.$$

Applying part (ii) of the theorem, we find that $\det(A) = \det(B) = \det(C)$. Applying part (v) of the theorem gives $\det(C) = 0$, so $\det(A) = 0$ follows.

We caution you to be sure to keep track of the changes in the determinant that arise whenever a row is multiplied by a constant. Since the second row operation is the most frequently used and does not change the determinant, it is easy to become sloppy and forget this.

Remark. Most scientific calculators are capable of computing determinants, and this has greatly reduced the need for computing determinants by row operations. Once a matrix has been entered into the calculator, the *det* command will do the rest. However, the relationship between the determinant function and row operations is still important to us because it enables us to understand the most important properties of the determinant. These results are given in the next section. In the meantime, you are encouraged to learn how to compute determinants on your calculator and computer, and you should use your machine to check your answers to the problems below.

Problems

In problems 1 and 2 compute the determinants of the following matrices in three different ways. Use a cofactor expansion, use row operations, and then try to find the quickest method by mixing these techniques. In this third approach you strive to get zeros quickly by row operations and then use cofactors to obtain the determinant of the resulting matrix. Finally, check your answers using a calculator or computer, if available.

1. (a) $\begin{pmatrix} 2 & 3 & 9 \\ 1 & 1 & 2 \\ 1 & 3 & 1 \end{pmatrix}$ (b) $\begin{pmatrix} 1 & 2 & 1 \\ 4 & 7 & 8 \\ 1 & 4 & 1 \end{pmatrix}$ (c) $\begin{pmatrix} 1 & 4 & 7 \\ 2 & 5 & 8 \\ 3 & 6 & 9 \end{pmatrix}$

   (d) $\begin{pmatrix} 3 & 0 & 2 & 3 \\ 0 & 2 & 4 & 0 \\ 1 & 0 & 1 & 1 \\ 1 & 0 & 1 & 0 \end{pmatrix}$ (e) $\begin{pmatrix} 2 & 0 & 2 & 0 \\ 0 & 1 & 0 & 1 \\ 2 & 0 & 3 & 0 \\ 0 & 1 & 0 & 2 \end{pmatrix}$ (f) $\begin{pmatrix} 2 & 1 & 0 & 0 \\ 1 & 2 & 1 & 0 \\ 0 & 1 & 2 & 1 \\ 0 & 0 & 1 & 2 \end{pmatrix}$

2. (a) $\begin{pmatrix} 1 & 5 & 0 \\ 1 & 1 & 1 \\ 1 & 2 & 1 \end{pmatrix}$ (b) $\begin{pmatrix} 1 & 2 & 1 \\ 4 & 8 & 3 \\ 5 & 10 & 1 \end{pmatrix}$ (c) $\begin{pmatrix} 3 & 2 & 1 \\ 1 & 2 & 3 \\ 3 & 3 & 3 \end{pmatrix}$

   (d) $\begin{pmatrix} 1 & 1 & 1 & 1 \\ 0 & 2 & 0 & 2 \\ 0 & 2 & 0 & 3 \\ 1 & 1 & 1 & 0 \end{pmatrix}$ (e) $\begin{pmatrix} 1 & 0 & 0 & 0 \\ 7 & 1 & 7 & 5 \\ 9 & 0 & 2 & 0 \\ 4 & 0 & 5 & 1 \end{pmatrix}$ (f) $\begin{pmatrix} 2 & 0 & 1 & 0 \\ 0 & 0 & 0 & 1 \\ 4 & 3 & 1 & 2 \\ 2 & 0 & 0 & 3 \end{pmatrix}$

3. Suppose that A is an $n \times n$ matrix with more than $n^2 - n$ entries that are 0. Show that $\det(A) = 0$.
4. If A and B are 2×2 matrices, do you think that $\det(A + B) \neq \det(A) + \det(B)$? Give a proof if you think it is true, and give a counterexample if you think it is false.
5. Verify by direct calculation that $\det(AB) = \det(A)\det(B)$ for 2×2 matrices.
6. Show that the 3×3 Vandermonde determinant

$$\det \begin{pmatrix} 1 & a & a^2 \\ 1 & b & b^2 \\ 1 & c & c^2 \end{pmatrix}$$

   is given by $(b - a)(c - a)(c - b)$. When is the Vandermonde matrix invertible?
7. A skew-symmetric matrix is a matrix A for which $A^t = -A$. Suppose that A is an $n \times n$ skew-symmetric matrix where n is odd. Show $\det(A) = 0$. Is this true when n is even?
8. Use a calculator or computer to evaluate the determinants of each of the matrices in Prob. 1 in Sec. 4.2. Try to find a relationship between the determinants of these matrices and your observations about the matrix powers.

Group Project: Determinants and Row Operations

(a) Verify each of the five assertions of Theorem 14 in the 2×2 and 3×3 cases by explicit computation, using the formulas for the determinant.

(b) Now verify each of the five assertions of Theorem 14 in the 3×3 case by using various cofactor expansions and the fact that the theorem is true in the 2×2 case. From this, explain why you believe the result is true in general.

4.4 Properties and Applications of the Determinant

In this section we detail two important properties of the determinant. The first is the fact that the nonvanishing of the determinant is equivalent to invertibility. Recall that the question of invertibility motivated our original discussion of the 2×2 and 3×3 determinants. We also show that the determinant of a product is the product of determinants of the factors. This famous result is understood by the use of elementary matrices and *not* by a long calculation. These results are followed by two applications of the cofactor expansion of the determinant. The first is a computation of the inverse matrix in terms of determinants. While not always practical from a computational point of view, this description of the inverse has important theoretical uses. The second application, Cramer's rule, describes the unique solution to $A\vec{X} = \vec{b}$ whenever A is invertible.

The Determinant and Invertibility

We recall from Sec. 3.3 that a square matrix A is invertible precisely in case A can be row-reduced to I_n. This shows, together with Theorem 14, that A is invertible if and only if $\det(A) \neq 0$. Furthermore, these row operations also show that A is invertible if and only if $rk(A) = n$. Finally, we note by Theorem 7 (ii) that the condition $rk(A) = n$ is equivalent to the homogeneous system $A\vec{X} = \vec{0}$ having only the trivial solution. All of these equivalent forms of invertibility are summarized in the following theorem.

Theorem 15. *Suppose that A is an $n \times n$ matrix. Then the following are equivalent:*

(a) $\det(A) \neq 0$;

(b) *A is invertible;*

(c) $rk(A) = n$;

(d) *The homogeneous system $A\vec{X} = \vec{0}$ has only the trivial solution.*

Here are some examples.

Examples. Compute the determinants of the following matrices and determine if they are invertible.

(a) $A = \begin{pmatrix} 2 & 3 \\ 5 & 7 \end{pmatrix}$ (b) $B = \begin{pmatrix} 1 & 2 & 0 \\ 5 & 1 & 0 \\ 3 & 3 & 1 \end{pmatrix}$ (c) $C = \begin{pmatrix} 3 & 1 & 0 & 1 \\ 2 & 0 & 0 & 0 \\ 2 & 2 & 1 & 0 \\ 0 & 1 & 0 & 1 \end{pmatrix}$

Solution. (a) $\det(A) = 2 \cdot 7 - 5 \cdot 3 = -1$, and A is therefore invertible.

(b) $\det(B) = 1 \cdot 1 \cdot 1 - 1 \cdot 0 \cdot 3 + 2 \cdot 3 \cdot 0 - 2 \cdot 5 \cdot 1 + 0 \cdot 5 \cdot 3 - 0 \cdot 1 \cdot 3 = 1 - 0 + 0 - 10 + 0 + 0 = -9$ using the formula given in the subsection on 3×3 determinants. Consequently, B is invertible.

(c) The sequence of row operations

$$\begin{pmatrix} 3 & 1 & 0 & 1 \\ 2 & 0 & 0 & 0 \\ 2 & 2 & 1 & 0 \\ 0 & 1 & 0 & 1 \end{pmatrix} \mapsto \begin{pmatrix} 0 & 1 & 0 & 1 \\ 2 & 0 & 0 & 0 \\ 2 & 2 & 1 & 0 \\ 0 & 1 & 0 & 1 \end{pmatrix} \mapsto \begin{pmatrix} 0 & 0 & 0 & 0 \\ 2 & 0 & 0 & 0 \\ 2 & 2 & 1 & 0 \\ 0 & 1 & 0 & 1 \end{pmatrix}$$

shows that $\det(C) = 0$ and that C cannot be invertible.

The Determinant of a Product

We noted earlier in this chapter that row operations can be obtained by multiplication by elementary matrices. We later determined how row operations affect the determinant. Putting these two results together gives us the following result, the verification of which is checked by direct calculation.

Lemma. *Suppose that E is an $n \times n$ elementary matrix and that A is an arbitrary $n \times n$ matrix. Then* $\det(EA) = \det(E)\det(A)$.

Next we recall from the final corollary in Sec. 4.1 that any invertible matrix can be expressed as a product of elementary matrices. The lemma can now be applied iteratively to obtain the following theorem.

Theorem 16. *If A and B are $n \times n$ matrices, then*

$$\det(AB) = \det(A)\det(B).$$

Proof. If A is invertible, then by the last corollary in Sec. 4.1, $A = E_s E_{s-1} \cdots E_1$ where $E_1, E_2, \ldots, E_s$ are elementary matrices. Applying the lemma repeatedly gives

$$\begin{aligned}
\det(AB) &= \det(E_sE_{s-1}\cdots E_1B) \\
&= \det(E_s)\det(E_{s-1}\cdots E_1B) \\
&= \cdots = \det(E_s)\det(E_{s-1})\cdots\det(E_1)\det(B) \\
&= \det(E_sE_{s-1})\det(E_{s-2})\cdots\det(E_1)\det(B) \\
&= \cdots = \det(E_sE_{s-1}\cdots E_1)\det(B) \\
&= \det(A)\det(B).
\end{aligned}$$

If A is not invertible, then by Theorem 15 $\det(A) = 0$. Suppose that C is the reduced row-echelon form of A. Then by Theorem 11 there is an invertible matrix M for which $MA = C$. As C has a row of zeros, the definition of matrix multiplication shows that CB has a row of zeros. Since $CB = (MA)B = M(AB)$, we see that AB is row equivalent to a matrix with a row of zeros. It follows that AB cannot be invertible either. Hence, $0 = \det(AB) = 0\det(B) = \det(A)\det(B)$, as desired. □

Suppose that A is an invertible matrix. Then we know that $AA^{-1} = I$ is the identity matrix. The theorem shows that $\det(A)\det(A^{-1}) = \det(AA^{-1}) = \det(I) = 1$. In other words, $\det(A^{-1})$ is the reciprocal of $\det(A)$. We emphasize this as the next corollary.

Corollary. *If A is an invertible matrix, then*

$$\det(A^{-1}) = \frac{1}{\det(A)}.$$

The Adjoint Matrix

We now turn to the applications mentioned earlier. Recall that in the subsection on inverting square matrices in Sec. 3.3 we gave a direct formula for the inverse of a 2×2 matrix. We found that if

$$A = \begin{pmatrix} a & b \\ c & d \end{pmatrix} \quad \text{then} \quad A^{-1} = \frac{1}{ad - bc}\begin{pmatrix} d & -b \\ -c & a \end{pmatrix}.$$

We recognize the denominator $ad - bc$ as the determinant of A. The matrix

$$\begin{pmatrix} d & -b \\ -c & a \end{pmatrix}$$

is a special matrix known as the *adjoint* of A, which is denoted $\mathrm{adj}(A)$. Observe that the product $A\,\mathrm{adj}(A)$ is $\det(A)I_2$. Our next task is to generalize this observation.

Definition. Let $A = (a_{ij})$ be an $n \times n$ matrix. We define the *adjoint* of A, denoted $\mathrm{adj}(A)$, to be the $n \times n$ matrix whose jith entry is $(-1)^{i+j}\det(A(i \mid j))$. In other words,

$$\mathrm{adj}(A)(j, i) = (-1)^{i+j}\det(A(i \mid j)).$$

Look closely at the uses of i and j in the definition of the adjoint. Note that it specifies the jith entry of $\mathrm{adj}(A)$ in terms of $\det(A(i \mid j))$. This reversal of i and j is *not* a typographical error and plays a key role in the proof of Theorem 17. Using the adjoint, we have the following elegant description of the inverse of an invertible matrix.

Theorem 17. *For any invertible matrix A,*

$$A^{-1} = \frac{1}{\det(A)}\mathrm{adj}(A).$$

Proof. We directly compute $A[\det(A)^{-1} \cdot \mathrm{adj}(A)]$. Since this product is $\det(A)^{-1} \cdot A \cdot \mathrm{adj}(A)$, we must show that $A \cdot \mathrm{adj}(A)$ is $\det(A)I_n$. As the ith row of A is $\begin{pmatrix} a_{i1} & a_{i2} & \cdots & a_{in} \end{pmatrix}$ and the jth column of $\mathrm{adj}(A)$ is

$$\begin{pmatrix} (-1)^{j+1}\det(A(j \mid 1)) \\ (-1)^{j+2}\det(A(j \mid 2)) \\ \vdots \\ (-1)^{j+n}\det(A(j \mid n)) \end{pmatrix},$$

we see using the definition of matrix multiplication that the ijth entry of the product $A \cdot \mathrm{adj}(A)$ is

$$\sum_{k=1}^{n} a_{ik}(-1)^{i+k}\det(A(j \mid k)).$$

Observe that if $i = j$, this is precisely the cofactor expansion $\mathcal{DET}_{n,j}(A) = \det(A)$. In case $i \neq j$, this sum is $\mathcal{DET}_{n,i}(A_i^j)$, where A_i^j is the matrix obtained from A by replacing the jth row of A with the ith row of A. Since A_i^j has two identical rows, we know $\mathcal{DET}_{n,i}(A_i^j) = \det(A_i^j) = 0$. From this we see $A \cdot \mathrm{adj}(A) = \det(A)I_n$, as required. □

For example, consider the matrix

$$B = \begin{pmatrix} 4 & 2 & 1 & 0 \\ 0 & 3 & 2 & 0 \\ 0 & 1 & 4 & 0 \\ 2 & 1 & 0 & 1 \end{pmatrix}.$$

Direct computation of the adjoint of B gives

$$\mathrm{adj}(B) = \begin{pmatrix} 10 & 7 & 1 & 0 \\ 0 & 16 & -8 & 0 \\ 0 & -4 & 12 & 0 \\ 20 & -2 & 6 & 40 \end{pmatrix}.$$

Since $\det(B) = 40$, the theorem gives

$$B^{-1} = \frac{1}{40}\begin{pmatrix} 10 & 7 & 1 & 0 \\ 0 & 16 & -8 & 0 \\ 0 & -4 & 12 & 0 \\ 20 & -2 & 6 & 40 \end{pmatrix}.$$

Cramer's Rule

The adjoint inversion formula has the following nice application.

Theorem 18 (Cramer's rule). *Assume that $A = (a_{ij})$ is an $n \times n$ invertible matrix. Let $\vec{b} = (b_i)$ be a column of n constants. We denote by $A(i)$ the $n \times n$ matrix obtained from A by replacing the ith column of A by $\vec{b}$. Then the system of equations $A\vec{X} = \vec{b}$ has the unique solution given by*

$$X_i = \frac{\det(A(i))}{\det(A)}.$$

Proof. We know that $A\vec{X} = \vec{b}$ has the solution $\vec{X} = A^{-1}\vec{b} = \det(A)^{-1}\mathrm{adj}(A)\vec{b}$. Therefore, it suffices to show that the ith row entry in the column matrix $\mathrm{adj}(A)\vec{b}$ is precisely $\det(A(i))$. The definition shows that the ith row of $\mathrm{adj}(A)$ is

$$\left((-1)^{i+1}\det(A(1 \mid i)) \quad (-1)^{i+2}\det(A(2 \mid i)) \quad \cdots \quad (-1)^{i+n}\det(A(n \mid i))\right).$$

Multiplying matrices, we see that the ith entry of $\mathrm{adj}(A)\vec{b}$ is

$$\sum_{k=1}^{n}(-1)^{i+k}\det(A(k \mid i))b_k = \sum_{k=1}^{n}(-1)^{i+k}b_k\det(A(k \mid i)).$$

But this is the cofactor expansion $\mathcal{DET}_i^n(A(i))$, which equals $\det(A(i))$. □

For example, Cramer's rule shows that the matrix equation

$$\begin{pmatrix} 1 & 2 & 4 \\ 3 & 4 & 1 \\ 2 & 1 & 0 \end{pmatrix}\begin{pmatrix} X \\ Y \\ Z \end{pmatrix} = \begin{pmatrix} 1 \\ 2 \\ 1 \end{pmatrix}$$

has the unique solution given by

$$X = \frac{\det\begin{pmatrix} 1 & 2 & 4 \\ 2 & 4 & 1 \\ 1 & 1 & 0 \end{pmatrix}}{\det\begin{pmatrix} 1 & 2 & 4 \\ 3 & 4 & 1 \\ 2 & 1 & 0 \end{pmatrix}}, \quad Y = \frac{\det\begin{pmatrix} 1 & 1 & 4 \\ 3 & 2 & 1 \\ 2 & 1 & 0 \end{pmatrix}}{\det\begin{pmatrix} 1 & 2 & 4 \\ 3 & 4 & 1 \\ 2 & 1 & 0 \end{pmatrix}}, \quad Z = \frac{\det\begin{pmatrix} 1 & 2 & 1 \\ 3 & 4 & 2 \\ 2 & 1 & 1 \end{pmatrix}}{\det\begin{pmatrix} 1 & 2 & 4 \\ 3 & 4 & 1 \\ 2 & 1 & 0 \end{pmatrix}}.$$

We find that $X = \frac{-7}{-17} = \frac{7}{17}$, $Y = \frac{-3}{-17} = \frac{3}{17}$, and $Z = \frac{-1}{-17} = \frac{1}{17}$.

Problems

1. Use the adjoint formula to invert the following matrices. Check your results using a calculator or computer, if available. Otherwise, check your answer by multiplying to see that you get the identity.

   (a) $\begin{pmatrix} -1 & 0 & -1 \\ -3 & 1 & 1 \\ 0 & 2 & -1 \end{pmatrix}$ (b) $\begin{pmatrix} 1 & -1 & 1 \\ 2 & 0 & 1 \\ 3 & 4 & -2 \end{pmatrix}$

2. Use Cramer's rule to solve the following systems of equations. Be sure to check your answers by plugging them back into the system.

   (a) $\begin{pmatrix} -1 & 0 & 1 \\ 0 & 1 & 1 \\ 0 & 1 & -1 \end{pmatrix}\begin{pmatrix} X \\ Y \\ Z \end{pmatrix} = \begin{pmatrix} 3 \\ 1 \\ 4 \end{pmatrix}$

   (b) $\begin{pmatrix} 1 & -1 & 1 \\ 2 & 0 & 1 \\ 3 & 4 & -2 \end{pmatrix}\begin{pmatrix} X \\ Y \\ Z \end{pmatrix} = \begin{pmatrix} 1 \\ 1 \\ 1 \end{pmatrix}$

3. Compute the determinant of the following matrices by any manner you like.

   (a) $\begin{pmatrix} 2 & 3 & 4 & 1 \\ 1 & 3 & 2 & 3 \\ 0 & 0 & 2 & 4 \\ 0 & 0 & 4 & 9 \end{pmatrix}$ (b) $\begin{pmatrix} 1 & 2 & 3 & 0 \\ 5 & 6 & 7 & 0 \\ 9 & 10 & 11 & 0 \\ 13 & 14 & 15 & 16 \end{pmatrix}$

   (c) $\begin{pmatrix} 0 & 2 & -3 & 0 & 0 \\ 1 & 2 & 3 & 0 & 0 \\ 0 & 3 & 2 & 0 & 0 \\ 3 & 6 & 5 & -9 & 2 \\ 2 & 8 & 0 & 1 & 1 \end{pmatrix}$ (d) $\begin{pmatrix} 0 & 0 & 1 & 1 \\ 0 & 0 & 2 & 1 \\ 1 & 1 & 0 & 0 \\ 1 & 2 & 0 & 0 \end{pmatrix}$

4. Two matrices A and B are said to *anticommute* if $AB = -BA$. Suppose that n is odd and A and B are $n \times n$ real matrices that anticommute. Show that one of A or B is not invertible. Show this can fail if n is even.

5. Suppose that $\det(A^n) = 1$ for some natural number n. What can you say about $\det(A)$?

6. (a) Show that for any two $n \times n$ matrices A and B, $\det(AB) = \det(BA)$.
   (b) If P is an invertible $n \times n$ matrix and if A is another $n \times n$ matrix, show that $\det(PAP^{-1}) = \det(A)$.

Group Project: Properties of the Adjoint

(a) Suppose that A is an $n \times n$ matrix and $rk(A) < n - 1$. Show that adj(A) is the matrix of all zeros. (Hint: The rank of any submatrix of A is also less than $n - 1$.)

(b) If A is an $n \times n$ matrix and $rk(A) = n - 1$, show that $rk(\text{adj}(A)) = 1$.

(c) Show that if A is an $n \times n$ matrix, then $\det(\text{adj}(A)) = \det(A)^{n-1}$.

(d) Show that for any $n \times n$ matrix A, $\text{adj}(A)^t = \text{adj}(A^t)$.

(e) Show that for any $n \times n$ matrix A, $\text{adj}(\text{adj}(A)) = \det(A)^{n-2}A$.

Group Project: Partitioned Matrices

If A is an $n \times m$ matrix, it can be partitioned into smaller matrices. For example, if $1 \leq r < n$ and if $1 \leq s < m$, then A could be viewed as a 2×2 partitioned matrix with upper left entry an $r \times s$ matrix, upper right entry an $r \times (m - s)$ matrix, lower left entry an $(n - r) \times s$ matrix, and lower right entry an $(n - r) \times (m - s)$ matrix. Of course, A could be partitioned in many other ways too.

(a) Make the notion of a partitioned matrix precise. Then make sure that your definition of partitioned is set up so that if all dimensions of the entry matrices are correct, then the product of two partitioned matrices can be found by using normal matrix multiplication applied to the blocks. Illustrate this process of "block multiplication" with some examples.

(b) Show that if A is an $n \times n$ partitioned matrix that is block upper triangular with square blocks on the diagonal, then $\det(A)$ is the product of the determinants of the diagonal blocks of A.

4.5 The LU-Decomposition

When applied to specific examples, the fact that row operations arise as multiplication by elementary matrices can give quite a bit of information. For example, suppose that A is an $n \times n$ matrix. Suppose also that Gaussian elimination row-reduces A to an upper triangular matrix U *and* that no row interchanges were used during the elimination process. Then we can express this upper triangular matrix U as $U = E_1E_2 \cdots E_sA$, where the E_i are lower triangular elementary matrices. (The E_i are lower triangular since the only row operations used in reducing to U involving adding a multiple of a row to a row below it.) Since $E_1E_2 \cdots E_s$ is lower triangular, its inverse is also. If we denote $(E_1 E_2 \cdots E_s)^{-1} = L$, we find $A = LU$, which shows that A can be expressed as a lower triangular matrix times an upper triangular matrix.

As an example, we consider the matrix

$$A = \begin{pmatrix} 1 & 2 & 0 \\ 1 & 3 & 1 \\ 0 & 2 & 4 \end{pmatrix}.$$

The sequence of row operations

$$\begin{pmatrix} 1 & 2 & 0 \\ 1 & 3 & 1 \\ 0 & 2 & 4 \end{pmatrix} \mapsto \begin{pmatrix} 1 & 2 & 0 \\ 0 & 1 & 1 \\ 0 & 2 & 4 \end{pmatrix} \mapsto \begin{pmatrix} 1 & 2 & 0 \\ 0 & 1 & 1 \\ 0 & 0 & 2 \end{pmatrix}$$

row-reduces A to an upper triangular matrix. This sequence of operations corresponds to multiplying A by the elementary matrices

$$\begin{pmatrix} 1 & 0 & 0 \\ 0 & 1 & 0 \\ 0 & -2 & 1 \end{pmatrix} \quad \text{and} \quad \begin{pmatrix} 1 & 0 & 0 \\ -1 & 1 & 0 \\ 0 & 0 & 1 \end{pmatrix}.$$

This shows

$$U = \begin{pmatrix} 1 & 2 & 0 \\ 0 & 1 & 1 \\ 0 & 0 & 2 \end{pmatrix} = \begin{pmatrix} 1 & 0 & 0 \\ 0 & 1 & 0 \\ 0 & -2 & 1 \end{pmatrix} \begin{pmatrix} 1 & 0 & 0 \\ -1 & 1 & 0 \\ 0 & 0 & 1 \end{pmatrix} \begin{pmatrix} 1 & 2 & 0 \\ 1 & 3 & 1 \\ 0 & 2 & 4 \end{pmatrix}$$
$$= \begin{pmatrix} 1 & 0 & 0 \\ -1 & 1 & 0 \\ 2 & -2 & 1 \end{pmatrix} \begin{pmatrix} 1 & 2 & 0 \\ 1 & 3 & 1 \\ 0 & 2 & 4 \end{pmatrix} = LA.$$

Inverting the lower triangular matrix L obtains the LU-decomposition for the original matrix A as $A = L^{-1}U$, that is

$$A = \begin{pmatrix} 1 & 0 & 0 \\ 1 & 1 & 0 \\ 0 & 2 & 1 \end{pmatrix} \begin{pmatrix} 1 & 2 & 0 \\ 0 & 1 & 1 \\ 0 & 0 & 2 \end{pmatrix}.$$

In this process, the matrix L can a bit messy to calculate. However, L^{-1} is readily found without calculating L first. The jith entry of L^{-1} is equal to $-m(j, i)$, where $m(j, i)$ is the "multiplier" used to zero out the jith entry during the row operations. For example, in our calculation above, we multiplied our first row by -1 and added it to the second to eliminate the 2,1 entry, and then we multiplied the second row by -2 and added it to the third in order to eliminate the 3,2 entry. Observe that the 2,1 entry of L^{-1} is 1 and the 3,2 entry L^{-1} is 2. You are asked to verify this more generally in the second group project at the end of this section.

We restate the preceding observation as the next theorem.

Theorem 19. *Suppose that Gaussian elimination transforms an $n \times n$ matrix A into an upper triangular matrix without using any row interchanges. Then A can be written as a product $A = LU$, where L is lower triangular and U is upper triangular.*

Permutation Matrices

The LU-decomposition just described is useful for a variety of reasons. Unfortunately, it is not always possible to row-reduce an arbitrary matrix to an upper triangular matrix without using row interchanges. In order to obtain LU-decompositions in the general case, we need to use permutation matrices.

Definition. A permutation matrix is a matrix obtained from the identity matrix by a sequence of row interchanges.

For example, the matrices

$$P_1 = \begin{pmatrix} 1 & 0 & 0 \\ 0 & 0 & 1 \\ 0 & 1 & 0 \end{pmatrix} \quad \text{and} \quad P_2 = \begin{pmatrix} 0 & 1 & 0 \\ 0 & 0 & 1 \\ 1 & 0 & 0 \end{pmatrix}$$

are each 3×3 permutation matrices. P_1 is obtained from I_3 by one row interchange (the second and third rows), and P_2 is obtained from I_3 by two row interchanges (an interchange of the first and second rows followed by an interchange of the second and third rows). The next fact is analogous to Theorem 10 established for elementary matrices. This can be easily checked with a few calculations.

Theorem 20. *Suppose that P is an $n \times n$ permutation matrix and A is an $n \times m$ matrix. Then the matrix PA is the matrix obtained from A by applying the row interchanges that were used to obtain P from I_n.*

The *PA* = *LU* Decomposition

Returning to our problem of a finding general LU-decompositions, we note that if A is an $n \times n$ matrix, then Gaussian elimination shows that A can be row-reduced to an upper triangular matrix U. However, one may have to use row interchanges in the process. Suppose that no row interchanges were required until the beginning of the ith stage of the reduction. In this case, one does not have a nonzero leading entry in the ith row of the matrix at that stage and a row below the ith row must be interchanged with the ith. Now assume this row interchange had been accomplished *before* the

Gaussian elimination ever started. Then at each reduction stage prior to the ith, the matrices would look the same, except that the ith row would already have been exchanged with the selected row below. When we reach the ith stage we can continue the elimination without an interchange since the newly selected ith row has a nonzero lead coefficient.

Here is an example to illustrate the point just made. Consider

$$A = \begin{pmatrix} 1 & 2 & 3 \\ 0 & 0 & 2 \\ 1 & 3 & 4 \end{pmatrix}.$$

We are forced to perform a row interchange in the second stage of our Gaussian elimination, as shown next.

$$A = \begin{pmatrix} 1 & 2 & 3 \\ 0 & 0 & 2 \\ 1 & 3 & 4 \end{pmatrix} \mapsto \begin{pmatrix} 1 & 2 & 3 \\ 0 & 0 & 2 \\ 0 & 1 & 1 \end{pmatrix} \mapsto \begin{pmatrix} 1 & 2 & 3 \\ 0 & 1 & 1 \\ 0 & 0 & 2 \end{pmatrix}$$

Suppose, however, we premultiplied A by the permutation matrix that interchanges the second and third rows, and then applied Gaussian elimination. We would find

$$PA = \begin{pmatrix} 1 & 0 & 0 \\ 0 & 0 & 1 \\ 0 & 1 & 0 \end{pmatrix} \begin{pmatrix} 1 & 2 & 3 \\ 0 & 0 & 2 \\ 1 & 3 & 4 \end{pmatrix} = \begin{pmatrix} 1 & 2 & 3 \\ 1 & 3 & 4 \\ 0 & 0 & 2 \end{pmatrix} \mapsto \begin{pmatrix} 1 & 2 & 3 \\ 0 & 1 & 1 \\ 0 & 0 & 2 \end{pmatrix}.$$

This process shows that PA does have an LU-decomposition (even though A does not). In fact, we can write

$$PA = \begin{pmatrix} 1 & 0 & 0 \\ 1 & 1 & 0 \\ 0 & 0 & 1 \end{pmatrix} \begin{pmatrix} 1 & 2 & 3 \\ 0 & 1 & 1 \\ 0 & 0 & 2 \end{pmatrix} = LU.$$

(Here, the matrix U is the result of our elimination and the matrix L comes from the inverse of the row operation used in the reduction.)

These ideas show that if one correctly exchanges rows in a matrix A *before* Gaussian elimination is applied, it is possible to start with a matrix that can be reduced to upper triangular form without row exchanges. The matrix resulting from the initial row exchanges will have the form PA, and the matrix PA will have an LU-decomposition. This is summarized in the next theorem.

Theorem 21. *If A is a $n \times n$ matrix, then there is a permutation matrix P, an upper triangular matrix U, and a lower triangular matrix L such that $PA = LU$.*

Example. Suppose that we desired to find the $PA = LU$ representation where

$$A = \begin{pmatrix} 1 & 4 & 2 & 3 \\ 1 & 2 & 1 & 0 \\ 2 & 6 & 3 & 1 \\ 0 & 0 & 1 & 4 \end{pmatrix}.$$

The first two stages of row operations on A give

$$A \mapsto \begin{pmatrix} 1 & 4 & 2 & 3 \\ 0 & -2 & -1 & -3 \\ 0 & -2 & -1 & -5 \\ 0 & 0 & 1 & 4 \end{pmatrix} \mapsto \begin{pmatrix} 1 & 4 & 2 & 3 \\ 0 & -2 & -1 & -3 \\ 0 & 0 & 0 & -2 \\ 0 & 0 & 1 & 4 \end{pmatrix}.$$

We thus need to interchange the third and fourth rows in order to obtain an upper triangular matrix. If we set

$$P = \begin{pmatrix} 1 & 0 & 0 & 0 \\ 0 & 1 & 0 & 0 \\ 0 & 0 & 0 & 1 \\ 0 & 0 & 1 & 0 \end{pmatrix},$$

then we row-reduce

$$PA = \begin{pmatrix} 1 & 4 & 2 & 3 \\ 1 & 2 & 1 & 0 \\ 0 & 0 & 1 & 4 \\ 2 & 6 & 3 & 1 \end{pmatrix}$$

to find its LU-decomposition. In order to keep track of the lower diagonal matrix that arises in our row reduction, we apply our elementary operations to the identity matrix augmented by the matrix PA as follows:

$$\left(\begin{array}{cccc|cccc} 1 & 0 & 0 & 0 & 1 & 4 & 2 & 3 \\ 0 & 1 & 0 & 0 & 1 & 2 & 1 & 0 \\ 0 & 0 & 1 & 0 & 0 & 0 & 1 & 4 \\ 0 & 0 & 0 & 1 & 2 & 6 & 3 & 1 \end{array}\right)$$

$$\mapsto \left(\begin{array}{cccc|cccc} 1 & 0 & 0 & 0 & 1 & 4 & 2 & 3 \\ -1 & 1 & 0 & 0 & 0 & -2 & -1 & -3 \\ 0 & 0 & 1 & 0 & 0 & 0 & 1 & 4 \\ -2 & 0 & 0 & 1 & 0 & -2 & -1 & -5 \end{array}\right)$$

$$\mapsto \left(\begin{array}{cccc|cccc} 1 & 0 & 0 & 0 & 1 & 4 & 2 & 3 \\ -1 & 1 & 0 & 0 & 0 & -2 & -1 & -3 \\ 0 & 0 & 1 & 0 & 0 & 0 & 1 & 4 \\ -1 & -1 & 0 & 1 & 0 & 0 & 0 & -2 \end{array}\right).$$

This calculation shows that $L^{-1}(PA) = U$, where

$$L^{-1} = \begin{pmatrix} 1 & 0 & 0 & 0 \\ -1 & 1 & 0 & 0 \\ 0 & 0 & 1 & 0 \\ -1 & -1 & 0 & 1 \end{pmatrix} \quad \text{and} \quad U = \begin{pmatrix} 1 & 4 & 2 & 3 \\ 0 & -2 & -1 & -3 \\ 0 & 0 & 1 & 4 \\ 0 & 0 & 0 & -2 \end{pmatrix}.$$

One readily computes that

$$L = \begin{pmatrix} 1 & 0 & 0 & 0 \\ 1 & 1 & 0 & 0 \\ 0 & 0 & 1 & 0 \\ 2 & 1 & 0 & 1 \end{pmatrix}.$$

We have found all the matrices in the expression $PA = LU$. (Also note that the off-diagonal entries of L can be found using the remarks before Theorem 19. The three row operations were subtracting the first row from the second row, subtracting twice the first row from the fourth row, and subtracting the second row from the fourth row. Therefore the off-diagonal entries of L are 1, 2, and 1.)

A Tridiagonal Matrix, Its *LU*-Decomposition, and Its Determinant

We next consider a 5×5 tridiagonal matrix. These special types of matrices are important in the study of some difference equations that are discrete versions of differential equations. (We will investigate one in detail in Chap. 7.) For now we will determine its LU-decomposition, and from that its determinant.

We consider

$$T = \begin{pmatrix} 2 & -1 & 0 & 0 & 0 \\ -1 & 2 & -1 & 0 & 0 \\ 0 & -1 & 2 & -1 & 0 \\ 0 & 0 & -1 & 2 & -1 \\ 0 & 0 & 0 & -1 & 2 \end{pmatrix}.$$

The process of reducing T to an upper triangular matrix looks as follows:

$$T = \begin{pmatrix} 2 & -1 & 0 & 0 & 0 \\ -1 & 2 & -1 & 0 & 0 \\ 0 & -1 & 2 & -1 & 0 \\ 0 & 0 & -1 & 2 & -1 \\ 0 & 0 & 0 & -1 & 2 \end{pmatrix} \mapsto \begin{pmatrix} 2 & -1 & 0 & 0 & 0 \\ 0 & \frac{3}{2} & -1 & 0 & 0 \\ 0 & -1 & 2 & -1 & 0 \\ 0 & 0 & -1 & 2 & -1 \\ 0 & 0 & 0 & -1 & 2 \end{pmatrix}$$

$$\mapsto \begin{pmatrix} 2 & -1 & 0 & 0 & 0 \\ 0 & \frac{3}{2} & -1 & 0 & 0 \\ 0 & 0 & \frac{4}{3} & -1 & 0 \\ 0 & 0 & -1 & 2 & -1 \\ 0 & 0 & 0 & -1 & 2 \end{pmatrix} \mapsto \begin{pmatrix} 2 & -1 & 0 & 0 & 0 \\ 0 & \frac{3}{2} & -1 & 0 & 0 \\ 0 & 0 & \frac{4}{3} & -1 & 0 \\ 0 & 0 & 0 & \frac{5}{4} & -1 \\ 0 & 0 & 0 & -1 & 2 \end{pmatrix}$$

$$\mapsto \begin{pmatrix} 2 & -1 & 0 & 0 & 0 \\ 0 & \frac{3}{2} & -1 & 0 & 0 \\ 0 & 0 & \frac{4}{3} & -1 & 0 \\ 0 & 0 & 0 & \frac{5}{4} & -1 \\ 0 & 0 & 0 & 0 & \frac{6}{5} \end{pmatrix}.$$

In the sequence of operations, we multiplied the first row by $\frac{1}{2}$ and added it to the second, multiplied the second row by $\frac{2}{3}$ and added it to the third, multiplied the third row by $\frac{3}{4}$ and added it to the fourth, and multiplied the fourth row by $\frac{4}{5}$ and added it to the fifth. These operations describe L^{-1}, and remembering to invert them in obtaining L shows that T has the LU-decomposition

$$T = \begin{pmatrix} 1 & 0 & 0 & 0 & 0 \\ -\frac{1}{2} & 1 & 0 & 0 & 0 \\ 0 & -\frac{2}{3} & 1 & 0 & 0 \\ 0 & 0 & -\frac{3}{4} & 1 & 0 \\ 0 & 0 & 0 & -\frac{4}{5} & 1 \end{pmatrix} \begin{pmatrix} 2 & -1 & 0 & 0 & 0 \\ 0 & \frac{3}{2} & -1 & 0 & 0 \\ 0 & 0 & \frac{4}{3} & -1 & 0 \\ 0 & 0 & 0 & \frac{5}{4} & -1 \\ 0 & 0 & 0 & 0 & \frac{6}{5} \end{pmatrix} = LU.$$

We observe that that both L and U are bidiagonal matrices. Furthermore, since the determinant of T is the product of the determinants of L and U, we find that $\det(T) = 6$. The ideas behind this calculation are more general, and you are asked to think about them in the first group project in the next problems.

Problems

1. Find an LU-decomposition for the following matrices. What is the determinant in each case?

   (a) $\begin{pmatrix} 1 & 3 \\ 2 & 7 \end{pmatrix}$ (b) $\begin{pmatrix} 1 & 1 & 0 \\ 0 & 1 & 1 \\ 1 & 1 & 1 \end{pmatrix}$ (c) $\begin{pmatrix} 1 & 2 & 3 \\ 1 & 0 & 1 \\ 0 & 1 & 0 \end{pmatrix}$

2. Find an LU-decomposition for the following matrices. What is the determinant in each case?

   (a) $\begin{pmatrix} 2 & 1 \\ 1 & 4 \end{pmatrix}$ (b) $\begin{pmatrix} 1 & 1 & 0 \\ 1 & 0 & 1 \\ 0 & 1 & 1 \end{pmatrix}$ (c) $\begin{pmatrix} 1 & 2 & 0 \\ 0 & 2 & 1 \\ 0 & 1 & 0 \end{pmatrix}$

3. Find a $PA = LU$-decomposition where A is

   (a) $\begin{pmatrix} 0 & 3 \\ 2 & 7 \end{pmatrix}$ (b) $\begin{pmatrix} 0 & 1 & 0 \\ 0 & 1 & 1 \\ 1 & 1 & 0 \end{pmatrix}$ (c) $\begin{pmatrix} 0 & 1 & 3 \\ 1 & 0 & 1 \\ 1 & 0 & 0 \end{pmatrix}$.

4. Find a $PA = LU$-decomposition where A is

   (a) $\begin{pmatrix} 0 & 1 \\ 1 & 4 \end{pmatrix}$ (b) $\begin{pmatrix} 0 & 1 & 0 \\ 0 & 0 & 1 \\ 1 & 1 & 0 \end{pmatrix}$ (c) $\begin{pmatrix} 1 & 1 & 3 \\ 1 & 1 & 1 \\ 0 & 2 & 0 \end{pmatrix}$.

Group Project: More Tridiagonal Matrices

(a) Generalize the ideas in the last part of Sec. 4.5 above to find the *LU*-decomposition and determinant of a 5×5 tridiagonal matrix whose entries along each diagonal are the same.

(b) Suppose T is a tridiagonal matrix such as just considered in part (a). Use the *LU*-decomposition for T to give a speedy solution to $T\vec{X} = \vec{v}$ for any column of five entries $\vec{v}$.

(c) Many calculators and computers have the ability to find *LU*-decompositions for you. Try both parts (a) and (b) for a tridiagonal 8×8 matrix using your calculator or computer.

Group Project: Multipliers and the *LU*-Decomposition

In this project you will explain why the remark above Theorem 19 is true. Suppose that A is an $n \times m$ matrix and suppose that $E_{n,n-1} \cdots E_{2,1}A = U$ represents a sequence of elementary row operations. Here, each $E_{i,j}$ is the elementary matrix representing the row operation that adds $-m_{ij} \cdot (i\text{th row})$ to the jth row in order to eliminate the ijth entry.

(a) Show that one can represent

$$E_{i,j} = \begin{pmatrix} 1 & 0 & \cdots & 0 \\ \vdots & -m_{ij} & \cdots & \vdots \\ 0 & \cdots & 0 & 1 \end{pmatrix}$$

and that

$$E_{i,j}^{-1} = \begin{pmatrix} 1 & 0 & \cdots & 0 \\ \vdots & m_{ij} & \cdots & \vdots \\ 0 & \cdots & 0 & 1 \end{pmatrix}.$$

(b) Now show that

$$E_{2,1}^{-1} \cdots E_{2,1}^{-1} = \begin{pmatrix} 1 & 0 & \cdots & 0 \\ m_{21} & \cdots & \ddots & \vdots \\ \cdots & \cdots & \ddots & 0 \\ m_{n1} & \cdots & m_{n,n-1} & 1 \end{pmatrix}.$$

In other words, when no row exchanges are required, one obtains $A = LU$, where L is a lower triangular matrix whose entries below the diagonal are simply the "multipliers" used in the Gaussian elimination.

CHAPTER 5

KEY CONCEPTS OF LINEAR ALGEBRA IN $\mathbf{R}^n$

5.1 Linear Combinations and Subspaces

In this chapter we introduce the basic concepts of linear algebra. These ideas, namely linear independence, span, basis, and dimension, are crucial to all applications of this subject.

Equations for Studying a Network Flow Problem

In many problems where the methods of linear algebra are useful, it is necessary to analyze a large number of linear equations or conditions. In fact, at times so many equations arise that the computations look hopeless. Of course, one can use computers to study such problems, but again a large batch of equations can consume programming time. For this reason it becomes crucial to eliminate as many equations as possible that carry redundant information about the problem. The idea of redundancy of equations is made precise using the notion of a linear combination.

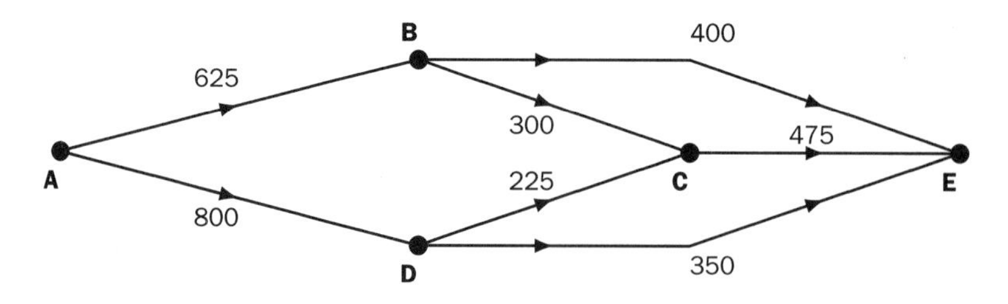

Fig. 5.1. A freeway system with seven sections between *A* and *E*

We illustrate the idea of linear combinations by considering a network flow problem and the various equations it gives. In Fig. 5.1 we have a graph with five vertices A, B, C, D, and E. We shall suppose that the graph represents possible freeway routes from point A to point E through a city. We shall assume that each segment in the graph has the carrying capacity indicated in cars per minute. Our problem is to determine the total carrying capacity between A and E of the freeway system (we will only consider one direction). We should also determine what the rate of flow should be along each freeway section in order to obtain the maximum capacity, which is important if we want to regulate traffic during peak hours.

In order to study this freeway network, we will work with seven variables, one to represent the traffic flow along each of the seven freeway segments. We let X_{AB} denote the flow along the segment $\overrightarrow{AB}$, X_{AC} denote the flow along the segment $\overrightarrow{AC}$, and so forth. Then, since the total flow, denoted f, is given by the total number of cars leaving A and arriving at E, we have $f = X_{AB} + X_{AD}$ and $f = X_{BE} + X_{CE} + X_{DE}$. We write these first two equations as

$$X_{AB} + X_{AD} - f = 0, \tag{1}$$

$$X_{BE} + X_{CE} + X_{DE} - f = 0. \tag{2}$$

Also, since the flow into B equals the flow out of B, and similarly for C and D, we obtain

$$X_{BC} + X_{BE} - X_{AB} = 0, \tag{3}$$

$$X_{DC} + X_{DE} - X_{AD} = 0, \tag{4}$$

$$X_{BC} + X_{DC} - X_{CE} = 0. \tag{5}$$

However, these are not all the possible equations. For example, we also have

$$X_{AB} + X_{DC} + X_{DE} - f = 0, \tag{6}$$

since the total flow equals the flow along $\overrightarrow{AB}$ plus the total along both $\overrightarrow{DC}$ and $\overrightarrow{DE}$. Likewise, the reader can check that

$$X_{BE} + X_{BC} + X_{DC} + X_{DE} - f = 0, \tag{7}$$

$$X_{AD} + X_{BC} + X_{BE} - f = 0. \tag{8}$$

At this point we have produced eight equations among our seven variables. We also have the seven inequalities $X_{AB} \leq 625$, $X_{AD} \leq 800$, $X_{BE} \leq 400$, $X_{BC} \leq 300$, $X_{DC} \leq 225$, $X_{DE} \leq 350$, and $X_{CE} \leq 475$. Which equations are crucial and which can be ignored? If we look carefully, we can see that one equation for each vertex of our graph suffices in the study of this network. For example, the first five equations given can be used to derive the last three. To see this, first note that adding equations (1) and (4) gives (6)

$$\begin{aligned} 0 &= (X_{AB} + X_{AD} - f) + (X_{DC} + X_{DE} - X_{AD}) \\ &= X_{AB} + X_{DC} + X_{DE} - f. \end{aligned}$$

Similarly, adding equations (1) and (3) gives (8), and adding equations (2) and (5) gives (7).

Our expressions show that these three equations are *linear combinations* of the first five. This means that our original five equations contain the same amount of information as all eight. Consequently, any problem involving our freeway network can be solved using the information in five equations only. Note also that the information in the third, fourth, and fifth equations can be obtained from the first, second, sixth, seventh, and eighth equations, by reversing the calculations just given. This observation will be used later in solving our freeway optimization question.

The Maximal Freeway Flow

Let's return to the question of finding the optimal traffic flow in our freeway network. The reader may recognize that the problem is one of linear optimization with constraints. However, instead of applying the simplex algorithm (see Sec. 3.5), we will analyze the situation directly, and our solution will lead us to some new ideas. Since $f = X_{AB} + X_{AD}$ and we have $X_{AB} \leq 625$ and $X_{AD} \leq 800$, we see that $f \leq 625 + 800 = 1425$. Similarly, as $f = X_{BE} + X_{CE} + X_{DE}$ and as $X_{BE} \leq 400$, $X_{DE} \leq 350$, and $X_{CE} \leq 475$, we know that $f \leq 400 + 350 + 475 = 1225$. At this point, we know the flow is at most 1225 cars per minute. But can this be the maximal flow? To determine this we must use the information contained in five equations. Instead of using the third, fourth, and fifth equations in our list, we shall utilize the last three,

$$f = X_{AB} + X_{DC} + X_{DE},$$

$$f = X_{BE} + X_{BC} + X_{DC} + X_{DE},$$

$$f = X_{AD} + X_{BC} + X_{BE},$$

since these three equations determine values for f, the quantity we want to maximize. These equations, together with our inequalities $X_{AB} \leq 625$, $X_{AD} \leq 800$, $X_{BE} \leq 400$, $X_{BC} \leq 300$, $X_{DC} \leq 225$, $X_{DE} \leq 350$, and $X_{CE} \leq 475$ show that

$$f \leq 625 + 225 + 350 = 1200,$$

$$f \leq 400 + 300 + 225 + 350 = 1275,$$

$$f \leq 800 + 300 + 400 = 1500.$$

We now see that a maximum of 1200 cars per minute can flow through our freeway network. Since we have utilized a full set equations in this analysis, this is the maximal flow.

We can find the flows through each freeway section during maximal flow by first using the flow values given by the fifth equation when $f = 1200$. This shows $X_{AB} = 625$, $X_{DC} = 225$, and $X_{DE} = 350$. The fourth equation now shows that $X_{AD} = X_{DC} + X_{DE} = 575$. To find the values of X_{BE}, X_{BC}, and X_{CE}, we can apply the second, third, and fifth equations together with values just obtained to get the system

$$X_{BE} + X_{BC} = 625$$

$$-X_{BC} + X_{CE} = 225$$

$$X_{BE} + X_{CE} = 1200 - 350 = 850.$$

The third equation in this system can be obtained by adding the first two, that is, it is a linear combination of them. This means that although our optimal flow is 1200 for the entire network, this flow does not have a unique value for each variable. For example, if we allow $X_{BE} = 400$ (its maximal capacity), then we find $X_{BC} = 225$ and $X_{CE} = 450$, all of which are allowable values. This gives a set of flow values that maximize the total freeway network flow. On the other hand, if we set $X_{CE} = 475$ (its maximal capacity), then we find $X_{BC} = 250$ and $X_{BE} = 375$, again all of which are allowable values. This gives another set of flow values that maximize the total freeway network flow. In general, any solution to the first two equations of our latter system that satisfy our variable constraints will give a maximal set of flow values. For an extension of the techniques initiated in this problem, see the group project at the end of this section.

In the preceding analysis, we inspected the network to develop equations and inequalities with which we experimented until we found the possible solutions. However, our procedure was not efficiently organized, and at times the reader may have wondered why we used the equations we did. In this chapter we will study two basic notions from linear algebra, namely linear combinations and linear independence, which help clarify how to solve efficiently problems like our freeway flow. Also, as you can imagine, for larger networks (imagine studying the Los Angeles freeway system!) the situation is similar but becomes incredibly messy. So it becomes important to utilize matrices to facilitate computation. The dependence of equations on other equations can be calculated using row operations.

Linear Combinations of Vectors

In this section, n will be a fixed natural number. An element of $\mathbf{R}^n$ will be called an *n-vector*. However, in most discussions, the n is unimportant, so we will usually refer to n-vectors simply as vectors. It proves to be convenient to represent elements of $\mathbf{R}^n$ as columns, that is, as column vectors. This is the notation we adopted in Chap. 2 when we discussed geometric vectors. We remarked at that time that there were good reasons for the column notation. These reasons will unfold in this chapter.

We begin with the definition of a linear combination of a collection of vectors.

Definition. Suppose $\vec{v}_1, \vec{v}_2, \ldots, \vec{v}_n$ are vectors and $r_1, r_2, \ldots, r_n$ are real numbers. The vector

$$\vec{w} = r_1\vec{v}_1 + r_2\vec{v}_2 + \cdots + r_n\vec{v}_n$$

is called a *linear combination* of $\vec{v}_1, \vec{v}_2, \ldots, \vec{v}_n$.

For example, the equation

$$\begin{pmatrix} 1 \\ 4 \\ 1 \\ 1 \end{pmatrix} + 2\begin{pmatrix} 3 \\ 1 \\ 2 \\ 0 \end{pmatrix} - \begin{pmatrix} 0 \\ 0 \\ 1 \\ 0 \end{pmatrix} = \begin{pmatrix} 7 \\ 6 \\ 4 \\ 1 \end{pmatrix}$$

shows that the vector

$$\begin{pmatrix} 7 \\ 6 \\ 4 \\ 1 \end{pmatrix}$$

is a linear combination of

$$\begin{pmatrix} 1 \\ 4 \\ 1 \\ 1 \end{pmatrix}, \quad \begin{pmatrix} 3 \\ 1 \\ 2 \\ 0 \end{pmatrix}, \quad \text{and} \quad \begin{pmatrix} 0 \\ 0 \\ 1 \\ 0 \end{pmatrix}.$$

We also note that $\vec{0} = 0\vec{v}_1 + 0\vec{v}_2 + \cdots + 0\vec{v}_n$, which means the zero vector is always a linear combination of any nonempty set of vectors.

The Span

It is important to study the set of all linear combinations of a collection of vectors $\{r_1\vec{v}_1 + r_2\vec{v}_2 + \cdots + r_n\vec{v}_n \mid r_1, r_2, \ldots, r_n \in \mathbf{R}\}$. It is called the span, as is explained next.

Definition. The set of all linear combinations of $\vec{v}_1, \vec{v}_2, \ldots, \vec{v}_n$ is called the *span* of $\vec{v}_1, \vec{v}_2, \ldots, \vec{v}_n$ and is denoted by $\operatorname{span}\{\vec{v}_1, \vec{v}_2, \ldots, \vec{v}_n\}$.

The span of any set of nonzero vectors, $\operatorname{span}\{\vec{v}_1, \vec{v}_2, \ldots, \vec{v}_n\}$, is always an infinite collection of vectors. Visually, the span of a single nonzero vector in $\mathbf{R}^3$ is a line, since it is the set of all multiples of that vector. The span of two vectors in $\mathbf{R}^3$ that point in different directions is the plane that contains them.

We note that the vector

$$\begin{pmatrix} 2 \\ 1 \\ 4 \end{pmatrix}$$

does *not* lie in the span of

$$\vec{v}_1 = \begin{pmatrix} 2 \\ 1 \\ 0 \end{pmatrix} \quad \text{and} \quad \vec{v}_2 = \begin{pmatrix} 3 \\ 3 \\ 0 \end{pmatrix},$$

because any linear combination of these two vectors cannot have a nonzero third entry. Another way to see this is to try to solve the vector equation

$$r\begin{pmatrix} 2 \\ 1 \\ 0 \end{pmatrix} + s\begin{pmatrix} 3 \\ 3 \\ 0 \end{pmatrix} = \begin{pmatrix} 2 \\ 1 \\ 4 \end{pmatrix}.$$

We obtain the inconsistent system

$$\begin{pmatrix} 2 & 3 \\ 1 & 3 \\ 0 & 0 \end{pmatrix}\begin{pmatrix} r \\ s \end{pmatrix} = \begin{pmatrix} 2 \\ 1 \\ 4 \end{pmatrix}.$$

The span of the two vectors, $\operatorname{span}\{\vec{v}_1, \vec{v}_2\}$, is the collection of all vectors of the form

$$r\begin{pmatrix} 2 \\ 1 \\ 0 \end{pmatrix} + s\begin{pmatrix} 3 \\ 3 \\ 0 \end{pmatrix} \quad \text{where } r, s \in \mathbf{R}.$$

This span can be expressed in terms of the two parameters r and s as

$$\left\{ \begin{pmatrix} 2r+3s \\ r+3s \\ 0 \end{pmatrix} \mid r,s \in \mathbf{R} \right\}.$$

Testing for the Span

In the previous example we saw that the nonexistence of a linear combination was a consequence of the inconsistency of a system of linear equations. More generally, the question of determining if an n-vector is a linear combination of a set of other n-vectors can always be reduced to the problem of determining if a system of equations has a solution.

For example, suppose we need to know if the vectors

$$\vec{w}_1 = \begin{pmatrix} 2 \\ 4 \\ 3 \end{pmatrix}, \quad \vec{w}_2 = \begin{pmatrix} 2 \\ 5 \\ 3 \end{pmatrix}$$

are linear combinations of the vectors

$$\vec{v}_1 = \begin{pmatrix} 1 \\ 3 \\ 2 \end{pmatrix}, \quad \vec{v}_2 = \begin{pmatrix} 1 \\ 1 \\ 1 \end{pmatrix}, \quad \vec{v}_3 = \begin{pmatrix} 1 \\ 5 \\ 3 \end{pmatrix} \quad \in \mathbf{R}^3.$$

We consider the two systems of equations

$$\begin{pmatrix} 1 & 1 & 1 \\ 3 & 1 & 5 \\ 2 & 1 & 3 \end{pmatrix} \begin{pmatrix} X \\ Y \\ Z \end{pmatrix} = \begin{pmatrix} 2 \\ 4 \\ 3 \end{pmatrix} \quad \text{and} \quad \begin{pmatrix} 1 & 1 & 1 \\ 3 & 1 & 5 \\ 2 & 1 & 3 \end{pmatrix} \begin{pmatrix} X \\ Y \\ Z \end{pmatrix} = \begin{pmatrix} 2 \\ 5 \\ 3 \end{pmatrix}.$$

Observe that the columns of the coefficient matrices are the vectors $\vec{v}_1$, $\vec{v}_2$, and $\vec{v}_3$. Recall from Sec. 4.1 that in any matrix product AB the columns of the product are linear combinations of the columns of A. This means that $\vec{w}_1$ is a linear combination of $\vec{v}_1$, $\vec{v}_2$, and $\vec{v}_3$ if the first system has a solution, and $\vec{w}_2$ is such if the second system has a solution. The first system has the solution $X = 1, Y = 1, Z = 0$, while the second system has no solutions. Hence we can write $\vec{w}_1 = 1 \cdot \vec{v}_1 + 1 \cdot \vec{v}_2 + 0\vec{v}_3$.

We record the idea of what we just did in the next theorem.

Theorem 22. *Suppose $\{\vec{v}_1, \vec{v}_2, \ldots, \vec{v}_m\}$ is a set of n-vectors. Let A be the $n \times m$ matrix $A = (\vec{v}_1 \quad \vec{v}_2 \quad \cdots \quad \vec{v}_m)$ whose columns are the $\vec{v}_i$, and let $\vec{X}$ denote a column of m variables. Then $\vec{w} \in \mathbf{R}^n$ is a linear combination of $\vec{v}_1, \vec{v}_2, \ldots, \vec{v}_m$ if and only if the equation $A\vec{X} = \vec{w}$ has a solution.*

Proof. The equation $A\vec{X} = \vec{w}$ is

$$\begin{pmatrix} \vec{v}_1 & \vec{v}_2 & \cdots & \vec{v}_m \end{pmatrix} \begin{pmatrix} X_1 \\ X_2 \\ \vdots \\ X_m \end{pmatrix} = \vec{w},$$

which has a solution if and only if

$$X_1\vec{v}_1 + X_2\vec{v}_2 + \ldots + X_m\vec{v}_m = \vec{w}.$$

Thus, the existence of a solution to $A\vec{X} = \vec{w}$ is equivalent to being able to express $\vec{w}$ as a linear combination of $\vec{v}_1, \vec{v}_2, \ldots, \vec{v}_m$. □

The span of a set of vectors can be found using Gaussian elimination. If $\vec{v}_1, \vec{v}_2$, and $\vec{v}_3$ are the three vectors considered above, then the theorem says that

$$\begin{pmatrix} 1 & 1 & 1 \\ 3 & 1 & 5 \\ 2 & 1 & 3 \end{pmatrix} \begin{pmatrix} X \\ Y \\ Z \end{pmatrix} = \begin{pmatrix} a \\ b \\ c \end{pmatrix}$$

has a solution precisely when

$$\begin{pmatrix} a \\ b \\ c \end{pmatrix} \in \operatorname{span}\{\vec{v}_1, \vec{v}_2, \vec{v}_3\}.$$

Row-reducing the augmented matrix gives

$$\left(\begin{array}{ccc|c} 1 & 1 & 1 & a \\ 3 & 1 & 5 & b \\ 2 & 1 & 3 & c \end{array}\right) \mapsto \left(\begin{array}{ccc|c} 1 & 1 & 1 & a \\ 0 & -2 & 2 & b-3a \\ 0 & -1 & 1 & c-2a \end{array}\right) \mapsto \left(\begin{array}{ccc|c} 1 & 1 & 1 & a \\ 0 & 1 & -1 & -\frac{1}{2}b + \frac{3}{2}a \\ 0 & 0 & 0 & c - \frac{1}{2}b - \frac{1}{2}a \end{array}\right).$$

We find that our vector lies in the span whenever $2c = a + b$. In other words, $\operatorname{span}\{\vec{v}_1, \vec{v}_2, \vec{v}_3\} = \{(a, b, \frac{1}{2}(a+b) \mid a, b \in \mathbf{R}\}$.

Subspaces of $\mathbf{R}^n$

Subsets of vectors that arise as the span of a collection of vectors are extremely important in linear algebra. They are called *subspaces*.

Definition. A *subspace* of $\mathbf{R}^n$ is a subset $V \subseteq \mathbf{R}^n$ of n-vectors such that

(i) $\vec{0} \in V$;

(ii) If $\vec{v} \in V$ and $a \in \mathbf{R}$, then $a\vec{v} \in V$;

(iii) If $\vec{v}_1, \vec{v}_2 \in V$, then $(\vec{v}_1 + \vec{v}_2) \in V$.

Equivalently, a subspace of $\mathbf{R}^n$ is a nonempty set of n-vectors that is closed under the operations of scalar multiplication and addition.

$\mathbf{R}^n$ is a subspace of itself as is the subset $\{\vec{0}\}$ consisting of a single vector. Also, the xy-plane in $\mathbf{R}^3$, namely $\{(x, y, 0) \mid x, y \in \mathbf{R}\}$, is a subspace of $\mathbf{R}^3$ because scalar multiples and sums of vectors in the xy-plane still lie in the xy-plane. The xy-plane is the span of the vectors $(1, 0, 0)$ and $(0, 1, 0)$. The same reasoning applies to show that the xz-plane and the yz-plane are subspaces of $\mathbf{R}^3$.

At times we will use the phrase "vector space" to mean "subspace of $\mathbf{R}^n$" (some n). Often we will not specify the integer n, but of course this will become necessary before we do any concrete calculations. There is a more general notion of a vector space, which includes the subspaces of $\mathbf{R}^n$. In this text we will always work with subspaces of $\mathbf{R}^n$, although the results we state for vector spaces will also be true in the more general situation. We shall use capitals such as V, W, ... to denote vector spaces.

The span of any set of vectors in $\mathbf{R}^n$ is a subspace of $\mathbf{R}^n$. Whenever $V = \text{span}\{\vec{v}_1, \vec{v}_2, \ldots, \vec{v}_m\}$, we say that $\vec{v}_1, \vec{v}_2, \ldots, \vec{v}_m$ span V, or we say that $\{\vec{v}_1, \vec{v}_2, \ldots, \vec{v}_m\}$ is a spanning set for V.

Theorem 23. *The span of any set of vectors,* $\text{span}\{\vec{v}_1, \vec{v}_2, \ldots, \vec{v}_m\}$, *is a subspace.*

Proof. First note that $\vec{0} = 0\vec{v}_1 + 0\vec{v}_2 + \cdots + 0\vec{v}_m$, so $\vec{0}$ lies in this set. Consider a vector $\vec{w} = a_1\vec{v}_1 + a_2\vec{v}_2 + \cdots + a_m\vec{v}_m$ in the span. Then for any real number b we have $b\vec{w} = b(a_1\vec{v}_1 + a_2\vec{v}_2 + \cdots + a_m\vec{v}_m) = ba_1\vec{v}_1 + ba_2\vec{v}_2 + \cdots + ba_m\vec{v}_m$, which is also a linear combination of $\vec{v}_1, \vec{v}_2, \ldots, \vec{v}_m$. This shows closure under scalar multiplication. Next consider $\vec{w}_1 = a_1\vec{v}_1 + a_2\vec{v}_2 + \cdots + a_m\vec{v}_m$ and $\vec{w}_2 = b_1\vec{v}_1 + b_2\vec{v}_2 + \cdots + b_m\vec{v}_m$. Then, as $\vec{w}_1 + \vec{w}_2 = (a_1 + b_1)\vec{v}_1 + (a_2 + b_2)\vec{v}_2 + \cdots + (a_m + b_m)\vec{v}_m$, we see that $\vec{w}_1 + \vec{w}_2$ is also a linear combination of $\vec{v}_1, \vec{v}_2, \ldots, \vec{v}_m$. This shows that $\text{span}\{\vec{v}_1, \vec{v}_2, \ldots, \vec{v}_m\}$ is a vector space. □

Theorem 5 said that the set of solutions to a homogeneous system of equations is closed under vector addition and scalar multiplication. This means that such solutions form a subspace.

Corollary. *The set of solutions to a system of homogeneous equations is a subspace.*

For example,

$$W = \left\{ \begin{pmatrix} 4t \\ -2t \\ t \end{pmatrix} \mid t \in \mathbf{R} \right\}$$

is a vector space. One way to see this is to note that by back-substitution W is the set of solutions to

$$\begin{aligned} X \qquad - 4Z &= 0 \\ Y + 2Z &= 0. \end{aligned}$$

The corollary shows that the subset of all vectors $(x, y, z) \in \mathbf{R}^3$ for which $x + y - z = 0$ is a vector space since it is the set of solutions to a homogeneous system. However, the collection of all (x, y, z) for which $x + y - z = 1$ is *not* a vector space. Note for example that it does not contain $\vec{0}$.

Suppose that $V \subseteq \mathbf{R}^1$ is a subspace and $r \in V$ is nonzero. Then $kr \in V$ for all real numbers $k \in \mathbf{R}$. But kr can be any real number, so $V = \mathbf{R}$ follows. This shows that the only subspaces of $\mathbf{R}$ are $\{0\}$ and $\mathbf{R}$. A subspace of $\mathbf{R}^2$ is either $\{(0, 0)\}$, a line through $(0, 0)$, or all of $\mathbf{R}^2$. To see this, note that all scalar multiples of a nonzero vector in $\mathbf{R}^2$ form the line through $(0, 0)$ and that vector, and the set of all linear combinations of any two noncollinear vectors in $\mathbf{R}^2$ is all of $\mathbf{R}^2$. By similar reasoning, a vector subspace of $\mathbf{R}^3$ must be $\{(0, 0, 0)\}$, a line through $(0, 0, 0)$, a plane through $(0, 0, 0)$, or all of $\mathbf{R}^3$.

Problems

1. Can you express the linear equation $3X + 4Y - 2Z + W = 0$ using a linear combination of $2X + Y - 2Z + 2W = 0$, $3Y + 2Z - W = 0$, and $2X + W = 0$?

2. Determine if the vector

$$\begin{pmatrix} 3 \\ 2 \\ 2 \end{pmatrix}$$

is a linear combination of the following:

(a) $\begin{pmatrix} 0 \\ 1 \\ 1 \end{pmatrix}, \begin{pmatrix} 2 \\ 0 \\ 0 \end{pmatrix}, \begin{pmatrix} 1 \\ 0 \\ 0 \end{pmatrix}$

(b) $\begin{pmatrix} 1 \\ 0 \\ 0 \end{pmatrix}, \begin{pmatrix} 1 \\ 2 \\ 2 \end{pmatrix}$

(c) $\begin{pmatrix} 0 \\ 1 \\ 1 \end{pmatrix}, \begin{pmatrix} 1 \\ 1 \\ 0 \end{pmatrix}, \begin{pmatrix} 0 \\ 1 \\ 0 \end{pmatrix}, \begin{pmatrix} 0 \\ 1 \\ 0 \end{pmatrix}$

3. Find an expression for the span of the following vectors using as few parameters as possible.

(a) $\begin{pmatrix} 2 \\ 1 \\ 2 \end{pmatrix}, \begin{pmatrix} 0 \\ 0 \\ 0 \end{pmatrix}, \begin{pmatrix} -6 \\ -3 \\ -6 \end{pmatrix}$ (b) $\begin{pmatrix} 1 \\ 1 \\ 1 \end{pmatrix}, \begin{pmatrix} 0 \\ 2 \\ 2 \end{pmatrix}, \begin{pmatrix} 1 \\ 0 \\ 1 \end{pmatrix}$

(c) $\begin{pmatrix} 2 \\ 1 \\ 0 \\ 1 \end{pmatrix}, \begin{pmatrix} 1 \\ 3 \\ 1 \\ 0 \end{pmatrix}$ (d) $\begin{pmatrix} 1 \\ 2 \end{pmatrix}, \begin{pmatrix} 0 \\ 0 \end{pmatrix}, \begin{pmatrix} 0 \\ 2 \end{pmatrix}, \begin{pmatrix} 1 \\ 3 \end{pmatrix}$

4. Which of the following sets of vectors are subspaces of $\mathbf{R}^3$? Give reasons.
   (a) $\{(1, 1, 1)\}$
   (b) $\{(0, 0, 0), (1, 1, 1), (2, 2, 2), \ldots\}$
   (c) $\{r(1, 1, 1) + s(2, 4, 5) \mid r, s \in \mathbf{R}\}$
   (d) $\{(1, 1, 1) + s(2, 4, 5) \mid s \in \mathbf{R}\}$
   (e) $\{(t, 1 - t, 2t) \mid t \in \mathbf{R}\}$
   (f) $\{(t, t, 0) \mid t \in \mathbf{R}\}$
   (g) $\{(x, y, z) \mid x, y, z \in \mathbf{R} \quad \text{and} \quad x - z = 0\}$
   (h) $\{(x, y, z) \mid x, y, z \in \mathbf{R} \quad \text{and} \quad x - z = 1\}$
   (i) $\{(u, v, w) \in \mathbf{R}^3 \mid uvw = 0\}$
   (j) $\{(u, v, w) \in \mathbf{R}^3 \mid u^2 + v^2 + w^2 = 0\}$
5. Show that if a vector $\vec{v}$ is a linear combination of the vectors $\vec{v}_1, \vec{v}_2, \ldots, \vec{v}_n$, then $\text{span}\{\vec{v}, \vec{v}_1, \vec{v}_2, \ldots, \vec{v}_n\} = \text{span}\{\vec{v}_1, \vec{v}_2, \ldots, \vec{v}_n\}$.
6. Find a system of equations whose solutions are the span of

$$\begin{pmatrix} 1 \\ 0 \\ 2 \end{pmatrix} \quad \text{and} \quad \begin{pmatrix} 6 \\ 0 \\ 2 \end{pmatrix} \quad \text{inside} \quad \mathbf{R}^3.$$

7. Suppose that V and W are subspaces of $\mathbf{R}^n$. Under what conditions is $V \cap W$ a subspace? How about $V \cup W$?
8. Two solutions to the freeway optimization were found in the Maximal Freeway Flow subsection. There are others. Find a description of all of them.

Group Project: Subspace Properties

These two problems deal with sums of subspaces. If V and W are subspaces of $\mathbf{R}^n$, then we define $V + W = \{\vec{v} + \vec{w} \mid \vec{v} \in V, \vec{w} \in W\}$.

1. Suppose that V and W are subspaces of $\mathbf{R}^n$.
   (a) Show that $V + W$ is a subspace of $\mathbf{R}^n$.
   (b) Suppose that $V \cap W = \{\vec{0}\}$. Show that every $\vec{u} \in V + W$ can be expressed uniquely in the form $\vec{u} = \vec{v} + \vec{w}$, where $\vec{v} \in V$ and $\vec{w} \in W$. When this occurs, we say that V and W are *independent subspaces* and $V + W$ is a *direct sum*. In this case we denote $V + W$ by $V \oplus W$.

2. Suppose that U_1, U_2, and W are subspaces of $\mathbf{R}^n$.
   (a) Show that $(U_1 \cap W) + (U_2 \cap W) \subseteq (U_1 + U_2) \cap W$.
   (b) Give an example in $\mathbf{R}^2$ that shows equality need not hold in 2(a).

Group Project: The Max Flow-Min Cut Theorem and Integer Programming

In the traffic flow problem considered in this section, we found that the maximal flow was determined by a single equation (6). The five inequalities for f in this example were each obtained by totaling the maximal flow across a *cut.* Intuitively, a cut is obtained by cutting the graph into two pieces with a pair of scissors, where only the edges, not vertices, are cut, and where the beginning and ending vertices lie on different parts.

Suppose that one is considering a flow problem similar to the traffic flow problem. These problems are often called integer programming problems.

(a) Consider some cut of your network. Explain why the total flow of the network cannot exceed the sum of the capacities of the cut edges.

(b) Next explain why the maximal flow is the minimal sum of the capacities of cut edges among all possible cuts. This is called the *max flow-min cut theorem.* Hint: If you have a maximal flow, consider cuts along edges where the flow attains the capacity of the cut.

(c) Use the max flow-min cut theorem to determine the maximal flow of the following expansion of our earlier freeway network with capacities as indicated.

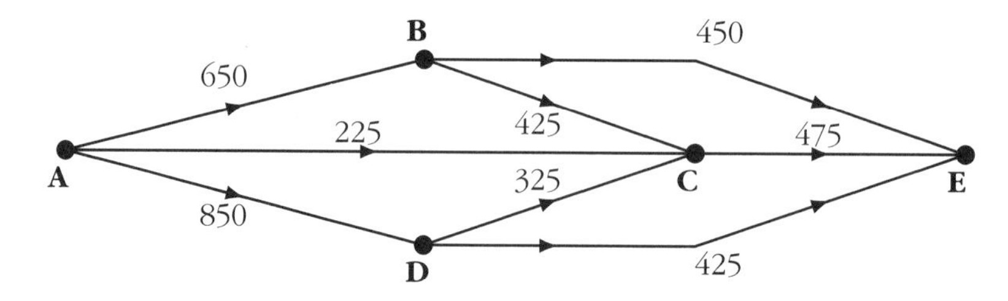

5.2 Linear Independence

The Stoichiometry and the Kinetic Description of the Formation of Hydrogen Bromide

In the study of chemical reactions, *stoichiometry* is the mathematics of keeping track of the chemical components of the system. Many chemical reactions are not single-step processes, but are instead a multistage process with many interrelated reactions occurring simultaneously. Even if the reactants and products of a reaction are completely understood, it is often quite difficult to determine the exact mechanism of the reaction, and many theories may be consistent with experimental observation. An important problem for chemists is to determine if a reaction could be the result of a proposed collection of intermediate reactions, and if so to determine how the intermediate reactions depend on one another. The dependence of component reactions in a reaction is important in understanding overall reaction mechanisms and rates. Using stoichiometry, we can algebraically analyze the dependence of reactions, associating each component reaction with a linear equation in which the variables represent a chemical species present in the system. The dependency of the reactions is reflected in the dependency of the associated linear equations.

We illustrate this by considering what is believed to be the kinetic description of the formation of hydrogen bromide. The overall reaction is that of hydrogen, H_2, and bromine, Br_2, combining to form hydrogen bromide, HBr, which is indicated by the expression

$$H_2 + Br_2 \longrightarrow 2HBr.$$

The 2 on the right was added to balance the equation by indicating that two hydrogen bromide molecules are produced from each pair of hydrogen and bromine molecules. Of course, one could have written $2H_2 + 2Br_2 \longrightarrow 4HBr$, or some other multiple of this equation, but the convention is to leave the expression in the simplest form possible.

It turns out, however, that in the formation of hydrogen bromide in this reaction, a number of other reactions actually take place, the net result of which is the above equation. This is called the *reaction mechanism.* Current theory suggests that the following reactions take place simultaneously in a reaction forming hydrogen bromide.[1]

$$\begin{gathered}
Br_2 \longrightarrow 2Br \\
H_2 + Br \longrightarrow HBr + H \\
H + Br_2 \longrightarrow HBr + Br \\
2Br \longrightarrow Br_2 \\
HBr + H \longrightarrow H_2 + Br
\end{gathered}$$

Note that the second and third reactions in this list produce HBr and that the fourth and fifth reactions are the reverse reactions of the first and second reactions.

We now assume that all five of these reactions are occurring simultaneously (with presumably different rates). Our task is to understand if and how the desired reaction $H_2 + Br_2 \longrightarrow 2HBr$ could arise as a result of the

[1]This mechanism has been suggested to explain the experimental observation due to Bodenstein and Lind [*Z. Phys. Chem.* **57**, 168 (1906)] that the rate of this reaction is proportional to the concentration of H_2 and to the square root of the concentration of Br_2.

five reactions. In order to understand this, we consider a system of linear equations. We denote the quantity of each chemical species by variables, setting $X_1 = Br_2$, $X_2 = Br$, $X_3 = H_2$, $X_4 = H$, $X_5 = HBr$. Then the original reaction $H_2 + Br_2 \to 2HBr$ is represented by the equation $X_1 + X_3 = 2X_5$. Our reaction mechanism leads to five additional equations, which we write as the homogeneous system

$$\begin{aligned}
-X_1 + X_2 \qquad\qquad &= 0 \qquad (1)\\
-X_2 - X_3 + X_4 + X_5 &= 0 \qquad (2)\\
-X_1 + X_2 \qquad - X_4 + X_5 &= 0 \qquad (3)\\
X_1 - 2X_2 \qquad\qquad &= 0 \qquad (4)\\
X_2 + X_3 - X_4 - X_5 &= 0 \qquad (5)
\end{aligned}$$

where, according to convention, the reactants have negative coefficients, and the products' coefficients are positive.

The matrix on the left below is the coefficient matrix of this system *with* a row on the bottom added to represent the equation, $-X_1 - X_3 + 2X_5 = 0$, of our overall reaction. Gaussian elimination applied to this matrix gives

$$\begin{pmatrix} -1 & 2 & 0 & 0 & 0 \\ 0 & -1 & -1 & 1 & 1 \\ -1 & 1 & 0 & -1 & 1 \\ 1 & -2 & 0 & 0 & 0 \\ 0 & 1 & 1 & -1 & -1 \\ -1 & 0 & -1 & 0 & 2 \end{pmatrix} \mapsto \begin{pmatrix} -1 & 2 & 0 & 0 & 0 \\ 0 & -1 & -1 & 1 & 1 \\ 0 & 0 & 1 & -2 & 0 \\ 0 & 0 & 0 & 0 & 0 \\ 0 & 0 & 0 & 0 & 0 \\ 0 & 0 & 0 & 0 & 0 \end{pmatrix}.$$

Since we have obtained a rank-three matrix, this shows that the three equations

$$\begin{aligned}
-X_1 + 2X_2 \qquad\qquad &= 0 \qquad (1)\\
-X_2 - X_3 + \quad X_4 + X_5 &= 0 \qquad (2)\\
X_3 + -2X_4 \qquad &= 0 \qquad (6)
\end{aligned}$$

are *independent* and that the overall reaction equation, $-X_1 - X_3 + 2X_5 = 0$, lies in their span. Note that equations (1) and (2) are our original reaction equations. Equation (6) was obtained in our row reduction by subtracting equations (1) and (2) from (3), but for reasons that will be clear momentarily, we will view it as the sum of equations (3), (4), and (5).

We may now analyze our overall reaction, $H_2 + Br_2 \to 2HBr$. Its equation, $-X_1 - X_3 + 2X_5 = 0$, can be obtained by adding equations (1), (2) twice, and (6) as

$$\begin{aligned}
0 &= (-X_1 + 2X_2) + 2(-X_2 - X_3 + X_4 + X_5) + (X_3 - 2X_4)\\
&= -X_1 - X_3 + 2X_5.
\end{aligned}$$

In particular, this shows that the overall reaction can arise as the outcome of the five reactions listed in the mechanism. Since equation (6) is the sum of equations (3), (4), and (5), we see that our overall reaction equation can be expressed as

$$\begin{aligned} 0 &= (-X_1 + 2X_2) + 2(-X_2 - X_3 + X_4 + X_5) \\ &\quad + (-X_1 + X_2 - X_4 + X_5) + (X_1 - 2X_2) + (X_2 + X_3 - X_4 - X_5) \\ &= -X_1 - X_3 + 2X_5. \end{aligned}$$

This expression is a sum of positive multiples of each component equation, and it therefore tells us how the reactions in the mechanism could combine to produce the overall reaction. We see that we can view the overall reaction as the following sequence of events:

(a) a bromine molecule disassociates: $Br_2 \to 2Br$;
(b) add two hydrogen molecules: $2H_2 + 2Br \to 2HBr + 2H$;
(c) a free hydrogen finds a second bromine molecule: $H + Br_2 \to HBr + Br$;
(d) a second free hydrogen reacts with a hydrogen bromide molecule: $HBr + H \to H_2 + Br$, returning a hydrogen molecule and a free bromine;
(e) two bromine atoms find each other: $2Br \to Br_2$, returning another bromine molecule to the system.

Note that although two H_2 molecules and two Br_2 molecules participated in the reaction mechanism, the overall effect was simply $H_2 + Br_2 \to 2HBr$.

Subspaces Associated with Matrices

Determining the span of the rows of the matrix considered in our analysis of the hydrogen bromide stoichiometry was crucial to explaining the reaction mechanism. We need terminology for such spans.

Definition. Suppose that A is an $m \times n$ matrix. The subspace of $\mathbf{R}^n$ spanned by the rows of A is called the *row space* of A. The subspace of $\mathbf{R}^m$ spanned by the columns of A is called the *column space* of A. The subspace of $\mathbf{R}^n$ of all solutions to $A\vec{X} = \vec{0}$ is called the *null space* of A. The row space and column space of A are abbreviated as row(A) and col(A). We denote the null space by ker(A), where "ker" is short for "kernel."

Determining the Row Space

For an arbitrary matrix A it is usually quite difficult to determine any properties of its row space just by looking at the matrix. In order to find the set of solutions to a homogeneous system of equations (which is finding the null space), one can apply Gaussian elimination to obtain a row equivalent (reduced), row-echelon matrix. In order to study the row space of a matrix, one can again use Gaussian elimination.

Recall from Sec. 4.1 that the rows of a matrix product PB are always linear combinations of the rows of B. This means that any linear combinations of the rows of PB must be a linear combination of the rows of B. In other words, the row space of PB is contained in the row space of B. Furthermore, if P is invertible, then as $B = P^{-1}(PB)$, this same observation shows that the row space of B is contained in the row space of PB. This shows that whenever P is invertible, both B and PB have the same row space.

This observation is extremely useful. Recall that all row operations are given by multiplication by an invertible matrix. So if A and B are row equivalent matrices, then A and B have the same row space. This is restated as the next theorem.

Theorem 24. *Suppose that A and B are row equivalent matrices. Then* $\mathrm{row}(A) = \mathrm{row}(B)$.

For example, the row reduction in our analysis in Sec. 5.1 gives

$$\begin{pmatrix} -1 & 2 & 0 & 0 & 0 \\ 0 & -1 & -1 & 1 & 1 \\ -1 & 1 & 0 & -1 & 1 \\ 1 & -2 & 0 & 0 & 0 \\ 0 & 1 & 1 & -1 & -1 \end{pmatrix} \mapsto \begin{pmatrix} 1 & -2 & 0 & 0 & 0 \\ 0 & 1 & 1 & -1 & -1 \\ 0 & 0 & 1 & -2 & 0 \\ 0 & 0 & 0 & 0 & 0 \\ 0 & 0 & 0 & 0 & 0 \end{pmatrix}.$$

The second matrix is the reduced row-echelon form of the first. From this we see that $(1, -2, 0, 0, 0)$, $(0, 1, 1, -1, -1)$, and $(0, 0, 1, -2, 0)$ form a spanning set for the row space of each matrix.

Observe that we can have three vectors in this spanning set because the rank of the matrix is 3. In fact, the rank is the minimum possible number of spanning vectors for the row space. This will discussed in detail shortly.

The Null Space and Gaussian Elimination

We developed Gaussian elimination in order to solve systems of equations. Since the null space of a matrix is precisely the set of solutions to the associated homogeneous system of equations, and since row operations do not change the set of solutions to a system of equations, we obtain the next theorem.

Theorem 25. *Suppose that A and B are row equivalent matrices. Then* $\ker(A) = \ker(B)$.

For example, consider the matrix

$$C = \begin{pmatrix} 1 & 0 & 1 & 0 \\ 2 & 3 & 4 & 5 \\ 4 & 3 & 6 & 5 \end{pmatrix}.$$

Row-reducing gives

$$\begin{pmatrix} 1 & 0 & 1 & 0 \\ 2 & 3 & 4 & 5 \\ 4 & 3 & 6 & 5 \end{pmatrix} \mapsto \begin{pmatrix} 1 & 0 & 1 & 0 \\ 0 & 3 & 2 & 5 \\ 0 & 3 & 2 & 5 \end{pmatrix}$$
$$\mapsto \begin{pmatrix} 1 & 0 & 1 & 0 \\ 0 & 3 & 2 & 5 \\ 0 & 0 & 0 & 0 \end{pmatrix}.$$

The null space is the span of two vectors,

$$\text{span}\left\{ \begin{pmatrix} 0 \\ 5 \\ 0 \\ -3 \end{pmatrix}, \begin{pmatrix} 3 \\ 2 \\ -3 \\ 0 \end{pmatrix} \right\}.$$

To see this, note that the last two variables in the homogeneous system associated with the matrix at the end of our row reduction are free. For convenience we can choose $W = -3$, $Z = 0$ to get the first vector, and $W = 0$, $Z = -3$ to get the second. (The elimination also shows that the row space of C is the span of two vectors, $\text{span}\{(1, 0, 1, 0), (0, 3, 2, 5)\}$.)

Linear Independence

The matrix considered in our discussion of hydrogen bromide had a row space spanned by its five rows, while a row-equivalent matrix has the same row space with a spanning set of three vectors. Is it possible that two vectors could suffice? The answer is no, and the key concept that enables us to answer this is that of *linear independence.*

The reader must study this definition carefully. In particular, note the "if . . . , then" nature of this definition. There is a great tendency among students (at least initially) to oversimplify the meaning of linear independence.

Definition. A set of vectors $\{\vec{v}_1, \vec{v}_2, \ldots, \vec{v}_n\}$ is said to be *linearly independent* if whenever $r_1, r_2, \ldots, r_n$ are real numbers and $r_1\vec{v}_1 + r_2\vec{v}_2 + \cdots + r_n\vec{v}_n = \vec{0}$, then $r_1 = 0, r_2 = 0, \ldots, r_n = 0$. In other words, the only way to express $\vec{0}$ as a linear combination of $\vec{v}_1, \vec{v}_2, \ldots, \vec{v}_n$ is to use only zeros as coefficients. If $\{\vec{v}_1, \vec{v}_2, \ldots, \vec{v}_n\}$ is not linearly independent, we say that $\{\vec{v}_1, \vec{v}_2, \ldots, \vec{v}_n\}$ is *linearly dependent.*

According to the definition, $\{\vec{v}_1, \vec{v}_2\}$ is linearly dependent if one can find $r_1, r_2 \in \mathbf{R}$ (not both 0) so that $r_1\vec{v}_1 + r_2\vec{v}_2 = \vec{0}$. If $r_1 \neq 0$, we find that $\vec{v}_1 = -\frac{r_2}{r_1}\vec{v}_2$. In case $r_1 = 0$, we find that $\vec{v}_2 = \vec{0}$. This shows that whenever a set of two vectors is linearly dependent one of the vectors is a multiple of the other.

Examples of Linear Independence

Note that no one of the vectors

$$\begin{pmatrix} 1 \\ 0 \end{pmatrix}, \begin{pmatrix} 0 \\ 1 \end{pmatrix}, \begin{pmatrix} 1 \\ 1 \end{pmatrix}$$

is a multiple of another. However, the set of these three vectors is linearly dependent since

$$\begin{pmatrix} 1 \\ 0 \end{pmatrix} + \begin{pmatrix} 0 \\ 1 \end{pmatrix} - \begin{pmatrix} 1 \\ 1 \end{pmatrix} = \begin{pmatrix} 0 \\ 0 \end{pmatrix}.$$

This example shows that you cannot hastily generalize to more than two vectors our observation that a set of two vectors is linearly independent if and only if one is a multiple of the other.

The definition of linear independence is an "if ..., then" statement. This means that if you want to see whether a collection of vectors is linearly independent, you must first make an assumption and then check a second condition. For example, to show that the vectors

$$\begin{pmatrix} 1 \\ 0 \\ 1 \end{pmatrix}, \begin{pmatrix} 1 \\ 1 \\ 1 \end{pmatrix}, \begin{pmatrix} 0 \\ 1 \\ 1 \end{pmatrix}$$

are linearly independent in $\mathbf{R}^3$, we assume that the equation

$$a\begin{pmatrix} 0 \\ 0 \\ 1 \end{pmatrix} + b\begin{pmatrix} 1 \\ 0 \\ 1 \end{pmatrix} + c\begin{pmatrix} 0 \\ 1 \\ 0 \end{pmatrix} = \begin{pmatrix} 0 \\ 0 \\ 0 \end{pmatrix}$$

is true. These equations say that $b = 0$, $c = 0$, and $a + b = 0$. Of course, from this we find $a = 0$ as well. This is the desired conclusion in the definition of linear independence, so we have shown what is required.

The zero vector can never be part of a linearly independent set of vectors. For example, the set of vectors

$$\left\{\begin{pmatrix} 0 \\ 0 \end{pmatrix}, \begin{pmatrix} 1 \\ 1 \end{pmatrix}\right\}$$

is *not* linearly independent since one can express

$$1\begin{pmatrix} 0 \\ 0 \end{pmatrix} + 0\begin{pmatrix} 1 \\ 1 \end{pmatrix} = \begin{pmatrix} 0 \\ 0 \end{pmatrix}.$$

Linear Independence and Gaussian Elimination

Recall from Sec. 4.1 that if A is a matrix and if $\vec{v}$ is a column vector, then the column $A\vec{v}$ is a linear combination of the columns of A. This observation enables us to test for linear independence by calculating matrix rank.

Theorem 26. *Suppose $\vec{v}_1, \vec{v}_2, \ldots, \vec{v}_m$ is a collection of n-vectors. Let $A = (\vec{v}_1 \quad \vec{v}_2 \quad \cdots \quad \vec{v}_m)$ be the $n \times m$ matrix whose columns are the $\vec{v}_i$. Then $\{\vec{v}_1, \vec{v}_2, \ldots, \vec{v}_m\}$ is linearly independent if and only if* $\mathrm{rk}(A) = m$.

Proof. Expanding the matrix equation $A\vec{X} = \vec{0}$ according to the definition of multiplication, we find

$$(\vec{v}_1 \quad \vec{v}_2 \quad \cdots \quad \vec{v}_m)\begin{pmatrix} X_1 \\ X_2 \\ \vdots \\ X_m \end{pmatrix} = \vec{0}$$

if and only if

$$X_1\vec{v}_1 + X_2\vec{v}_2 + \cdots + X_m\vec{v}_m = \vec{0}.$$

This shows that the linear independence of $\{\vec{v}_1, \vec{v}_2, \ldots, \vec{v}_m\}$ is equivalent to the system $A\vec{X} = \vec{0}$ having the unique solution $X_1 = 0, X_2 = 0, \ldots, X_m = 0$. According to the Theorem 7(ii), this occurs if and only if $\mathrm{rk}(A) = m$. This proves the theorem. □

As an example we consider the vectors

$$\begin{pmatrix} -1 \\ 0 \\ -1 \\ 1 \\ 0 \end{pmatrix}, \begin{pmatrix} 2 \\ -1 \\ 1 \\ -2 \\ 1 \end{pmatrix}, \begin{pmatrix} 0 \\ -1 \\ 0 \\ 0 \\ 1 \end{pmatrix}, \begin{pmatrix} 0 \\ 1 \\ -1 \\ 0 \\ -1 \end{pmatrix}, \begin{pmatrix} 0 \\ 1 \\ 1 \\ 0 \\ -1 \end{pmatrix},$$

which are the columns of our matrix from the beginning of this section. Our row reduction

$$\begin{pmatrix} -1 & 2 & 0 & 0 & 0 \\ 0 & -1 & -1 & 1 & 1 \\ -1 & 1 & 0 & -1 & 1 \\ 1 & -2 & 0 & 0 & 0 \\ 0 & 1 & 1 & -1 & -1 \end{pmatrix} \mapsto \begin{pmatrix} 1 & -2 & 0 & 0 & 0 \\ 0 & 1 & 1 & -1 & -1 \\ 0 & 0 & 1 & -2 & 0 \\ 0 & 0 & 0 & 0 & 0 \\ 0 & 0 & 0 & 0 & 0 \end{pmatrix}$$

shows these matrices have rank 3, and therefore the set of the original five vectors is not linearly independent. However the first three vectors listed form a linearly independent set since the matrix consisting of the first three columns has rank 3.

Problems

1. Find a spanning set of as few vectors as possible for the row space, column space, and null space of each of the following matrices.

   (a) $\begin{pmatrix} 1 & 2 & 1 \\ 3 & 6 & 3 \\ 2 & 4 & 2 \end{pmatrix}$ (b) $\begin{pmatrix} 1 & 3 & 4 \\ 1 & 3 & 5 \end{pmatrix}$ (c) $\begin{pmatrix} 2 & 1 & 4 \\ 1 & 0 & 5 \\ 3 & 1 & 9 \end{pmatrix}$

2. Find a spanning set of as few vectors as possible for the row space, column space, and null space of each of the following matrices.

   (a) $\begin{pmatrix} 2 & 1 \\ 1 & 5 \\ 1 & 2 \end{pmatrix}$ (b) $\begin{pmatrix} 2 & 7 & 8 & 4 \\ 1 & 3 & 7 & 1 \\ 1 & 4 & 1 & 3 \end{pmatrix}$ (c) $\begin{pmatrix} 2 & 3 & 6 & 8 \\ 2 & 4 & 3 & 1 \end{pmatrix}$

3. Which of the following sets of vectors are linearly independent?

   (a) $\left\{ \begin{pmatrix} 1 \\ 1 \\ 7 \end{pmatrix}, \begin{pmatrix} 1 \\ 1 \\ 2 \end{pmatrix}, \begin{pmatrix} 1 \\ 1 \\ 2 \end{pmatrix} \right\}$

   (b) $\left\{ \begin{pmatrix} 8 \\ 2 \\ 4 \end{pmatrix}, \begin{pmatrix} 4 \\ 1 \\ 2 \end{pmatrix} \right\}$

   (c) $\left\{ \begin{pmatrix} 0 \\ 1 \\ 0 \end{pmatrix}, \begin{pmatrix} 1 \\ 1 \\ 1 \end{pmatrix}, \begin{pmatrix} 4 \\ 7 \\ 4 \end{pmatrix} \right\}$

   (d) $\left\{ \begin{pmatrix} 8 \\ 18 \\ 3 \end{pmatrix}, \begin{pmatrix} 10 \\ 99 \\ 41 \end{pmatrix}, \begin{pmatrix} 11 \\ 2 \\ 0 \end{pmatrix}, \begin{pmatrix} 13 \\ 41 \\ 66 \end{pmatrix} \right\}$

4. Show that two matrices, possibly of different size, with the same row space have the same rank. (Hint: Study their reduced echelon forms.)

5. For each of the following collections of vectors, find all subcollections of minimal size with the same span.

   (a) $\begin{pmatrix} 2 \\ 1 \\ 1 \end{pmatrix}, \begin{pmatrix} 0 \\ 1 \\ 2 \end{pmatrix}, \begin{pmatrix} 4 \\ 4 \\ 6 \end{pmatrix}$

   (b) $\begin{pmatrix} 1 \\ 2 \\ 3 \end{pmatrix}, \begin{pmatrix} 3 \\ 2 \\ 1 \end{pmatrix}, \begin{pmatrix} 2 \\ 2 \\ 2 \end{pmatrix}, \begin{pmatrix} 1 \\ 1 \\ 1 \end{pmatrix}$

   (c) $\begin{pmatrix} 1 \\ 3 \\ 2 \end{pmatrix}, \begin{pmatrix} 0 \\ 2 \\ 6 \end{pmatrix}, \begin{pmatrix} 4 \\ 0 \\ 1 \end{pmatrix}$

   (d) $\begin{pmatrix} 0 \\ 0 \\ 0 \\ 0 \end{pmatrix}, \begin{pmatrix} 1 \\ 1 \\ 1 \\ 1 \end{pmatrix}, \begin{pmatrix} 1 \\ 0 \\ 0 \\ 1 \end{pmatrix}, \begin{pmatrix} 0 \\ 3 \\ 3 \\ 0 \end{pmatrix}, \begin{pmatrix} 2 \\ 1 \\ 1 \\ 2 \end{pmatrix}$

6. Assume that the set $\{\vec{v}_1, \vec{v}_2, \vec{v}_3\}$ is linearly independent in some vector space. Which of the following sets are linearly independent?

(a) $\{\vec{v}_1 + \vec{v}_2 + \vec{v}_3, \vec{v}_2 + \vec{v}_3\}$
(b) $\{\vec{0}, \vec{v}_1, \vec{v}_2 + \vec{v}_3\}$
(c) $\{9\vec{v}_1, 5\vec{v}_2, 6\vec{v}_1 - 7\vec{v}_2\}$
(d) $\{\vec{v}_1 + \vec{v}_2, \vec{v}_1 + \vec{v}_3, \vec{v}_2 - \vec{v}_3\}$

7. Show that a subset of any linearly independent set of vectors is linearly independent.
8. (a) Can you find three vectors in $\mathbf{R}^3$ that are linearly dependent, any two of which are linearly independent? If so, give such; if not, show why.
   (b) What happens to part (a) if you try it in $\mathbf{R}^2$?
9. Assume that $\vec{v}_1, \vec{v}_2, \ldots, \vec{v}_n$ are vectors in some vector space that satisfy the conditions $\vec{v}_1 \neq \vec{0}$, $\vec{v}_2 \notin \text{span}\{\vec{v}_1\}$, $\vec{v}_3 \notin \text{span}\{\vec{v}_1, \vec{v}_2\}, \ldots, \vec{v}_n \notin \text{span}\{\vec{v}_1, \vec{v}_2, \ldots, \vec{v}_{n-1}\}$. Show that $\{\vec{v}_1, \vec{v}_2, \ldots, \vec{v}_n\}$ is linearly independent.
10. Suppose that U and V are subspaces of $\mathbf{R}^n$ and that the subsets $\{\vec{u}_1, \vec{u}_2, \ldots, \vec{u}_s\} \subset U$ and $\{\vec{v}_1, \vec{v}_2 \ldots, \vec{v}_t\} \subset V$ are linearly independent. If $U \cap V = \{\vec{0}\}$, show that $\{\vec{u}_1, \vec{u}_2, \ldots, \vec{u}_s, \vec{v}_1, \vec{v}_2 \ldots, \vec{v}_t\}$ is linearly independent.
11. Assume that $\{\vec{v}_1, \vec{v}_2, \ldots, \vec{v}_n\}$ is a linearly independent subset of vectors in $\mathbf{R}^m$. Suppose that A is an $m \times m$ invertible matrix. Show that $\{A\vec{v}_1, A\vec{v}_2, \ldots, A\vec{v}_n\}$ is linearly independent.
12. Find conditions on the real numbers r, s, t which guarantee that

$$\left\{ \begin{pmatrix} 1 \\ 1 \\ r \end{pmatrix}, \begin{pmatrix} 1 \\ 0 \\ s \end{pmatrix}, \begin{pmatrix} -2 \\ -2 \\ t \end{pmatrix} \right\}$$

is linearly independent.

Group Project: How Many Linearly Independent Vectors Are Possible?

(a) Suppose that $\{\vec{v}_1, \vec{v}_2, \ldots, \vec{v}_m\}$ is linearly dependent in $\mathbf{R}^p$. Show there exists some j with $1 \leq j \leq m$ such that $\vec{v}_j$ is a linear combination of $\vec{v}_1, \vec{v}_2, \ldots, \vec{v}_{j-1}$. As a hint, write out a linear combination that is $\vec{0}$ and look for the last nonzero coefficient.

(b) Let V be a subspace of $\mathbf{R}^p$. Suppose that $\{\vec{w}_1, \vec{w}_2, \ldots, \vec{w}_m\}$ is linearly independent and each of $\vec{w}_1, \vec{v}_2, \ldots, \vec{w}_m$ is a linear combination of $\vec{v}_1, \vec{v}_2, \ldots, \vec{v}_n$. Show that $m \leq n$. In order to do this, one approach is to assume that $m > n$ and derive a contradiction. Consider the equation $X_1\vec{w}_1 + X_2\vec{w}_2 + \cdots + X_m\vec{w}_m = \vec{0}$ in the variables $X_1, X_2, \ldots, X_m$. Substitute expressions for the $\vec{w}_i$ in terms of the $\vec{v}_j$ in this equation and obtain

$$\begin{aligned} \vec{0} = {} & X_1(a_{11}\vec{v}_1 + a_{21}\vec{v}_2 + \cdots + a_{n1}\vec{v}_n) \\ & + X_2(a_{12}\vec{v}_1 + a_{22}\vec{v}_2 + \cdots + a_{n2}\vec{v}_n) \\ & + \cdots + X_m(a_{1m}\vec{v}_1 + a_{2m}\vec{v}_2 + \cdots + a_{nm}\vec{v}_n) \end{aligned}$$

$$= (a_{11}X_1 + a_{12}X_2 + \cdots + a_{1m}X_m)\vec{v}_1$$
$$+ (a_{21}X_1 + a_{22}X_2 + \cdots + a_{2m}X_m)\vec{v}_2$$
$$+ \cdots + (a_{n1}X_1 + a_{n2}X_2 + \cdots + a_{nm}X_m)\vec{v}_n.$$

What do you know about such systems?

5.3 Basis and Dimension

The Hydrogen Bromide Reaction Again

In the last section we saw how the study of linear combinations of equations is used in stoichiometry. At that time we worked with linear equations that represented the various reactions that occur in formation of hydrogen bromide. In our next investigation we will simplify the notation used in Sec. 5.2 and represent these linear equations by the row vectors whose ith entry are the coefficients of X_i. So, for example, our reaction $H_2 + Br \to HBr + H$, which was represented by the equation $-X_2 - X_3 + X_4 + X_5 = 0$, is now represented by the row vector $(0, -1, -1, 1, 1)$.

We now consider the question of whether it is possible to construct a simpler reaction mechanism involving only four of the component reactions considered earlier. We ask if $H_2 + Br_2 \to 2HBr$ could arise from a combination of the following four reactions.

$$Br_2 \to 2Br$$
$$H_2 + Br \to HBr + H$$
$$2Br \to Br_2$$
$$HBr + H \to H_2 + Br$$

We have deleted the third reaction $H + Br_2 \to HBr + Br$ from our earlier mechanism. It is conceivable that these reactions could suffice in the production of hydrogen bromide, because hydrogen bromide is a product of the second reaction, and hydrogen and bromine are the only reactants needed.

To solve this question, we must find out if the vector $(-1, 0, -1, 0, 2)$ representing our overall reaction $H_2 + Br_2 \to 2HBr$ is a linear combination of the vectors representing our four reactions. If we write the four vectors as columns, Theorem 22 shows that this is true if and only if the system of equations represented by the augmented matrix

$$\left(\begin{array}{cccc|c} -1 & 0 & 0 & 1 & -1 \\ 2 & -1 & 1 & 2 & 0 \\ 0 & -1 & 1 & 0 & -1 \\ 0 & 1 & -1 & 0 & 0 \\ 0 & 1 & -1 & 0 & 2 \end{array}\right)$$

has a solution. Inspection of the bottom two equations shows that this system is inconsistent. So, no matter how much we had hoped, *the overall reaction* $H_2 + Br_2 \to 2HBr$ *cannot be explained by the four listed reactions.*

Our earlier calculations showed that the linearly independent set of three vectors $\{(-1, 2, 0, 0, 0), (0, -1, -1, 1, 1), (0, 0, -1, 2, 0)\}$ contained the vector $(-1, 0, -1, 0, 2)$ in its span. But this vector does not lie in the span of the two vectors $(-1, 2, 0, 0, 0)$ and $(0, -1, -1, 1, 1)$. This information was crucial in helping us understand possible reaction mechanism. Collections of vectors that span a vector space and are linearly independent are extremely important. They are called bases.

Basis

Consider the row space $V = \text{row}(A)$ of the reduced row-echelon matrix

$$A = \begin{pmatrix} 1 & 0 & 3 & 0 \\ 0 & 1 & 5 & 0 \\ 0 & 0 & 0 & 1 \end{pmatrix}.$$

Every vector in V is, by definition, a linear combination of the rows of A. By considering such linear combinations, we can express

$$\begin{aligned} V &= \{r(1, 0, 3, 0) + s(0, 1, 5, 0) + t(0, 0, 0, 1) \mid r, s, t \in \mathbf{R}\} \\ &= \{(r, s, 3r + 5s, t) \mid r, s, t \in \mathbf{R}\}. \end{aligned}$$

Inspecting this description of V carefully, we observe that every element of V can be expressed *uniquely* as a linear combination $r(1, 0, 3, 0) + s(0, 1, 5, 0) + t(0, 0, 0, 1)$ of the rows of A. This crucial property of the rows of A motivates the following definition.

Definition. A set of vectors $\{\vec{v}_1, \vec{v}_2, \ldots, \vec{v}_n\}$ in a vector space V is called a *basis* for V if every vector $\vec{w} \in V$ can be expressed *uniquely* as a linear combination

$$\vec{w} = a_1\vec{v}_1 + a_2\vec{v}_2 + \cdots a_n\vec{v}_n$$

for real numbers $a_1, a_2, \ldots, a_n$.

The simplest example of a basis is the following. We denote by $\vec{e}_i = (0, \ldots, 0, 1, 0 \ldots, 0)$ the vector in $\mathbf{R}^n$ containing all zeros except for a one in the ith entry. It is easy to see that $\{\vec{e}_1, \vec{e}_2, \ldots, \vec{e}_n\}$ is a basis for $\mathbf{R}^n$, because any element $(a_1, a_2, \ldots, a_n) \in \mathbf{R}^n$ can be written uniquely as the linear combination

$$(a_1, a_2, \ldots, a_n) = a_1\vec{e}_1 + a_2\vec{e}_2 + \cdots + a_n\vec{e}_n.$$

$\{\vec{e}_1, \vec{e}_2, \ldots, \vec{e}_n\}$ is called the *standard basis* for $\mathbf{R}^n$.

A Basis Is Linearly Independent and Spanning

Suppose that $\{\vec{v}_1, \vec{v}_2, \ldots, \vec{v}_n\}$ is a basis for a vector space V. Then, since every vector in V is a linear combination of $\{\vec{v}_1, \vec{v}_2, \ldots, \vec{v}_n\}$, these vectors span V. Further, by our uniqueness assumption, the only way to express $\vec{0}$ as a linear combination of $\{\vec{v}_1, \vec{v}_2, \ldots, \vec{v}_n\}$ is $\vec{0} = 0\vec{v}_1 + 0\vec{v}_2 + \cdots + 0\vec{v}_n$. This shows that $\{\vec{v}_1, \vec{v}_2, \ldots, \vec{v}_n\}$ is linearly independent.

The converse of this is also true. Suppose that $\{\vec{v}_1, \vec{v}_2, \ldots, \vec{v}_n\}$ is linearly independent and spans a vector space V. Then every vector $\vec{v} \in V$ can be written as a linear combination

$$\vec{w} = a_1\vec{v}_1 + a_2\vec{v}_2 + \cdots + a_n\vec{v}_n$$

for real numbers $a_1, a_2, \ldots, a_n$. We ask if this representation is unique. Suppose there were another representation, that is, assume we have

$$\vec{w} = b_1\vec{v}_1 + b_2\vec{v}_2 + \cdots + b_n\vec{v}_n$$

for real numbers $b_1, b_2, \ldots, b_n$. Subtracting gives

$$\vec{0} = (a_1 - b_1)\vec{v}_1 + (a_2 - b_2)\vec{v}_2 + \cdots + (a_n - b_n)\vec{v}_n.$$

The linear independence of our basis vectors $\{\vec{v}_1, \vec{v}_2, \ldots, \vec{v}_n\}$ shows that $(a_1 - b_1) = 0, (a_2 - b_2) = 0, \ldots, (a_n - b_n) = 0$, and the uniqueness of our representation for $\vec{v}$ follows.

We have established the next theorem. Often this characterization is used for the definition of bases.

Theorem 27. *A set of vectors $\{\vec{v}_1, \vec{v}_2, \ldots, \vec{v}_n\}$ is a basis for a vector space V if and only if both of these conditions hold:*
(i) *$\{\vec{v}_1, \vec{v}_2, \ldots, \vec{v}_n\}$ is linearly independent;*
(ii) *$\text{span}\{\vec{v}_1, \vec{v}_2, \ldots, \vec{v}_n\} = V$.*

Finding Bases

Suppose that V is the span of a set of m vectors in $\mathbf{R}^n$. How do we find a basis for V? We can write these vectors as the rows of an $m \times n$ matrix and row-reduce to a row-echelon matrix. The nonzero rows of this row-echelon matrix will be a basis for V, because the row space of the row-echelon matrix is the same as that of the original matrix and the nonzero rows of a row-echelon matrix are linearly independent.

For example, if we start with the matrix

$$B = \begin{pmatrix} 1 & 1 & 8 & 2 \\ 0 & 1 & 5 & 3 \\ 1 & 0 & 3 & 0 \\ 1 & 1 & 8 & 5 \end{pmatrix}$$

and then apply Gaussian elimination, we find that its reduced row-echelon form is

$$R = \begin{pmatrix} 1 & 0 & 3 & 0 \\ 0 & 1 & 5 & 0 \\ 0 & 0 & 0 & 1 \\ 0 & 0 & 0 & 0 \end{pmatrix}.$$

The three nonzero rows of R are the three rows of the matrix A considered just before in the Basis subsection. This shows that the row space of B is same as the row space of A and that $\{(1, 0, 3, 0), (0, 1, 5, 0), (0, 0, 0, 1)\}$ is a basis for this vector space.

This observation shows even more. We see that for any $m \times n$ matrix M, the row space V of M has a basis with $r = \text{rk}(M)$ elements. This basis consists of the nonzero rows of R, where R is the reduced row-echelon form of M. Observe that the nonzero rows of any reduced row-echelon matrix are linearly independent because of the condition on the leading ones. Moreover, *it turns out that the number of elements in any two bases of a vector space is always the same.* To help understand this, suppose the row space V of M had a basis with s vectors where $s \leq r$. Let N be the $m \times n$ matrix with these s vectors as top rows and other rows zero, and let R' be the reduced echelon form of N. Since $\text{row}(N) = V = \text{row}(M)$, we must have $\text{row}(R) = \text{row}(R')$. Since both R and R' are reduced row-echelon matrices with the same row space, a row-by-row comparison starting at the top rows shows they must be the same. We conclude that $s = r$. (This is essentially the uniqueness of the reduced echelon form given in Theorem 4. See the group project at the end of this section for further discussion.)

The number of elements in any basis of a vector space is called its *dimension*. This shows that the row space of a matrix M is r-dimensional, where $r = \text{rk}(M)$.

Definition. Suppose that V is a vector space and has a basis with n elements for some natural number n. Then we say that V is *n-dimensional.*

For example, the dimension of the row space of matrix B just given is 3. Summarizing all of this information, we give the next results.

Theorem 28. *Suppose that $\{\vec{v}_1, \vec{v}_2, \ldots, \vec{v}_n\}$ and $\{\vec{w}_1, \vec{w}_2, \ldots, \vec{w}_m\}$ are both bases of a vector space V. Then $n = m$; that is, any two bases of a vector space have the same number of elements.*

Corollary. *Let A be an $m \times n$ matrix and suppose that R is a row-echelon form of A. A basis for the row space of A is the collection of nonzero rows of R. Therefore, the dimension of the row space of A is* $\text{rk}(A)$.

Examples of Bases

In view of the existence of the standard basis for $\mathbf{R}^n$, we see that (as expected!) the dimension of $\mathbf{R}^n$ is n. By convention the zero vector space $\{\vec{0}\}$ has as a basis the empty set $\varnothing$, which has zero elements. Thus, $\{\vec{0}\}$ is a zero-dimensional vector space. In general, any nonzero vector space has infinitely many different bases. However, all bases of a particular vector space have the same number of elements.

For example, the set of vectors

$$\left\{\begin{pmatrix}2\\1\end{pmatrix}, \begin{pmatrix}1\\1\end{pmatrix}\right\}$$

is a basis for $\mathbf{R}^2$. They are linearly independent by Theorem 25 since

$$\begin{pmatrix}2 & 1\\1 & 2\end{pmatrix}$$

is a rank-2 matrix. Further, any system of equations of the form

$$X\begin{pmatrix}2\\1\end{pmatrix} + Y\begin{pmatrix}1\\2\end{pmatrix} = \begin{pmatrix}a\\b\end{pmatrix}$$

can always be solved whenever a and b are real numbers. Hence the two vectors span $\mathbf{R}^2$.

The set of vectors

$$\left\{\begin{pmatrix}1\\-3\\4\end{pmatrix}, \begin{pmatrix}2\\2\\-5\end{pmatrix}, \begin{pmatrix}0\\-8\\13\end{pmatrix}\right\}$$

is not a basis for $\mathbf{R}^3$ since

$$2\begin{pmatrix}1\\-3\\4\end{pmatrix} - \begin{pmatrix}2\\2\\-5\end{pmatrix} - \begin{pmatrix}0\\-8\\13\end{pmatrix} = \begin{pmatrix}0\\0\\0\end{pmatrix}$$

shows they are not linearly independent.

The set of vectors

$$\left\{\begin{pmatrix}1\\0\\0\end{pmatrix}, \begin{pmatrix}\pi\\1\\0\end{pmatrix}, \begin{pmatrix}e\\\sqrt{2}\\1\end{pmatrix}\right\}$$

is a basis for $\mathbf{R}^3$. They are linearly independent and span $\mathbf{R}^3$.

A Basis for Our Network Flow Problem

In Sec. 5.1 we studied the problem of optimizing the flow in a freeway network. We found eight equations that gave useful information about the freeway system, but we also discovered that five equations contained all the information needed to solve our problem. We might wonder if fewer equations could suffice. To find out, we apply what we have learned about bases. We have a collection of eight equations in seven unknowns. If we order the variables of our equations as X_{AB}, X_{AD}, X_{BE}, X_{BC}, X_{DC}, X_{DE}, and X_{CE}, then our eight equations (see the list in Sec. 5.1) are represented by the augmented matrix

$$\left(\begin{array}{ccccccc|c} 1 & 1 & 0 & 0 & 0 & 0 & 0 & f \\ 0 & 0 & 1 & 0 & 0 & 1 & 1 & f \\ -1 & 0 & 1 & 1 & 0 & 0 & 0 & 0 \\ 0 & -1 & 0 & 0 & 1 & 1 & 0 & 0 \\ 0 & 0 & 0 & 1 & 1 & 0 & -1 & 0 \\ 1 & 0 & 0 & 0 & 1 & 1 & 0 & f \\ 0 & 0 & 1 & 1 & 1 & 1 & 0 & f \\ 0 & 1 & 1 & 1 & 0 & 0 & 0 & f \end{array}\right).$$

This matrix can be row-reduced to

$$\left(\begin{array}{ccccccc|c} 1 & 1 & 0 & 0 & 0 & 0 & 0 & f \\ 0 & 1 & 0 & 0 & -1 & -1 & 0 & 0 \\ 0 & 0 & 1 & 0 & 0 & 1 & 1 & f \\ 0 & 0 & 0 & 1 & 1 & 0 & -1 & 0 \\ 0 & 0 & 0 & 0 & 1 & -1 & -2 & -f \\ 0 & 0 & 0 & 0 & 0 & 0 & 0 & 0 \\ 0 & 0 & 0 & 0 & 0 & 0 & 0 & 0 \\ 0 & 0 & 0 & 0 & 0 & 0 & 0 & 0 \end{array}\right),$$

which is a rank-5 matrix. In particular, we see that any basis of the row space of our original matrix must have five elements. This means that five equations can summarize the information in our network flow (as we have seen), but it also means that *no fewer than five equations* can contain this same information.

Bases as Minimal Spanning Sets

A *minimal spanning set* is a collection of vectors that span a vector space V but for which any proper subset does *not* span V. (A proper subset of a set S is a subset different from S.) Suppose that $\{\vec{v}_1, \vec{v}_2, \ldots, \vec{v}_n\}$ spans a vector space V but is not a minimal spanning set. Then some subcollection, say $\{\vec{v}_1, \vec{v}_2, \ldots, \vec{v}_{n-1}\}$ also spans V. This means that $\vec{v}_n$ is in the span of $\{\vec{v}_1, \vec{v}_2, \ldots, \vec{v}_{n-1}\}$, and so we can write $\vec{v}_n = k_1\vec{v}_1 + k_2\vec{v}_2 + \cdots + k_{n-1}\vec{v}_{n-1}$ for some real numbers $k_1, k_2, \ldots, k_{n-1}$. Since $k_1\vec{v}_1 + k_2\vec{v}_2 + \cdots + k_{n-1}\vec{v}_{n-1} - \vec{v}_n = \vec{0}$, we find that $\{\vec{v}_1, \vec{v}_2, \ldots, \vec{v}_n\}$ is *not* linearly independent.

Since bases for vector spaces are linearly independent, and since they are spanning sets, this discussion shows that bases are minimal spanning sets. In fact, every minimal spanning set is a basis. For example, the set of the first five network flow equations in our example is a minimal spanning set and as we have seen is also a basis.

Theorem 29. *Suppose $\vec{v}_1, \vec{v}_2, \ldots, \vec{v}_n$ are elements of a vector space V. Then $\{\vec{v}_1, \vec{v}_2, \ldots, \vec{v}_n\}$ is a basis for V if and only if $\{\vec{v}_1, \vec{v}_2, \ldots, \vec{v}_n\}$ is a minimal spanning set for V.*

Proof. Suppose that $\{\vec{v}_1, \vec{v}_2, \ldots, \vec{v}_n\}$ is a minimal spanning set for V. We must show that $\{\vec{v}_1, \vec{v}_2, \ldots, \vec{v}_n\}$ is linearly independent. If not, suppose that $k_1\vec{v}_1 + k_2\vec{v}_2 + \cdots + k_{n-1}\vec{v}_{n-1} + k_n\vec{v}_n = \vec{0}$, where we assume that $k_n \neq 0$. Then we can write $\vec{v}_n = -k_n^{-1}(k_1\vec{v}_1 + k_2\vec{v}_2 + \cdots + k_{n-1}\vec{v}_{n-1})$, which shows that $\vec{v}_n$ lies in the span of $\vec{v}_1, \vec{v}_2, \ldots, \vec{v}_{n-1}$. But now we see that $\vec{v}_1, \vec{v}_2, \ldots, \vec{v}_n$ is not a minimal spanning set after all, which is a contradiction. This shows that minimal spanning sets are linearly independent and hence are bases. □

Recall that in Sec. 3.4 it was shown that the solutions to the equation $A\vec{X} = \vec{0}$ in n variables can be described by $n - \text{rk}(A)$ parameters. Suppose that $R\vec{X} = \vec{0}$ is a reduced echelon form for a system with $s = n - \text{rk}(A)$ free variables. Denote by $\vec{w}_i$ the solution to $R\vec{X} = \vec{0}$ for which the ith free variable is assigned 1 and all other free variables are assigned 0. Then back-substitution shows that the collection $\{\vec{w}_1, \ldots, \vec{w}_s\}$ is a spanning set for the solution space. In fact, we claim that the collection $\{\vec{w}_1, \ldots, \vec{w}_s\}$ is a minimal spanning set for the solution space, because if $\vec{w}_j$ were left out then any linear combination of the remaining $\vec{w}_i$ would assign the value 0 to the jth free variable. In particular, the solution $\vec{w}_j$ does not lie in the span of the remaining $\vec{w}_i$. Thus no proper subcollection of $\{\vec{w}_1, \ldots, \vec{w}_s\}$ can span. This discussion establishes the following result.

Theorem 30. *Suppose that A is an $m \times n$ matrix. Then* $\ker(A)$ *has a minimal spanning set containing $n - \text{rk}(A)$ vectors.*

The Dimension Theorem

Up to this point, we have studied bases of vector spaces by looking at row spaces of matrices and using the reduced echelon form to find a basis. But bases for the null space of a matrix are also important since they are solution sets to homogeneous systems of linear equations. Our next objective is to understand the dimension theorem. This theorem describes the fundamental relation between the rank of a matrix A and the dimension of its null space, $\ker(A)$. It is an important result about matrices that is also crucial to the study of linear transformations (taken up in Chap. 8).

We just noted that for any matrix A, the vector space $\ker(A)$ has a minimal spanning set with $n - \text{rk}(A)$ elements. This means that the dimension of

$\ker(A)$ is $n - \text{rk}(A)$. The dimension of $\ker(A)$ is called the *nullity* of A and is denoted by $\text{null}(A)$. We shall use this result often, and we will refer to it as the dimension theorem. Sometimes this result is referred to as the "rank plus nullity theorem."

Theorem 31 (Dimension Theorem). *Suppose A is an $m \times n$ matrix and* $\text{rk}(A) = r$. *Then* $\ker(A) = \{\vec{v} \mid A\vec{v} = \vec{0}\}$ *is a $(n-r)$-dimensional vector space. In other words,* $\text{rk}(A) + \text{null}(A) = n$.

As an example, consider the matrix

$$A = \begin{pmatrix} 3 & 0 & 2 & 2 \\ 6 & 1 & 0 & 0 \end{pmatrix}.$$

By inspection, the rows of A are linearly independent, so A has rank 2. Since A is 2×4, $\text{null}(A) = 2$. In particular, the homogeneous system of equations $A\vec{X} = \vec{0}$ has a two-dimensional vector space of solutions. Two linearly independent solutions can be found by inspection. For example, both

$$\begin{pmatrix} 0 \\ 0 \\ -1 \\ 1 \end{pmatrix} \quad \text{and} \quad \begin{pmatrix} 1 \\ -6 \\ -\frac{3}{2} \\ 0 \end{pmatrix}$$

are solutions. Consequently, we know that all the solutions to this homogeneous system are given by $X_1 = t$, $X_2 = -6t$, $X_3 = -u - \frac{3}{2}t$, and $X_4 = u$ for parameters t and u.

Bases for the Row Space, Column Space, and Null Space

Suppose that A is an $m \times n$ matrix and R is the reduced row-echelon form of A. Bases for the row space of A and for the null space were studied earlier in this chapter. However, what can be said about the column space of A? Can it be determined using the reduced row-echelon form? Of course, the column space of A is the same as the row space of A^t, so one could compute the reduced row-echelon form of A^t and find a basis for its row space. But is it possible to find a basis for A using R directly? It turns out one can.

Theorem 26 says that a set of columns of A are linearly independent precisely if the corresponding columns of R are linearly independent. Evidently, the columns of R containing the leading ones are linearly independent. This shows that the corresponding set of columns of A are linearly independent. Furthermore, Theorem 22 shows that the remaining columns of A are linear combinations of these columns. Thus we have found a basis for the column space of A.

We collect all this information in the next result.

Theorem 32 (Finding bases for the row, column, and null spaces). *Suppose that A is an $m \times n$ matrix and R is the reduced row-echelon form of A.*

(i) *A basis for the row space of A is the set of nonzero rows of R. Its dimension is* $\mathrm{rk}(A)$.

(ii) *A basis for the column space of A is the set of columns of A that correspond to the columns containing leading ones in R. The dimension of the column space of A is* $\mathrm{rk}(A)$.

(iii) *A basis for the null space of A can be found using R and back-substitution, with a basis element corresponding to each free variable. The dimension of the null space of A is* $n - \mathrm{rk}(A)$.

Remark. A nice consequence of parts (i) and (ii) of Theorem 32 is the fact that for any matrix A we have $\mathrm{rk}(A) = \mathrm{rk}(A^t)$, because the row space of A^t is the same as the column space of A.

As an example of each of these, we consider the matrix

$$M = \begin{pmatrix} 1 & 3 & 2 & 1 & 0 \\ 0 & 1 & 1 & 0 & 1 \\ 2 & 7 & 5 & 2 & 1 \end{pmatrix}.$$

The reduced echelon form of M is

$$\begin{pmatrix} 1 & 0 & -1 & 1 & -3 \\ 0 & 1 & 1 & 0 & 1 \\ 0 & 0 & 0 & 0 & 0 \end{pmatrix}.$$

Applying (i), we find that the row space of M is two-dimensional where a basis is the rows of the reduced echelon form of M, namely

$$\{(1, 0, -1, 1, -3), (0, 1, 1, 0, 1)\}.$$

If we apply part (ii) of the theorem, we find that the column space of M has as a basis the first and second columns of M, namely

$$\left\{ \begin{pmatrix} 1 \\ 0 \\ 2 \end{pmatrix}, \begin{pmatrix} 3 \\ 1 \\ 7 \end{pmatrix} \right\}.$$

Be sure to return to the original columns of M! Finally, part (iii) tells us that a basis for the null space of M will have $5 - 2 = 3$ vectors, that are obtained by back-substitution. One such basis is

$$\left\{ \begin{pmatrix} 3 \\ -1 \\ 0 \\ 0 \\ 1 \end{pmatrix}, \begin{pmatrix} -1 \\ 0 \\ 0 \\ 1 \\ 0 \end{pmatrix}, \begin{pmatrix} 1 \\ -1 \\ 1 \\ 0 \\ 0 \end{pmatrix} \right\}.$$

Problems

1. Find a basis for the set of solutions to the system of equations

$$\begin{pmatrix} 0 & 1 & 5 & 2 & 8 \\ 2 & 1 & 0 & 0 & 1 \\ 1 & 0 & 0 & 0 & 3 \end{pmatrix} \begin{pmatrix} X \\ Y \\ Z \\ W \\ V \end{pmatrix} = \begin{pmatrix} 0 \\ 0 \\ 0 \end{pmatrix}.$$

2. Consider the matrix

$$M = \begin{pmatrix} 1 & 0 & 1 & 2 & 1 \\ 0 & 1 & 2 & 0 & 1 \\ 1 & 1 & 3 & 0 & 2 \\ 2 & 1 & 4 & 1 & 3 \end{pmatrix}.$$

   (a) Find a basis for the row space of M.

   (b) Find a basis for the column space of M. Show that the leading one columns in the reduced row-echelon form of M are not a basis for $\mathrm{col}(M)$.

   (c) Find a basis for the null space of M.

3. For each of the following matrices find a basis for the row space, the column space, and the null space.

   (a) $\begin{pmatrix} 1 & 1 & 3 \\ 3 & 0 & 2 \end{pmatrix}$ (b) $\begin{pmatrix} 1 & 1 \\ 0 & 1 \\ 2 & 2 \\ 1 & 3 \end{pmatrix}$

   (c) $\begin{pmatrix} 1 & 0 & 1 \\ 1 & 1 & 4 \\ 0 & 1 & 4 \end{pmatrix}$ (d) $\begin{pmatrix} 0 & 1 & 2 \\ 2 & 0 & 1 \\ 2 & 2 & 3 \\ 4 & 2 & 4 \end{pmatrix}$

4. For each of the following matrices find a basis for the row space, the column space, and the null space.

   (a) $\begin{pmatrix} 1 & 3 & 2 \\ 2 & 0 & 1 \\ 5 & 5 & 5 \end{pmatrix}$ (b) $\begin{pmatrix} 1 & 4 & 0 \\ 5 & 0 & 5 \\ 1 & 8 & -1 \end{pmatrix}$

   (c) $\begin{pmatrix} 2 & 1 & 1 & 1 \\ 1 & 1 & 2 & 2 \\ 1 & 2 & 3 & 0 \end{pmatrix}$ (d) $\begin{pmatrix} 1 & 1 \\ 0 & 1 \\ 1 & 0 \\ 2 & 0 \end{pmatrix}$

5. Find a basis for

   (a) $V = \mathit{span}\left\{ \begin{pmatrix} 1 \\ 2 \\ 3 \end{pmatrix}, \begin{pmatrix} 4 \\ 5 \\ 6 \end{pmatrix}, \begin{pmatrix} 7 \\ 8 \\ 9 \end{pmatrix} \right\}$

(b) $W = span\left\{\begin{pmatrix}1\\0\\1\end{pmatrix}, \begin{pmatrix}2\\4\\8\end{pmatrix}, \begin{pmatrix}0\\4\\6\end{pmatrix}, \begin{pmatrix}7\\2\\10\end{pmatrix}\right\}$.

6. Find a basis for

(a) $V = span\left\{\begin{pmatrix}1\\1\\1\\1\end{pmatrix}, \begin{pmatrix}1\\2\\3\\4\end{pmatrix}, \begin{pmatrix}4\\3\\2\\1\end{pmatrix}, \begin{pmatrix}9\\8\\7\\6\end{pmatrix}\right\}$;

(b) $W = span\left\{\begin{pmatrix}1\\4\\7\\2\end{pmatrix}, \begin{pmatrix}2\\2\\2\\6\end{pmatrix}, \begin{pmatrix}3\\6\\9\\8\end{pmatrix}, \begin{pmatrix}1\\1\\1\\1\end{pmatrix}\right\}$.

7. Do you believe that two matrices of the same size and rank are row equivalent? Why or why not?

8. Find all possible subsets of the vectors

$$\left\{\begin{pmatrix}1\\2\\1\end{pmatrix}, \begin{pmatrix}0\\1\\1\end{pmatrix}, \begin{pmatrix}1\\4\\3\end{pmatrix}, \begin{pmatrix}1\\0\\2\end{pmatrix}\right\}$$

that are a basis for their span.

9. Show that each standard basis vector $\vec{e}_1, \vec{e}_2, \vec{e}_3$ is a linear combination of the vectors in $\{(1,0,1),(0,1,1),(1,1,1)\}$. Explain why this is sufficient to show that the set $\{(1,0,1),(0,1,1),(1,1,1)\}$ is a basis for $\mathbf{R}^3$.

10. Suppose that $\{\vec{v}_1, \vec{v}_2, \ldots, \vec{v}_n\}$ is a basis of a vector space V. Assuming that n is odd, show that $\{\vec{v}_1 + \vec{v}_2, \vec{v}_2 + \vec{v}_3, \ldots, \vec{v}_{n-1} + \vec{v}_n, \vec{v}_n + \vec{v}_1\}$ is also basis of V. What happens if n is even?

Group Project: The Uniqueness of the Reduced Row-Echelon Form

Explain why the reduced row-echelon form R of a matrix A is unique using steps (a) and (b) as an outline:

(a) Show that the columns containing leading ones in R are characterized by being precisely those columns that are not linear combinations of the preceding columns.

(b) Show the other columns are completely determined by how they are expressed as a linear combination of the columns that precede them.

Explain how the uniqueness of the reduced echelon form can be used to show that the number of elements in a basis of a vector space is uniquely determined. (These ideas are indicated in the paragraph prior to the Theorem 27.)

Group Project: Extending Linearly Independent Sets to Bases

Suppose that $\{\vec{v}_1, \vec{v}_2, \ldots, \vec{v}_r\}$ is a linearly independent subset of $\mathbf{R}^n$. Explain why there exists $n - r$ vectors $\vec{v}_{r+1}, \vec{v}_{r+2}, \ldots, \vec{v}_n$ for which $\{\vec{v}_1, \vec{v}_2, \ldots, \vec{v}_n\}$ is a basis for $\mathbf{R}^n$. (Hint: Consider the vectors $\vec{e}_1, \vec{e}_2, \ldots, \vec{e}_n$ from the standard basis of $\mathbf{R}^n$. Consider adding these vectors one at a time to your set to fill it out to a basis. When do you add $\vec{e}_i$, and when do you pass it up?)

CHAPTER 6

MORE VECTOR GEOMETRY

6.1 The Dot Product

In this section we study the norm and the dot product in $\mathbf{R}^2$ and $\mathbf{R}^3$. The norm of a vector is its length. Closely related to the norm is the dot product. The dot product is a function of two vectors that is related not only to their length but to the angle between them as well. Both of these functions are crucial tools for using vectors to study the plane and the space around us. Many of the ideas in this chapter are extended to $\mathbf{R}^n$ in Chap. 9, where our intuitive understanding of geometry in $\mathbf{R}^2$ and $\mathbf{R}^3$ can be used (by analogy) to deepen our understanding of n-space.

Vector Length

The length of a vector in $\mathbf{R}^2$ was given in Sec. 2.1. The formula is a consequence of the Pythagorean theorem.

Definition. The vector

$$\vec{v} = \begin{pmatrix} r \\ s \end{pmatrix}$$

has length $\sqrt{r^2 + s^2}$. The length of a vector $\vec{v}$ is denoted $\|\vec{v}\|$ and is sometimes called the *norm* of $\vec{v}$.

For example, if

$$\vec{u} = \begin{pmatrix} 6 \\ -4 \end{pmatrix},$$

then the length of $\vec{u}$ is $\|\vec{u}\| = \sqrt{6^2 + (-4)^2} = \sqrt{36 + 16} = \sqrt{52} = 2\sqrt{13}$.

The situation in $\mathbf{R}^3$ is similar. Let $P = (p, q, r) \in \mathbf{R}^3$. Then the segment $\overline{OP}$ is the hypotenuse of the right triangle $\Delta\, OPQ$, where $Q = (p, q, 0)$. The length of the leg $\overline{PQ}$ is r, and using the length formula for the plane we see that the length of the leg $\overline{OQ}$ is $\sqrt{p^2 + q^2}$. The Pythagorean theorem now shows that the length of $\overline{OP}$ is $\sqrt{(\sqrt{p^2 + q^2})^2 + r^2} = \sqrt{p^2 + q^2 + r^2}$. We record this in the next definition.

Definition. The vector

$$\vec{v} = \begin{pmatrix} p \\ q \\ r \end{pmatrix}$$

has length $\sqrt{p^2 + q^2 + r^2}$. The length of a vector $\vec{v} \in \mathbf{R}^3$ is denoted $\|\vec{v}\|$ and is sometimes called the *norm* of $\vec{v}$ (see Fig. 6.1).

Norm Properties

Suppose that $\vec{v}$ is a vector in $\mathbf{R}^2$ or $\mathbf{R}^3$. The following conditions are satisfied by the norm.

Fig. 6.1. Length of a vector $\vec{v}$ in $\mathbf{R}^3$

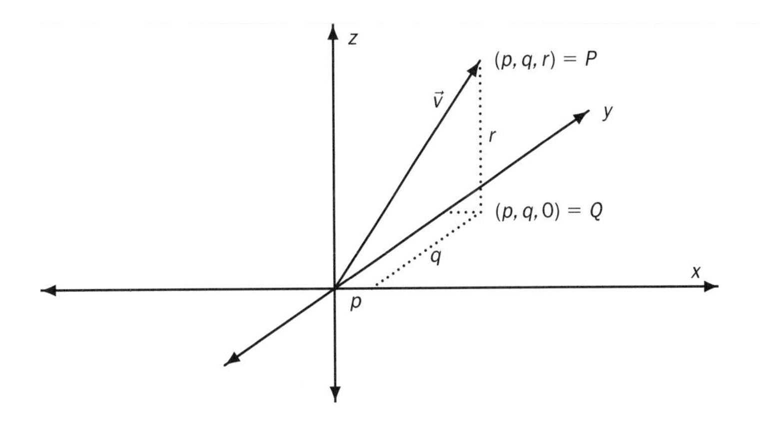

Theorem 33 (Norm properties).

(i) $\|\vec{v}\| \geq 0$ *for all* $\vec{v} \in V$, *and* $\|\vec{v}\| = 0$ *only if* $\vec{v} = \vec{0}$.

(ii) $\|k\vec{v}\| = |k| \cdot \|\vec{v}\|$ *for all real numbers* k *and* $v \in V$.

(iii) $\|\vec{u} + \vec{v}\| \leq \|\vec{u}\| + \|\vec{v}\|$ *for all* $\vec{u}, \vec{v} \in V$.

Property (i) is known as the positive-definite property of the norm. The point of this property is that the length of a vector is never negative and that the only vector with length zero is the zero vector. Property (ii) shows how scalar multiplication affects the length of a vector. This is what you should expect. Multiplying a vector by a real number r multiplies its length by $|r|$. (The absolute value is necessary because vector lengths cannot be negative.) Finally, property (iii) is known as the triangle inequality. This third property can be seen geometrically since it corresponds to the fact that the length of a side of a triangle is always less than the sum of the lengths of the other two sides. This is illustrated in Fig. 6.2.

The first two properties (i) and (ii) of Theorem 33 can be verified by some algebra using the definitions of the distance function. To check property (i) for $\mathbf{R}^3$ suppose that

$$\vec{v} = \begin{pmatrix} p \\ q \\ r \end{pmatrix}.$$

We first note that $p^2 + q^2 + r^2$ is always nonnegative and that it is zero *only* in case *each* of p, q and r are zero. Since $\sqrt{p^2 + q^2 + r^2}$ is always nonnegative and is zero only in case $p^2 + q^2 + r^2$ is zero, the property follows.

We next check property (ii) for $\mathbf{R}^3$. For this we consider

$$\vec{v} = \begin{pmatrix} p \\ q \\ r \end{pmatrix} \quad \text{and} \quad k\vec{v} = \begin{pmatrix} kp \\ kq \\ kr \end{pmatrix}.$$

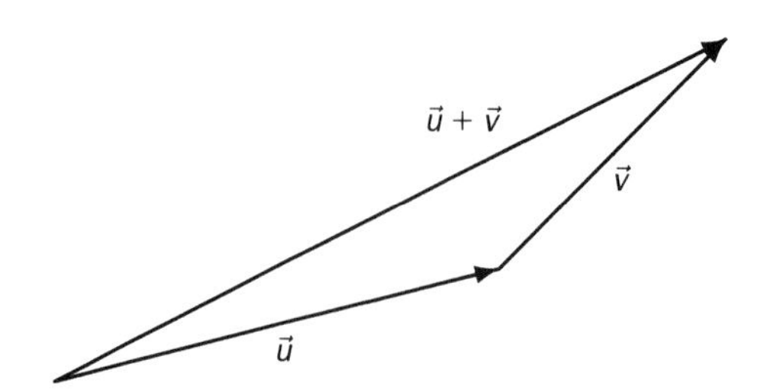

Fig. 6.2. The triangle inequality

Applying the definition of the norm, we find that

$$\begin{aligned}\|k\vec{v}\| &= \sqrt{(kp)^2 + (kq)^2 + (kr)^2} \\ &= \sqrt{k^2(p^2 + q^2 + r^2)} \\ &= |k|\sqrt{p^2 + q^2 + r^2} = |k| \cdot \|\vec{v}\|,\end{aligned}$$

which gives property (ii).

Suppose, for example, we wish to find the distance between the points $R = (2, -4, 1)$ and $S = (3, 7, -1)$ in $\mathbf{R}^3$. The distance between R and S is the length of the directed segment $\overrightarrow{RS}$. This directed segment specifies the vector

$$\begin{pmatrix} 3-2 \\ 7-(-4) \\ -1-1 \end{pmatrix} = \begin{pmatrix} 1 \\ 11 \\ -2 \end{pmatrix}$$

whose length is computed to be $\sqrt{1^2 + 11^2 + (-2)^2} = \sqrt{1 + 121 + 4} = \sqrt{126} = 3\sqrt{14}$.

The Dot Product

Closely related to the norm is dot product. We first give an algebraic definition and then explore the geometry.

Definition. (a) Suppose that

$$\vec{u} = \begin{pmatrix} a_1 \\ a_2 \end{pmatrix} \quad \text{and} \quad \vec{v} = \begin{pmatrix} b_1 \\ b_2 \end{pmatrix}$$

are two vectors in $\mathbf{R}^2$. The *dot product* of $\vec{u}$ and $\vec{v}$ is defined to be

$$\vec{u} \cdot \vec{v} = a_1 b_1 + a_2 b_2 .$$

(b) Suppose that

$$\vec{u} = \begin{pmatrix} a_1 \\ a_2 \\ a_3 \end{pmatrix} \quad \text{and} \quad \vec{v} = \begin{pmatrix} b_1 \\ b_2 \\ b_3 \end{pmatrix}$$

are two vectors in $\mathbf{R}^3$. The *dot product* of $\vec{u}$ and $\vec{v}$ is defined to be

$$\vec{u} \cdot \vec{v} = a_1 b_1 + a_2 b_2 + a_3 b_3 .$$

For example, consider the two vectors

$$\vec{u} = \begin{pmatrix} 2 \\ 5 \end{pmatrix} \quad \text{and} \quad \vec{v} = \begin{pmatrix} -1 \\ 3 \end{pmatrix}$$

in $\mathbf{R}^2$. Then $\vec{u} \cdot \vec{v} = 2(-1) + 5 \cdot 3 = -2 + 15 = 13$. Similarly, if

$$\vec{u} = \begin{pmatrix} 2 \\ 5 \\ 1 \end{pmatrix} \quad \text{and} \quad \vec{v} = \begin{pmatrix} 1 \\ -1 \\ 3 \end{pmatrix}$$

in $\mathbf{R}^3$, then $\vec{u} \cdot \vec{v} = 2 \cdot 1 + 5 \cdot (-1) + 1 \cdot 3 = 2 - 5 + 3 = 0$.

Often, the dot product of two vectors $\vec{u} \cdot \vec{v}$ is denoted by wedges as $\langle \vec{u}, \vec{v} \rangle$. This notation is useful because it reminds us that $\vec{u}$ and $\vec{v}$ are vectors. (Sometimes the dot in the dot product is confused with the multiplication of numbers.) We emphasize, however, that the dot product of two vectors is a real number (*not* a vector)! The bracket notation for the dot product is used in this next list of properties of the dot product.

Theorem 34 (Properties of the dot product).

(i) *Suppose that* $\vec{v}$ *is a vector. Then* $\vec{v} \cdot \vec{v} = \|\vec{v}\|^2$.

(ii) $\langle \vec{u}, \vec{v} \rangle = \langle \vec{v}, \vec{u} \rangle$ *for all vectors* $\vec{u}, \vec{v}$.

(iii) $\langle k\vec{u}, \vec{v} \rangle = k\langle \vec{u}, \vec{v} \rangle$ *for all* $k \in \mathbf{R}$ *and all vectors* $\vec{u}, \vec{v}$.

(iv) $\langle \vec{v}_1 + \vec{v}_2, \vec{u} \rangle = \langle \vec{v}_1, \vec{u} \rangle + \langle \vec{v}_2, \vec{u} \rangle$ *for all vectors* $\vec{u}, \vec{v}_1, \vec{v}_2$.

(v) $\langle \vec{v}, \vec{v} \rangle \geq 0$ *for all vectors* $\vec{v}$, *and* $\langle \vec{v}, \vec{v} \rangle = 0$ *only if* $\vec{v} = \vec{0}$.

To see (i), we consider the vectors

$$\vec{v} = \begin{pmatrix} a \\ b \end{pmatrix} \quad \text{and} \quad \vec{u} = \begin{pmatrix} p \\ q \\ r \end{pmatrix}$$

in $\mathbf{R}^2$ and $\mathbf{R}^3$. Then $\vec{v} \cdot \vec{v} = a \cdot a + b \cdot b = a^2 + b^2 = \|\vec{v}\|^2$ as required. Similarly, we find $\vec{u} \cdot \vec{u} = p \cdot p + q \cdot q + r \cdot r = p^2 + q^2 + r^2 = \|\vec{u}\|^2$ as required. The remaining properties (ii), (iii), (iv), and (v) can be checked by similar calculations.

Property (ii) of the dot product is known as *symmetry*, and (v) is referred to as the *positive-definite* property. Note that because of (ii), properties (iii) and (iv) hold as well on the opposite sides of the dot product bracket. Specifically, we have that

(ii′) $\langle \vec{u}, k\vec{v} \rangle = k\langle \vec{u}, \vec{v} \rangle$ for all $k \in \mathbf{R}$ and all vectors $\vec{u}, \vec{v}$.

(iii′) $\langle \vec{u}, \vec{v}_1 + \vec{v}_2 \rangle = \langle \vec{u}, \vec{v}_1 \rangle + \langle \vec{u}, \vec{v}_2 \rangle$ for all vectors $\vec{u}, \vec{v}_1, \vec{v}_2$.

Properties (iii), (iii′), (iv), and (iv′) taken together are called *bilinearity*.

Geometry of the Dot Product

We now turn to the geometry of the dot product. Our first key result is the following.

Theorem 35. *Suppose that $\vec{u}$ and $\vec{v}$ are nonzero vectors. Then*

$$\vec{u} \cdot \vec{v} = \|\vec{u}\| \, \|\vec{v}\| \cos(\theta),$$

where θ is the angle between the two vectors $\vec{u}$ and $\vec{v}$.

The key to understanding Theorem 35 is the law of cosines from basic trigonometry. The vectors $\vec{u}$ and $\vec{v}$ and the angle θ between them are drawn in Fig. 6.3. The triangle formed by these two vectors has sides of length $\|\vec{u}\|$, $\|\vec{v}\|$, with a third side of length $\|\vec{u} - \vec{v}\|$ opposite angle θ.

To prove the theorem, applying the law of cosines we have that

$$\|\vec{u} - \vec{v}\|^2 = \|\vec{u}\|^2 + \|\vec{v}\|^2 - 2\|\vec{u}\| \, \|\vec{v}\| \cos(\theta).$$

However, we know that $\|\vec{u}-\vec{v}\|^2 = (\vec{u}-\vec{v})\cdot(\vec{u}-\vec{v})$, $\|\vec{u}\|^2 = \vec{u}\cdot\vec{u}$, $\|\vec{v}\|^2 = \vec{v}\cdot\vec{v}$, and consequently we can rewrite our law of cosines as

$$(\vec{u} - \vec{v}) \cdot (\vec{u} - \vec{v}) = \vec{u} \cdot \vec{u} + \vec{v} \cdot \vec{v} - 2\|\vec{u}\| \, \|\vec{v}\| \cos(\theta).$$

Using the properties in Theorem 34, we find that $(\vec{u} - \vec{v}) \cdot (\vec{u} - \vec{v}) = \vec{u} \cdot (\vec{u} - \vec{v}) - \vec{v} \cdot (\vec{u} - \vec{v}) = \vec{u} \cdot \vec{u} - \vec{u} \cdot \vec{v} - \vec{v} \cdot \vec{u} + \vec{v} \cdot \vec{v}$, and consequently

$$\vec{u} \cdot \vec{u} + \vec{v} \cdot \vec{v} - 2\vec{u} \cdot \vec{v} = \vec{u} \cdot \vec{u} + \vec{v} \cdot \vec{v} - 2\|\vec{u}\| \, \|\vec{v}\| \cos(\theta).$$

Cancelling $\vec{u} \cdot \vec{u} + \vec{v} \cdot \vec{v}$ from both sides gives

$$-2\vec{u} \cdot \vec{v} = -2\|\vec{u}\| \, \|\vec{v}\| \cos(\theta).$$

The desired formula

$$\vec{u} \cdot \vec{v} = \|\vec{u}\| \, \|\vec{v}\| \cos(\theta)$$

follows from division by -2.

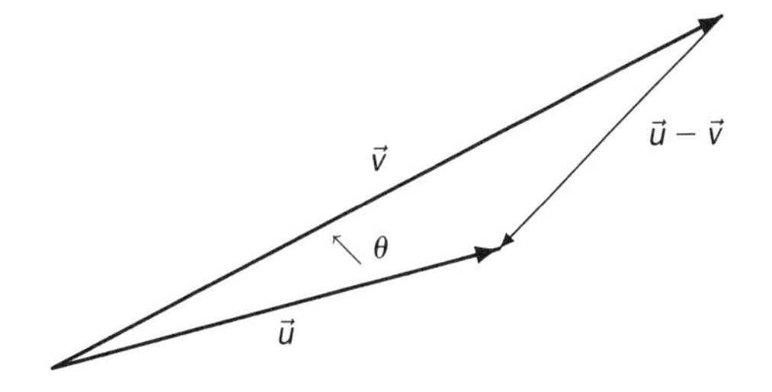

Fig. 6.3. Angle θ between vectors $\vec{u}$ and $\vec{v}$ and their triangle

For example, if

$$\vec{u} = \begin{pmatrix} 2 \\ 1 \\ 1 \end{pmatrix} \quad \text{and} \quad \vec{v} = \begin{pmatrix} 0 \\ 1 \\ 1 \end{pmatrix},$$

the theorem gives

$$\cos(\theta) = \frac{\vec{u} \cdot \vec{v}}{\|\vec{u}\| \, \|\vec{v}\|} = \frac{2}{\sqrt{6}\sqrt{2}} = \frac{2}{\sqrt{12}} \approx .577.$$

In this case we see that $\theta \approx .956$ radians ≈ 55 degrees.

Suppose that $\vec{u}$ and $\vec{v}$ are both nonzero vectors, yet $\vec{u} \cdot \vec{v} = 0$. Then Theorem 35 shows that $\cos(\theta) = 0$, where θ is the angle between $\vec{u}$ and $\vec{v}$. This means that $\theta = \frac{\pi}{2}$ radians $= 90$ degrees. In particular, $\vec{u}$ and $\vec{v}$ are perpendicular (which is often denoted by $\vec{u} \perp \vec{v}$). Perpendicular vectors are usually called *orthogonal*. We record this important application as a separate theorem.

Theorem 36. *Suppose that $\vec{u}$ and $\vec{v}$ are nonzero vectors. Then $\vec{u} \cdot \vec{v} = 0$ if and only if $\vec{u}$ and $\vec{v}$ are perpendicular.*

In $\mathbf{R}^3$ the set of vectors perpendicular to a given vector form a plane. This can be seen algebraically as follows. Suppose

$$\vec{u} = \begin{pmatrix} 1 \\ 2 \\ 1 \end{pmatrix}.$$

By direct calculation, if

$$\vec{v} = \begin{pmatrix} p \\ q \\ r \end{pmatrix}$$

and if $\vec{v} \perp \vec{u}$, we have $\vec{u} \cdot \vec{v} = p + 2q + r$. Therefore, $\vec{v}$ is perpendicular to $\vec{u}$ if and only if $p = -2q - r$. We see that the set of all vectors perpendicular to $\vec{u}$ are the vectors that form the plane $\mathcal{P} = \{(-2q - r, q, r) \mid q, r \in \mathbf{R}\}$.

The Tetrahedron and the Methane Bond Angle

Vector geometry is often quite useful in giving descriptions of three-dimensional objects. Most problems in the physical sciences and engineering would be difficult to describe without vectors. In chemistry, for example, it is important to be able to describe the locations of atoms in a molecule. This is usually done by viewing the atoms as vertices of a polyhedra whose edges can be described by vectors. We study an example of this next.

To start we study the geometry of the most basic polyhedra, the *tetrahedron* (see the Group Project in Sec. 1.1 for the definition and more discussion of polyhedra). A tetrahedron is a polyhedra whose edges are the six line segments between four points. We shall consider the situation where these four points (the vertices of the tetrahedron) are an equal distance from one another. Therefore each of the four faces of this tetrahedron will be an equilateral triangle. This kind of tetrahedron is important in basic chemistry because whenever four atoms are attached to a carbon atom in an organic molecule, they are are located on the vertices of a such a tetrahedron. For example, the carbon atom of the simple organic molecule methane, CH_4, is centered in this tetrahedron with vertices the hydrogen atoms. Such a tetrahedron as well as the same tetrahedron with its center is illustrated in Fig. 6.4.

An important problem is to compute the angle between a pair of line segments from the center point of this tetrahedron to its vertices. This angle is the bond angle between the hydrogen atoms that are attached to the carbon atom in the methane molecule.

In order to compute this angle, it is first necessary to obtain the coordinates of the vertices and center of our tetrahedron in $\mathbf{R}^3$. For this we shall choose an equilateral triangle in the xy-plane as one face of the tetrahedron. If we let $A = (0, 0, 0)$, $B = (2, 0, 0)$, and $C = (1, \sqrt{3}, 0)$, then ΔABC is an equilateral triangle on the xy-plane with sides of length 2. (The length 2 was chosen arbitrarily but also makes some calculations shorter.) The other vertex of our tetrahedron is a point $D = (a, b, c)$ that has distance 2 from the points A, B, and C. Hence, by the distance formula, we find

$$a^2 + b^2 + c^2 = 4, \ (a-2)^2 + b^2 + c^2 = 4, \ (a-1)^2 + (b - \sqrt{3})^2 + c^2 = 4.$$

Subtracting the first equation from the second gives $-4a+4 = 0$, so we find $a = 1$. Subtracting the first equation from the third gives $-2a+1-2\sqrt{3}b+3 = 0$, so since $a = 1$ we find that $b = \sqrt{\frac{1}{3}}$. Finally, substituting $a = 1$

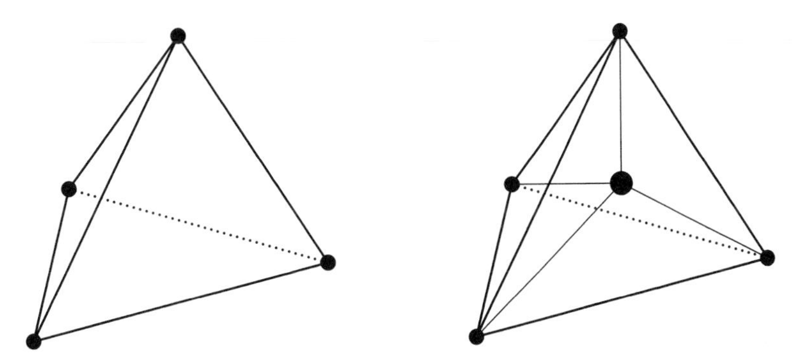

Fig. 6.4. A tetrahedron and the same tetrahedron with its center shown

and $b = \sqrt{\frac{1}{3}}$ into $a^2 + b^2 + c^2 = 4$ shows that $c^2 = \frac{8}{3}$. It follows that there are two points in $\mathbf{R}^3$ that are distance 2 from each of A, B, and C, namely $(1, \sqrt{\frac{1}{3}}, \pm\sqrt{\frac{8}{3}})$. For our calculations we will choose $D = (1, \sqrt{\frac{1}{3}}, \sqrt{\frac{8}{3}})$ as the fourth point of our tetrahedron (which will lie above the xy-plane).

Next we need to find the coordinates of the center point of the tetrahedron, which is the point equidistant from each of A, B, C, and D. Using a little visual geometry, we see that the x- and y- coordinates of this point must be the same as those of D, because a top view of the tetrahedron (looking straight down from above D) would hide the center point below D. Hence, if $P = (1, \sqrt{\frac{1}{3}}, p)$ is the center point, since P is equidistant from A and D we must have that

$$\|\overrightarrow{DP}\|^2 = \left(\sqrt{\frac{8}{3}} - p\right)^2 = \|\overrightarrow{AP}\|^2 = 1 + \frac{1}{3} + p^2.$$

This equation shows that $\frac{8}{3} - 2\sqrt{\frac{8}{3}}p = \frac{4}{3}$, which gives $p = \sqrt{\frac{1}{6}}$. This shows that the center point of our tetrahedron is $P = (1, \sqrt{\frac{1}{3}}, \sqrt{\frac{1}{6}})$.

We can now compute the methane bond angle. It is the angle between the vectors $\overrightarrow{PA}$ and $\overrightarrow{PD}$. Since we have computed the coordinates, we know that $\overrightarrow{PA} = (-1, -\sqrt{\frac{1}{3}}, -\sqrt{\frac{1}{6}})$ and $\overrightarrow{PD} = (0, 0, \sqrt{\frac{8}{3}} - \sqrt{\frac{1}{6}})$. We can also compute that $\|\overrightarrow{PA}\| = \sqrt{1 + \frac{1}{3} + \frac{1}{6}} = \sqrt{\frac{3}{2}}$ and $\|\overrightarrow{PD}\| = \sqrt{\frac{8}{3} - \frac{4}{3} + \frac{1}{6}} = \sqrt{\frac{3}{2}}$ (of course they are equal, as they should be). We denote by θ the angle between the vectors $\overrightarrow{PA}$ and $\overrightarrow{PD}$. Then

$$\cos(\theta) = \frac{\overrightarrow{PD} \cdot \overrightarrow{PA}}{\|\overrightarrow{PD}\|\,\|\overrightarrow{PA}\|} = \frac{\left(\sqrt{\frac{8}{3}} - \sqrt{\frac{1}{6}}\right)\left(-\sqrt{\frac{1}{6}}\right)}{\frac{3}{2}} = \frac{-\frac{2}{3} + \frac{1}{6}}{\frac{3}{2}} = -\frac{1}{3}.$$

This shows that $\theta = \cos^{-1}(-\frac{1}{3})$ or $109°28'$, often referred to as the *tetrahedral angle*. The tetrahedral angle is particularly important in organic chemistry since it is the angle at which carbon singly bonds to other atoms (if there are no other geometric constraints).

Note. Diamonds are a form of pure carbon where each atom is bonded to its four nearest neighbors in such a way that these four are centered at the vertices of a tetrahedron. The distance between each atom and each of its four nearest neighbors is 1.53×10^{-8} cm. In contrast, graphite is also pure carbon, but the atoms are not symmetrically arranged and in fact are not equally spaced. The tetrahedral bonding of carbon atoms in diamonds is what makes diamonds so hard.

Problems

1. Consider the following vectors in $\mathbf{R}^2$.

$$\vec{u} = \begin{pmatrix} 1 \\ -1 \end{pmatrix}, \quad \vec{v} = \begin{pmatrix} 2 \\ 4 \end{pmatrix}, \quad \vec{w} = \begin{pmatrix} 0 \\ 3 \end{pmatrix}$$

Find the following:

(a) $\|\vec{u}\|$;

(b) $\|\vec{v} - \vec{w}\|$;

(c) $\vec{w} \cdot \vec{v}$;

(d) $(\vec{u} + \vec{w}) \cdot (\vec{u} - \vec{w})$.

2. Consider the following vectors in $\mathbf{R}^3$.

$$\vec{p} = \begin{pmatrix} 1 \\ 0 \\ -1 \end{pmatrix}, \quad \vec{q} = \begin{pmatrix} 2 \\ 1 \\ 1 \end{pmatrix}, \quad \vec{r} = \begin{pmatrix} 1 \\ -5 \\ 3 \end{pmatrix}$$

Find the following:

(a) $\|\vec{p}\|$;

(b) $\|\vec{p} + 2\vec{q}\|$;

(c) $\vec{q} \cdot \vec{r}$;

(d) $\vec{p} \cdot (\vec{p} - \vec{q})$.

3. Consider the points $P = (1, 2)$, $Q = (2, -5)$, and $R = (1, 1)$ in $\mathbf{R}^2$. Sketch the points and find:

(a) the distance between P and Q;

(b) the distance between P and R;

(c) the angle between the segments $\overline{PQ}$ and $\overline{PR}$.

4. Consider the points $U = (2, 0, 1)$, $V = (0, 1, 1)$, and $T = (-1, 4, 1)$ in $\mathbf{R}^3$. Sketch the points and find:

(a) the distance between U and V;

(b) the distance between U and T;

(c) the angle between the segments $\overline{UV}$ and $\overline{UT}$.

5. According to the parallelogram law for the dot product,

$$\|\vec{v} + \vec{w}\|^2 + \|\vec{v} - \vec{w}\|^2 = 2\|\vec{v}\|^2 + 2\|\vec{w}\|^2$$

for all vectors $\vec{v}, \vec{w}$ in $\mathbf{R}^2$ or $\mathbf{R}^3$. Derive the parallelogram law using the properties of the dot product given in Theorem 34.

6. (a) Give an algebraic proof of the triangle inequality.

(b) Prove properties (ii), (iii), (iv), and (v) of Theorem 34.

Group Project: More Properties of the Dot Product

Try to show the following using the properties of the dot product (as given in Theorem 34) instead of direct computation.

(a) Show for two vectors $\vec{v}, \vec{w}$ in $\mathbf{R}^2$ or $\mathbf{R}^3$ that $\|\vec{v} + \vec{w}\|^2 = \|\vec{v}\|^2 + \|\vec{w}\|^2$ if and only if $\vec{v}$ and $\vec{w}$ are perpendicular.

(b) Show for two vectors $\vec{u}, \vec{v}$ in $\mathbf{R}^2$ or $\mathbf{R}^3$ that

$$\|\vec{u} - \vec{v}\|^2 = \|\vec{u}\|^2 - 2\langle\vec{u}, \vec{v}\rangle + \|\vec{v}\|^2.$$

(c) Show for two vectors $\vec{u}, \vec{v}$ in $\mathbf{R}^2$ or $\mathbf{R}^3$ that

$$\langle\vec{u}, \vec{v}\rangle = \frac{1}{4}\|\vec{u} + \vec{v}\|^2 - \frac{1}{4}\|\vec{u} - \vec{v}\|^2.$$

This result is called the *polar identity*.

6.2 Angles and Projections

The Boat and Chair Forms of Cyclohexane

In Sec. 2.1 we looked at vectors describing the location of carbon atoms in benzene, C_6H_6, that lie on a hexagon in a plane. Cyclohexane, C_6H_{12}, is an organic compound also having six carbon atoms bound together in a loop, but it differs from benzene in that there are twelve hydrogen atoms in the molecule instead of six. Since each carbon is bonded to two other carbon and two hydrogen atoms, the carbon atoms must be bound at the tetrahedral angle calculated in Sec. 6.1. Because of this, it is impossible for the carbon atoms in cyclohexane to lie in a single plane.

In order to understand positions of the carbon atoms in cyclohexane, we will represent them by six points P_1, P_2, P_3, P_4, P_5, P_6, arranged in $\mathbf{R}^3$ so that the six angles, $\angle P_1P_2P_3$, $\angle P_2P_3P_4$, $\angle P_3P_4P_5$, $\angle P_4P_5P_6$, $\angle P_5P_6P_1$, and $\angle P_6P_1P_2$ are each tetrahedral. It turns out that there are two geometrically different possibilities for the arrangement of six points satisfying these conditions, and each of these corresponds to different forms of cyclohexane found in nature. We set $P_1 = (0, 0, 0)$ and $P_2 = (2, 0, 0)$ (so the distance between our carbon atoms is 2). It turns out that the points P_1, P_2, P_4, and P_5 end up on a single plane with $\overline{P_1P_2}$ parallel to $\overline{P_4P_5}$ (and with P_3 and P_6 not on that plane), so we let this plane be the xy-plane and we assume that $P_4 = (2, 2b, 0)$ and $P_5 = (0, 2b, 0)$. Here, b is some number not yet determined.

Next we must try to locate the point P_3. By the symmetry of this arrangement, its y-coordinate must be half that of P_4 and P_5, so we can assume that $P_3 = (a, b, c)$ (which shows why we used $2b$ in P_4 and P_5.) We have $\overrightarrow{P_1P_2} = (2, 0, 0)$, $\overrightarrow{P_2P_3} = (a - 2, b, c)$, $\overrightarrow{P_3P_4} = (2 - a, b, -c)$, and

$\overrightarrow{P_4P_5} = (-2, 0, 0)$. Our assumption that the angle $\angle P_1P_2P_3$ is tetrahedral means that its cosine is $-\frac{1}{3}$ and therefore

$$\frac{-1}{3} = \frac{\overrightarrow{P_2P_1} \cdot \overrightarrow{P_2P_3}}{\|\overrightarrow{P_2P_1}\| \, \|\overrightarrow{P_2P_3}\|} = \frac{-2(a-2)}{4}.$$

(Note that we have to use the vector $\overrightarrow{P_2P_1}$ instead of $\overrightarrow{P_1P_2}$.) Consequently, $a = \frac{8}{3}$. Since the angles $\angle P_2P_3P_4$ and $\angle P_3P_4P_5$ are also tetrahedral, we find that

$$\frac{-1}{3} = \frac{\overrightarrow{P_3P_2} \cdot \overrightarrow{P_3P_4}}{4} = \frac{(a-2)^2 - b^2 + c^2}{4}.$$

Since $a = \frac{8}{3}$, we obtain $-\frac{1}{3} = \frac{1}{4}(\frac{4}{9} - b^2 + c^2)$, which gives $-b^2 + c^2 = -\frac{16}{9}$. But we also assumed that $\|\overrightarrow{P_2P_3}\| = \|\overrightarrow{P_3P_4}\| = 4$, and therefore $\frac{4}{9} + b^2 + c^2 = 4$, which shows $b^2 + c^2 = \frac{32}{9}$. Adding these, we find that $2c^2 = \frac{16}{9}$, and so $c = \pm\frac{2\sqrt{2}}{3}$. Substituting the c value shows that $b^2 = \frac{24}{9}$, so we can take $b = \frac{2\sqrt{6}}{3}$.

We now have the coordinates of five of the six carbon atoms in our model of cyclohexane. These points are $P_1 = (0, 0, 0)$, $P_2 = (2, 0, 0)$, $P_3 = (\frac{8}{3}, \frac{2\sqrt{6}}{3}, \pm\frac{2\sqrt{2}}{3})$, $P_4 = (2, \frac{4\sqrt{6}}{3}, 0)$, and $P_5 = (0, \frac{4\sqrt{6}}{3}, 0)$. The point P_6 must have the same possible positions relative to P_1 and P_5 as P_3 does relative to P_2 and P_4 (except sticking out in the opposite direction, as illustrated in Fig. 6.5). This shows that $P_6 = (-\frac{2}{3}, \frac{2\sqrt{6}}{3}, \pm\frac{2\sqrt{2}}{3})$. We find that there are two geometrically distinct solutions to our structure for cyclohexane. The first is where the z-coordinates of P_3 and P_6 have the same sign, which is called the "boat form of cyclohexane," and the second is where the z-coordinates of P_3 and P_6 have opposite signs, which is called the "chair form of cyclohexane." In the boat form the points P_3 and P_6 lie on the same side of the plane containing P_1, P_2, P_4, and P_5, while in the chair they lie on opposite sides. Both of these geometric forms occur naturally in cyclohexane. The cyclohexane ring system is also part of the structure of

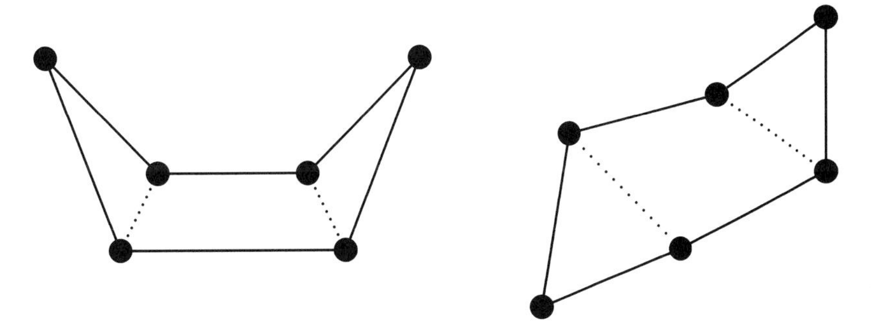

Fig. 6.5. The boat form (left) and the chair form (right) of cyclohexane

many other compounds such as terpenes, a class of natural products often found in essential oils. Familiar examples are camphor, used in medicine, and menthol, the flavor component of peppermint. These two forms of cyclohexane are pictured in Fig. 6.5.

Projections of Vectors

Recall the standard basis vectors $\vec{e}_1 = (1, 0, 0)$, $\vec{e}_2 = (0, 1, 0)$, and $\vec{e}_3 = (0, 0, 1)$ for $\mathbf{R}^3$. Clearly, if $\vec{v} = (a, b, c)$, we can write $\vec{v} = a\vec{e}_1 + b\vec{e}_2 + c\vec{e}_3$. This expression shows that $\vec{v}$ is the sum of its projections onto the standard basis vectors. Note that the three projections $a\vec{e}_1$, $b\vec{e}_2$, and $c\vec{e}_3$ are mutually orthogonal. It will be useful to generalize this. Fig. 6.6 shows the projection of a vector $\vec{u}$ onto a vector $\vec{v}$.

Consider two vectors $\vec{u}$ and $\vec{v}$. We denote by $\mathrm{proj}_{\vec{v}}(\vec{u})$ the projection of $\vec{u}$ onto $\vec{v}$ as illustrated in Fig. 6.6. If θ is the angle between these two vectors, then by trigonometry $\|\vec{u}\| \cos(\theta)$ is the length of this projection. Since $\vec{v}/\|\vec{v}\|$ is the vector of length 1 with the same direction as $\vec{v}$, the projection of $\vec{u}$ onto $\vec{v}$ is the vector

$$\mathrm{proj}_{\vec{v}}(\vec{u}) = \|\vec{u}\| \cos(\theta) \left(\frac{\vec{v}}{\|\vec{v}\|} \right) = \frac{\|\vec{u}\| \cos(\theta)}{\|\vec{v}\|} \vec{v}.$$

Since $\|\vec{v}\| = (\vec{v} \cdot \vec{v})^{1/2}$ and $\vec{u} \cdot \vec{v} = \|\vec{u}\| \, \|\vec{v}\| \cos(\theta)$, we find that

$$\begin{aligned} \mathrm{proj}_{\vec{v}}(\vec{u}) &= \frac{\|\vec{u}\| \, \|\vec{v}\| \cos(\theta)}{\|\vec{v}\| \, \|\vec{v}\|} \vec{v} \\ &= \frac{\vec{u} \cdot \vec{v}}{\|\vec{v}\|^2} \vec{v} = \frac{\vec{u} \cdot \vec{v}}{\vec{v} \cdot \vec{v}} \vec{v}. \end{aligned}$$

This calculation motivates the following definition.

Definition. If $\langle \; , \; \rangle$ denotes the dot product on either $\mathbf{R}^2$ or $\mathbf{R}^3$, we define the *orthogonal projection* of $\vec{u}$ onto $\vec{v}$ by

$$\mathrm{proj}_{\vec{v}}(\vec{u}) = \frac{\langle \vec{u}, \vec{v} \rangle}{\langle \vec{v}, \vec{v} \rangle} \vec{v}.$$

Observe in Fig. 6.6 that the vector $\vec{u} - \mathrm{proj}_{\vec{v}}(\vec{u})$ is perpendicular to the vector $\vec{v}$. This is exactly what we noted for the projections onto the standard

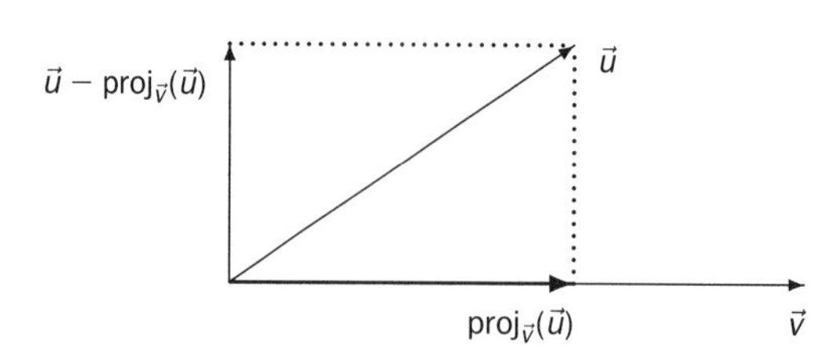

Fig. 6.6. Projection of vectors

basis vectors of $\mathbf{R}^3$. Using a few algebraic manipulations, we see this is true in general.

Theorem 37. *If both $\vec{u}$ and $\vec{v}$ are nonzero vectors, then $\vec{v}$ and $(\vec{u} - \text{proj}_{\vec{v}}(\vec{u}))$ are orthogonal.*

Proof. We check this in the following calculation, where we use the properties of the dot product given in Theorem 34. When interpreting this calculation, be sure to remember that $\langle \vec{u}, \vec{v} \rangle$ is a real number, not a vector:

$$\begin{aligned}
\langle \vec{v}, \vec{u} - \text{proj}_{\vec{v}}(\vec{u}) \rangle &= \langle \vec{v}, \vec{u} - \frac{\langle \vec{u}, \vec{v} \rangle}{\langle \vec{v}, \vec{v} \rangle} \vec{v} \rangle \\
&= \langle \vec{v}, \vec{u} \rangle - \langle \vec{v}, \frac{\langle \vec{u}, \vec{v} \rangle}{\langle \vec{v}, \vec{v} \rangle} \vec{v} \rangle \\
&= \langle \vec{v}, \vec{u} \rangle - \frac{\langle \vec{u}, \vec{v} \rangle}{\langle \vec{v}, \vec{v} \rangle} \langle \vec{v}, \vec{v} \rangle \\
&= \langle \vec{v}, \vec{u} \rangle - \langle \vec{u}, \vec{v} \rangle \\
&= 0.
\end{aligned}$$

□

More Applications

The dot product in $\mathbf{R}^2$ and $\mathbf{R}^3$ can be used to determine the equations of lines and planes.

Examples. (a) Suppose you are interested in the line ℓ_1 in the plane that is perpendicular to a given vector

$$\vec{v} = \begin{pmatrix} 2 \\ -3 \end{pmatrix}$$

and passes through $(0, 0)$. Then the geometric vectors that lie on ℓ are precisely those vectors whose dot product with $\vec{v}$ is zero. This shows that

$$\ell_1 = \{(x, y) \in \mathbf{R}^2 \mid 2x - 3y = 0\}.$$

If you are interested in the line ℓ_2 that is parallel to ℓ_1 but passes through the point $(5, 9)$ instead, simply translate the equation so that you take the dot product of $\vec{v}$ with the vector starting at $(5, 9)$ instead of $(0, 0)$. This results in

$$\begin{aligned}
\ell_2 &= \{(x, y) \in \mathbf{R}^2 \mid 2(x - 5) - 3(y - 9) = 0\} \\
&= \{(x, y) \in \mathbf{R}^2 \mid 2x - 3y = -17\}.
\end{aligned}$$

(b) Now suppose that you want to describe the plane $\mathcal{P}$ of all points perpendicular to the vector

$$\vec{v} = \begin{pmatrix} 2 \\ 3 \\ -1 \end{pmatrix}$$

in $\mathbf{R}^3$ and passing through $(3, 1, 1)$. The points on this plane are the endpoints of the geometric vectors that start at $(3, 1, 1)$ and have a direction perpendicular to $\vec{v}$. Using the dot product, we find that

$$\begin{aligned} \mathcal{P} &= \{(x, y, z) \in \mathbf{R}^3 \mid 2(x-3) + 3(y-1) - 1(z-1) = 0\} \\ &= \{(x, y, z) \in \mathbf{R}^3 \mid 2x + 3y - z = 8\}. \end{aligned}$$

(c) We can also use the theorem to compute the distance between a point P and a line ℓ. According to Euclidean geometry, this distance is the length of the line segment $\overline{PQ}$, where $\overline{PQ} \perp \ell$. For example, suppose $P = (5, 3)$ and ℓ is the line through the origin and $(4, -1)$. Let

$$\vec{v} = \begin{pmatrix} 4 \\ -1 \end{pmatrix} \quad \text{and} \quad \vec{u} = \begin{pmatrix} 5 \\ 3 \end{pmatrix}.$$

Then ℓ is the line spanned by $\vec{v}$. Using the geometry of the projection, we see that the line segment $\overline{PQ}$ has the same length $|\vec{u} - \operatorname{proj}_{\vec{v}}(\vec{u})|$. By a direct calculation we find $|\vec{u} - \operatorname{proj}_{\vec{v}}(\vec{u})| = \sqrt{17}$.

Rotations in the Plane

Recall that in Sec. 2.1 we were interested in six vectors that gave the edges of a regular hexagon. These six vectors were obtained by starting with the vector $(1, 0)$ and rotating it successively six times by an angle of $60°$. It turns out that all rotations in the plane can be computed as a matrix multiplication. In order to see this, recall that the coordinates of any point P on the unit circle in $\mathbf{R}^2$ can be expressed as $P = (\cos(\alpha), \sin(\alpha))$, where α denotes the angle the vector $(\cos(\alpha), \sin(\alpha))$ makes with the positive x-axis. Suppose the point P is rotated counterclockwise by an angle θ. Then we obtain the new point P_θ whose coordinates are $(\cos(\theta + \alpha), \sin(\theta + \alpha))$. We now rewrite this vector using the angle addition formulas from trigonometry and using matrix multiplication:

$$\begin{aligned} \begin{pmatrix} \cos(\theta+\alpha) \\ \sin(\theta+\alpha) \end{pmatrix} &= \begin{pmatrix} \cos(\theta)\cos(\alpha) - \sin(\theta)\sin(\alpha) \\ \sin(\theta)\cos(\alpha) + \cos(\theta)\sin(\alpha) \end{pmatrix} \\ &= \begin{pmatrix} \cos(\theta) & -\sin(\theta) \\ \sin(\theta) & \cos(\theta) \end{pmatrix} \begin{pmatrix} \cos(\alpha) \\ \sin(\alpha) \end{pmatrix}. \end{aligned}$$

This shows that the coordinates of the rotation of a point on the unit circle can be obtained by matrix multiplication.

The matrix

$$R_\theta = \begin{pmatrix} \cos(\theta) & -\sin(\theta) \\ \sin(\theta) & \cos(\theta) \end{pmatrix}$$

is the matrix that, when left-multiplied to any vector $\vec{v}$ in $\mathbf{R}^2$, rotates the vector $\vec{v}$ counterclockwise by the angle θ. For example, if $\theta = 60°$, we obtain

$$R_{60°} = \begin{pmatrix} \frac{1}{2} & -\frac{\sqrt{3}}{2} \\ \frac{\sqrt{3}}{2} & \frac{1}{2} \end{pmatrix}.$$

If we multiply the vector $(1, 0)$ repeatedly by $R_{60°}$ we obtain

$$R_{60°} \cdot \begin{pmatrix} 1 \\ 0 \end{pmatrix} = \begin{pmatrix} \frac{1}{2} \\ \frac{\sqrt{3}}{2} \end{pmatrix}, \qquad R_{60°} \cdot \begin{pmatrix} \frac{1}{2} \\ \frac{\sqrt{3}}{2} \end{pmatrix} = \begin{pmatrix} -\frac{1}{2} \\ \frac{\sqrt{3}}{2} \end{pmatrix},$$

$$R_{60°} \cdot \begin{pmatrix} -\frac{1}{2} \\ \frac{\sqrt{3}}{2} \end{pmatrix} = \begin{pmatrix} -1 \\ 0 \end{pmatrix}, \qquad R_{60°} \cdot \begin{pmatrix} -1 \\ 0 \end{pmatrix} = \begin{pmatrix} -\frac{1}{2} \\ -\frac{\sqrt{3}}{2} \end{pmatrix},$$

$$R_{60°} \cdot \begin{pmatrix} -\frac{1}{2} \\ -\frac{\sqrt{3}}{2} \end{pmatrix} = \begin{pmatrix} \frac{1}{2} \\ -\frac{\sqrt{3}}{2} \end{pmatrix}, \qquad R_{60°} \cdot \begin{pmatrix} \frac{1}{2} \\ -\frac{\sqrt{3}}{2} \end{pmatrix} = \begin{pmatrix} 1 \\ 0 \end{pmatrix},$$

which are the six unit vectors considered in Sec. 2.1.

Observe that if we take the dot product of successive vectors in our list we always obtain $\frac{1}{2}$. (For example, $(\frac{1}{2}, \frac{\sqrt{3}}{2}) \cdot (-\frac{1}{2}, \frac{\sqrt{3}}{2}) = -\frac{1}{4} + \frac{3}{4} = \frac{1}{2}$.) Since each of these vectors is a unit vector, Theorem 35 shows that the cosine of the angle between them is $\frac{1}{2}$. This verifies that the angle between these vectors is $60°$.

Schwarz's Inequality

We conclude this section with one final application of Theorem 36.

Theorem 38 (Schwarz's inequality). *Suppose that $\vec{u}$, and $\vec{v}$ are vectors. Then*

$$\vec{u} \cdot \vec{v} \leq \|\vec{u}\| \cdot \|\vec{v}\|.$$

Equality holds if and only if $\vec{v} = \vec{0}$ or $\vec{u} = k\vec{v}$ for some $k \geq 0$.

Proof. If $\vec{v} = \vec{0}$, the result is easy since both sides are zero. Thus we may assume that $\vec{v} \neq \vec{0}$. We know that $\vec{u} \cdot \vec{v} = \|\vec{u}\| \, \|\vec{v}\| \cos\theta$. In case $\vec{u}$ is not a multiple of $\vec{v}$, we know that $|\cos(\theta)| < 1$ and the result follows. In case $\vec{u}$ is a multiple of $\vec{v}$, then we know that θ is 0 or π and $|\cos(\theta)| = 1$, so we have equality as required. □

Problems

1. Consider the following vectors in $\mathbf{R}^2$.

$$\vec{u} = \begin{pmatrix} 1 \\ -1 \end{pmatrix}, \quad \vec{v} = \begin{pmatrix} 2 \\ 4 \end{pmatrix}, \quad \vec{w} = \begin{pmatrix} 0 \\ 3 \end{pmatrix}$$

   Find and sketch the following:
   (a) $\text{proj}_{\vec{v}}(\vec{u})$;
   (b) $\text{proj}_{\vec{u}}(\vec{v})$;
   (c) $\vec{w} - \text{proj}_{\vec{v}}(\vec{w})$.

2. Consider the following vectors in $\mathbf{R}^3$.

$$\vec{p} = \begin{pmatrix} 1 \\ 0 \\ -1 \end{pmatrix}, \quad \vec{q} = \begin{pmatrix} 2 \\ 1 \\ 1 \end{pmatrix}, \quad \vec{r} = \begin{pmatrix} 1 \\ -5 \\ 3 \end{pmatrix}$$

   Find and sketch the following:
   (a) $\text{proj}_{\vec{p}}(\vec{q})$;
   (b) $\text{proj}_{\vec{q}}(\vec{p})$;
   (c) $\vec{r} - \text{proj}_{\vec{q}}(\vec{r})$.

3. Consider the points $P = (1, 2)$, $Q = (2, -5)$, and $R = (1, 1)$ in $\mathbf{R}^2$. Find a parametric description of the following lines in $\mathbf{R}^2$, and sketch them:
   (a) the line perpendicular to the segment $\overline{PQ}$ and passing through $(0, 0)$;
   (b) the line perpendicular to the segment $\overline{PR}$ and passing through $(1, 1)$.

4. Consider the points $U = (2, 0, 1)$, $V = (0, 1, 1)$, and $T = (-1, 2, 1)$ in $\mathbf{R}^3$. Find a parametric description of the following lines in $\mathbf{R}^3$, and sketch them:
   (a) the plane perpendicular to the segment $\overline{UV}$ and passing through $(0, 0, 0)$;
   (b) the line perpendicular to the segment $\overline{VT}$ and passing through V.

5. Find the distances between the point P_1 and the other five points in the model of cyclohexane. Compare these distances to those of the carbon atoms in a benzene molecule. Explain your observations.

6. Let $\vec{v}$ be an arbitrary vector in $\mathbf{R}^2$ and let R_θ denote the rotation matrix given in Sec. 6.2. Compute the dot product $\langle \vec{v}, R_\theta \vec{v} \rangle$ by a direct calculation. Use Theorem 35 to show that the angle between $\vec{v}$ and $R_\theta \vec{v}$ is θ.

Group Project: More Geometry of Cyclohexane

In this project you will study some properties of the arrangement of the carbon atoms in cyclohexane.

(a) First construct models of the location of these six points in cyclohexane (both forms) whose coordinates were computed earlier. This can be done with a single piece of paper folded twice or with clay and toothpicks.

(b) Find the dihedral angle between the plane containing P_1, P_2, P_4, P_5 and the plane containing P_2, P_3, P_4 in both the chair and boat forms of cyclohexane modeled in this section. (See Ex. 4 in Sec. 6.3 ahead for a definition of dihedral angle if you are not familiar with it.)

(c) What other collections of four points in the two cyclohexane models lie in a common plane?

(d) Find coordinates of the twelve hydrogen atoms in our model of cyclohexane, where you assume that their distance is 1 from each carbon atom, and where each carbon atom has two hydrogen atoms bonded so that the four bonds to the carbon are all at tetrahedral angles. How close are these hydrogen atoms to one another? How well do they fit (you should look at your model to visualize their locations).

6.3 The Cross Product

Definition

There is a special product of vectors in $\mathbf{R}^3$ which gives another vector. This is called the cross product and it has important geometric properties that will be explored in this section.

Definition. Consider the vectors

$$\vec{u} = \begin{pmatrix} p_1 \\ q_1 \\ r_1 \end{pmatrix} \quad \text{and} \quad \vec{v} = \begin{pmatrix} p_2 \\ q_2 \\ r_2 \end{pmatrix} \in \mathbf{R}^3.$$

Then we define the *cross product* of $\vec{u}$ and $\vec{v}$ by

$$\vec{u} \times \vec{v} = \begin{pmatrix} q_1 r_2 - q_2 r_1 \\ p_2 r_1 - p_1 r_2 \\ p_1 q_2 - p_2 q_1 \end{pmatrix}.$$

We emphasize that the second entry of $\vec{u} \times \vec{v}$ is *not* $p_1 r_2 - p_2 r_1$ (it is the negative of this). The importance of this sign difference will become apparent shortly, but for now we must keep it straight for the purposes of computation. For example,

$$\begin{pmatrix} 2 \\ -1 \\ 1 \end{pmatrix} \times \begin{pmatrix} 3 \\ -2 \\ 4 \end{pmatrix} = \begin{pmatrix} (-1)4 - 1(-2) \\ 3 \cdot 1 - 2 \cdot 4 \\ 2(-2) - (-1)3 \end{pmatrix} = \begin{pmatrix} -2 \\ -5 \\ -1 \end{pmatrix}$$

and

$$\begin{pmatrix} 0 \\ 1 \\ 0 \end{pmatrix} \times \begin{pmatrix} 1 \\ 0 \\ 0 \end{pmatrix} = \begin{pmatrix} 0 \\ 0 \\ -1 \end{pmatrix}.$$

We recall that whenever $a, b, c, d \in \mathbf{R}$, the 2×2 determinant is given by

$$\det \begin{pmatrix} a & b \\ c & d \end{pmatrix} = ad - bc.$$

Using the determinant we can express the cross product as follows:

$$\begin{pmatrix} p_1 \\ q_1 \\ r_1 \end{pmatrix} \times \begin{pmatrix} p_2 \\ q_2 \\ r_2 \end{pmatrix} = \begin{pmatrix} \det \begin{pmatrix} q_1 & q_2 \\ r_1 & r_2 \end{pmatrix} \\ -\det \begin{pmatrix} p_1 & p_2 \\ r_1 & r_2 \end{pmatrix} \\ \det \begin{pmatrix} p_1 & p_2 \\ q_1 & q_2 \end{pmatrix} \end{pmatrix}.$$

Many students find this formulation of the cross product easier to remember since its entries can be viewed as the determinants of possible 2×2 matrices built from entries of the original two vectors. This is sometimes expressed as

$$\begin{pmatrix} p_1 \\ q_1 \\ r_1 \end{pmatrix} \times \begin{pmatrix} p_2 \\ q_2 \\ r_2 \end{pmatrix} = \det \begin{pmatrix} \mathbf{i} & \mathbf{j} & \mathbf{k} \\ p_1 & q_1 & r_1 \\ p_2 & q_2 & r_2 \end{pmatrix},$$

where we denote

$$\mathbf{i} = \begin{pmatrix} 1 \\ 0 \\ 0 \end{pmatrix}, \ \mathbf{j} = \begin{pmatrix} 0 \\ 1 \\ 0 \end{pmatrix}, \ \mathbf{k} = \begin{pmatrix} 0 \\ 0 \\ 1 \end{pmatrix}.$$

We point out that this latter expression for the cross product is a mnemonic device and is not really a determinant since it mixes vectors and scalars as entries.

Perpendicularity and the Cross Product

We next give the most important property of the cross product.

Theorem 39. *Whenever $\vec{u}$ and $\vec{v}$ are nonzero vectors in $\mathbf{R}^3$, and $\vec{v}$ is not a scalar multiple of $\vec{u}$, the vector $\vec{u} \times \vec{v}$ is nonzero and is perpendicular to both vectors $\vec{u}$ and $\vec{v}$.*

Proof. This theorem can be verified by direct computation. We suppose that

$$\vec{u} = \begin{pmatrix} p_1 \\ q_1 \\ r_1 \end{pmatrix} \quad \text{and} \quad \vec{v} = \begin{pmatrix} p_2 \\ q_2 \\ r_2 \end{pmatrix},$$

and so by definition

$$\vec{u} \times \vec{v} = \begin{pmatrix} q_1 r_2 - q_2 r_1 \\ p_2 r_1 - p_1 r_2 \\ p_1 q_2 - p_2 q_1 \end{pmatrix}.$$

Suppose that $\vec{u} \times \vec{v} = \vec{0}$. Since $\vec{u} \neq \vec{0}$, at least one of p_1, q_1, or r_1 is nonzero. If $q_1 \neq 0$, then $q_1 r_2 - q_2 r_1 = 0$ gives $r_2 = \frac{q_2}{q_1} r_1$ and $p_1 q_2 - p_2 q_1 = 0$ gives $p_2 = \frac{q_2}{q_1} p_1$. Of course, $q_2 = \frac{q_2}{q_1} q_1$, and this shows that $\vec{v} = \frac{q_2}{q_1} \vec{u}$. Hence $\vec{v}$ is a multiple of $\vec{u}$, contrary to the assumptions of the theorem. Similar reasoning shows that $\vec{v}$ is also a multiple of $\vec{u}$ if either p_1 or r_1 is nonzero. These contradictions show that $\vec{u} \times \vec{v} \neq \vec{0}$, giving the first part of the theorem.

We next compute the dot products:

$$\vec{u} \cdot (\vec{u} \times \vec{v}) = p_1(q_1 r_2 - q_2 r_1) + q_1(p_2 r_1 - p_1 r_2) + r_1(p_1 q_2 - p_2 q_1) = 0$$

and

$$\vec{v} \cdot (\vec{u} \times \vec{v}) = p_2(q_1 r_2 - q_2 r_1) + q_2(p_2 r_1 - p_1 r_2) + r_2(p_1 q_2 - p_2 q_1) = 0.$$

Since these dot products are zero, the result follows from Theorem 36. □

Theorem 39 will help us develop many applications of the cross product. As a start, we consider the problem of finding the equation of a plane through three noncollinear points. Using the three points, two vectors lying in the plane can be found. The cross product of these two vectors is a vector perpendicular to the plane, and therefore any vector in the plane has zero dot product with this vector. We illustrate this idea in the next example.

Example. Find the equation of the plane $\mathcal{P}$ through the three points $P = (1, -2, 3)$, $Q = (0, 1, 2)$, and $R = (-1, 2, 2) \in \mathbf{R}^3$.

Solution. The geometric vectors $\overrightarrow{PQ}$ and $\overrightarrow{PR}$ translated to the origin are

$$\vec{u} = \overrightarrow{PQ} = \begin{pmatrix} -1 \\ 3 \\ -1 \end{pmatrix} \quad \text{and} \quad \vec{v} = \overrightarrow{PR} = \begin{pmatrix} -2 \\ 4 \\ -1 \end{pmatrix}.$$

The cross product can be computed to be

$$\vec{u} \times \vec{v} = \begin{pmatrix} 1 \\ 1 \\ 2 \end{pmatrix}.$$

The plane $\mathcal{P}$ is perpendicular to $\vec{u} \times \vec{v}$ and passes through $(1, -2, 3)$. Hence, using the dot product we see the equation of $\mathcal{P}$ is

$$1(x-1) + 1(y+2) + 2(z-3) = 0.$$

Equivalently, the equation of $\mathcal{P}$ can be written as $x + y + 2z = 5$.

Properties of the Cross Product

We next give some additional properties of the cross product. Each of these properties can be verified by an algebraic computation.

Theorem 40. *Suppose that $\vec{u}$, $\vec{v}$, and $\vec{w}$ are vectors in $\mathbf{R}^3$ and that r is a real number. Then*

(i) $\vec{u} \times \vec{v} = -\vec{v} \times \vec{u}$;

(ii) $\vec{v} \times \vec{v} = \vec{0}$;

(iii) $r(\vec{u} \times \vec{v}) = (r\vec{u}) \times \vec{v} = \vec{u} \times (r\vec{v})$;

(iv) $\vec{u} \times (\vec{v} + \vec{w}) = (\vec{u} \times \vec{v}) + (\vec{u} \times \vec{w})$;

(v) $(\vec{u} + \vec{v}) \times \vec{w} = (\vec{u} \times \vec{w}) + (\vec{v} \times \vec{w})$.

Property (a) of the theorem can be seen by inspecting the definition of $\vec{u} \times \vec{v}$, because if the order of the vectors in the cross product is reversed, then the entries in the cross product move to the other side of the minus sign. Therefore, $\vec{u} \times \vec{v} = -\vec{v} \times \vec{u}$. This property is important. Note that for any two vectors $\vec{u}$ and $\vec{v}$, there are two directions that are perpendicular to both. Property (a) shows that both directions are given by the cross product; *however*, the vector you get depends on the order in which the cross product is taken. Which of these two directions arises when a given cross product is taken? The answer is given by the so-called right-hand rule. Suppose that the index finger of your right hand points in the direction of the vector $\vec{u}$ and the middle finger of your right hand points in the direction of the vector $\vec{v}$. Then if you manipulate your thumb to point in

Fig. 6.7. Right-hand rule

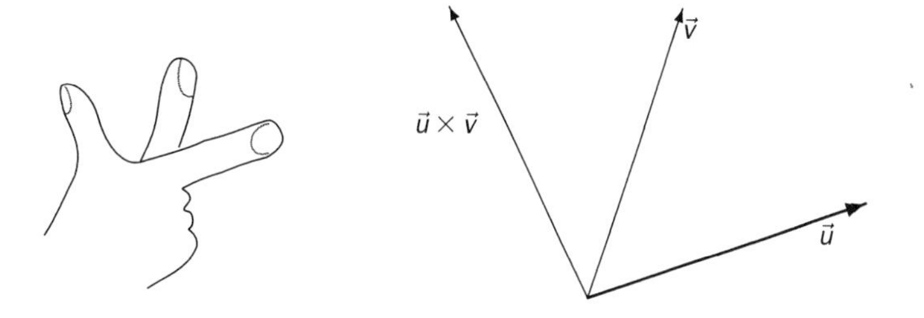

the direction perpendicular to the plane through $\vec{u}$ and $\vec{v}$, then it will point in the direction of $\vec{u} \times \vec{v}$ (see Fig. 6.7).

In the next calculation we illustrate how properties (b), (c), (d), and (e) of the theorem can be used in studying cross products. Suppose that

$$\vec{u} = \begin{pmatrix} 2 \\ 6 \\ 9 \end{pmatrix}, \quad \vec{v} = \begin{pmatrix} 1 \\ 2 \\ 3 \end{pmatrix}, \quad \text{and} \quad \mathbf{i} = \begin{pmatrix} 1 \\ 0 \\ 0 \end{pmatrix}.$$

Before computing $\vec{u} \times \vec{v}$ we note that $\vec{u} = 3\vec{v} - \mathbf{i}$. Hence, applying (e), then (c), and finally (b) yields

$$\begin{aligned} \vec{u} \times \vec{v} &= (3\vec{v} - \mathbf{i}) \times \vec{v} \\ &= 3(\vec{v} \times \vec{v}) + (-\mathbf{i} \times \vec{v}) \\ &= \begin{pmatrix} 0 \\ 0 \\ 0 \end{pmatrix} - \begin{pmatrix} 0 \\ -3 \\ 2 \end{pmatrix} = \begin{pmatrix} 0 \\ 3 \\ -2 \end{pmatrix}. \end{aligned}$$

The Length of the Cross Product

We have seen that the cross product of two vectors in $\mathbf{R}^3$ is a third vector perpendicular to the original two. Does the length of the cross product have any significance? The answer is yes and is explained in the next theorem.

Theorem 41. *Let $\vec{u}$ and $\vec{v}$ be vectors in $\mathbf{R}^3$. Then the length of $\vec{u} \times \vec{v}$ is*

$$|\vec{u} \times \vec{v}| = \|\vec{u}\|\,\|\vec{v}\|\,|\sin(\theta)|,$$

where θ is the angle between $\vec{u}$ and $\vec{v}$. This length is the same as the area of the parallelogram with adjacent edges $\vec{u}$ and $\vec{v}$, shown in Fig. 6.8.

Proof. Suppose that

$$\vec{u} = \begin{pmatrix} p_1 \\ q_1 \\ r_1 \end{pmatrix} \quad \text{and} \quad \vec{v} = \begin{pmatrix} p_2 \\ q_2 \\ r_2 \end{pmatrix}, \quad \text{so} \quad \vec{u} \times \vec{v} = \begin{pmatrix} q_1 r_2 - q_2 r_1 \\ p_2 r_1 - p_1 r_2 \\ p_1 q_2 - p_2 q_1 \end{pmatrix}.$$

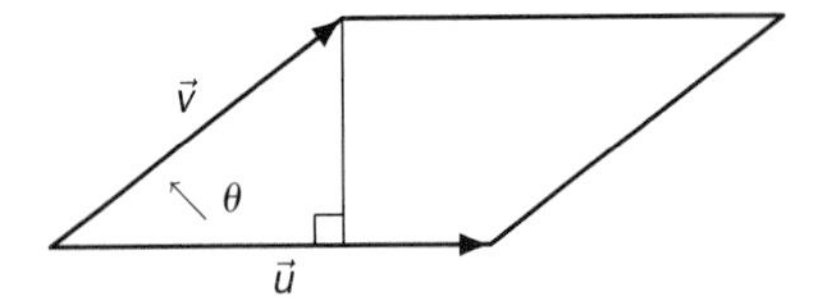

Fig. 6.8. Parallelogram with adjacent edges the vectors $\vec{u}$ and $\vec{v}$

According to the definitions (and a bit of algebraic manipulation),

$$\begin{aligned}
|\vec{u} \times \vec{v}|^2 &= (q_1 r_2 - q_2 r_1)^2 + (p_2 r_1 - p_1 r_2)^2 + (p_1 q_2 - p_2 q_1)^2 \\
&= (p_1^2 + q_1^2 + r_1^2)(p_2^2 + q_2^2 + r_2^2) - (p_1 p_2 + q_1 q_2 + r_2 r_2)^2 \\
&= \|\vec{u}\|^2 \|\vec{v}\|^2 - (\vec{u} \cdot \vec{v})^2 \\
&= \|\vec{u}\|^2 \|\vec{v}\|^2 - \|\vec{u}\|^2 \|\vec{v}\|^2 \cos^2(\theta) \\
&= \|\vec{u}\|^2 \|\vec{v}\|^2 (1 - \cos^2(\theta)) \\
&= \|\vec{u}\|^2 \|\vec{v}\|^2 \sin^2(\theta).
\end{aligned}$$

This gives the first result.

In order to understand the second statement, we look closely at the parallelogram with adjacent edges $\vec{u}$ and $\vec{v}$. Applying basic trigonometry we see that the length of the perpendicular segment from the edge $\vec{u}$ to the end point of $\vec{v}$ is $\|\vec{v}\| \sin(\theta)$ (see Fig. 6.8); in other words, $\|\vec{v}\| \sin(\theta)$ is the altitude of the triangle with sides $\vec{u}$ and $\vec{v}$ from the end point of $\vec{u}$ to the side $\vec{v}$. Since the area of a parallelogram is the length of the base given by $\vec{v}$ times this altitude, it follows that the area of the parallelogram in question is $\|\vec{u}\| \|\vec{v}\| \sin(\theta)$. □

Applications

We conclude this section with a sequence of problems applying the cross product and other ideas from this chapter in a variety of ways. There are several ways to solve some of these problems. The solutions presented here are chosen in order to give you an opportunity to visualize some of the concepts presented in the chapter. If you are familiar with other methods or can think of some other techniques—great! But be sure you understand the geometry behind your methods.

Example 1. Find the area of the triangle ΔPQR with end points $P = (2, 0, 1)$, $Q = (1, 1, 0)$, and $R = (-1, 3, 2)$ in $\mathbf{R}^3$.

Solution. The geometric vectors $\overrightarrow{PQ}$ and $\overrightarrow{PR}$ translated to the origin are

$$\vec{u} = \begin{pmatrix} -1 \\ 1 \\ -1 \end{pmatrix} \quad \text{and} \quad \vec{v} = \begin{pmatrix} -3 \\ 3 \\ 1 \end{pmatrix}.$$

Their cross product can be computed to be

$$\vec{u} \times \vec{v} = \begin{pmatrix} 4 \\ 4 \\ 0 \end{pmatrix}.$$

The length of the vector $\vec{u} \times \vec{v}$ is $\sqrt{16+16} = 4\sqrt{2}$. This shows that the area of the parallelogram with adjacent edges $\overline{PQ}$ and $\overline{PR}$ is $4\sqrt{2}$. The triangle ΔPQR has exactly half the area of this parallelogram. Therefore the area of ΔPQR is $2\sqrt{2}$.

Example 2. Find the distance between the point $Q = (3, 1, 2)$ in $\mathbf{R}^3$ and the plane $\mathcal{P}$ defined by the equation $2x - y + z = 13$.

Solution. The plane $\mathcal{P}$ is perpendicular to the vector

$$\vec{v} = \begin{pmatrix} 2 \\ -1 \\ 1 \end{pmatrix}.$$

(To see this look at the coefficients of the equation that defines $\mathcal{P}$.) Now consider the line ℓ in the direction of $\vec{v}$ and passing through Q. Parametrically, ℓ can be described as $\{(2t, -t, t) + (3, 1, 2) \mid t \in \mathbf{R}\} = \{(2t+3, -t+1, t+2) \mid t \in \mathbf{R}\}$. The intersection of ℓ and $\mathcal{P}$ is the point R on $\mathcal{P}$ given by $(2t+3, -t+1, t+2)$, where $2(2t+3) - (-t+1) + (t+2) = 13$. Solving for t, we find that $6t = 6$, or $t = 1$. This shows that $R = (5, 0, 3)$. Since ℓ is perpendicular to $\mathcal{P}$, it follows that the distance between Q and $\mathcal{P}$ is the distance between Q and R. Since $\overrightarrow{QR} = (2, -1, 1)$, this distance is $\sqrt{2^2 + (-1)^2 + 1^2} = \sqrt{6}$.

Example 3. Find a description of the line ℓ containing $(5, 5, 5)$ that is perpendicular to the plane $\mathcal{P}$ passing through $(0, 0, 0)$, $(1, 3, 2)$, and $(-1, 2, 1)$.

Solution. We must first find a direction vector for this line ℓ. Such a vector must be perpendicular to both

$$\vec{u} = \begin{pmatrix} 1 \\ 3 \\ 2 \end{pmatrix} \quad \text{and} \quad \vec{v} = \begin{pmatrix} -1 \\ 2 \\ 1 \end{pmatrix},$$

since these vectors lie on the plane $\mathcal{P}$. So we can take for our direction vector $\vec{u} \times \vec{v}$, which is readily computed to be

$$\vec{u} \times \vec{v} = \begin{pmatrix} -1 \\ -3 \\ 5 \end{pmatrix}.$$

Since ℓ passes through $(5, 5, 5)$, a parametric description for ℓ is $\ell = \{(-t+5, -3t+5, 5t+5) \mid t \in \mathbf{R}\}$.

Example 4. Suppose that $\mathcal{P}_1$ and $\mathcal{P}_2$ are nonparallel planes and that Q is a point on their line of intersection. Let $\vec{v}_1$ be a vector perpendicular to $\mathcal{P}_1$ at Q, and let $\vec{v}_2$ be a vector perpendicular to $\mathcal{P}_2$ at Q. The *dihedral angle* between $\mathcal{P}_1$ and $\mathcal{P}_2$ is defined to be the angle between the vectors $\vec{v}_1$ and $\vec{v}_2$.

Solution. Consider the planes described by the equations $2x - y - z = 3$ and $x + y + z = 7$. The vectors

$$\vec{v}_1 = \begin{pmatrix} 2 \\ -1 \\ -1 \end{pmatrix} \quad \text{and} \quad v_2 = \begin{pmatrix} 1 \\ 1 \\ 1 \end{pmatrix}$$

are perpendicular to these planes, respectively. Therefore, the dihedral angle θ between these planes satisfies

$$\cos(\theta) = \frac{\vec{v}_1 \cdot \vec{v}_2}{\|\vec{v}_1\| \, \|\vec{v}_2\|} = \frac{0}{\sqrt{6}\sqrt{3}} = 0.$$

This shows $\theta = \frac{\pi}{2}$ (radians) $= 90°$.

Problems

1. Consider the following vectors in $\mathbf{R}^3$.

$$\vec{p} = \begin{pmatrix} 1 \\ 0 \\ -1 \end{pmatrix}, \quad \vec{q} = \begin{pmatrix} 2 \\ 1 \\ 1 \end{pmatrix}, \quad \vec{r} = \begin{pmatrix} 1 \\ -5 \\ 3 \end{pmatrix}$$

   Find the following:
   (a) $\vec{p} \times \vec{q}$;
   (b) $\vec{q} \times \vec{r}$;
   (c) $\vec{p} \cdot (\vec{q} \times \vec{r})$;
   (d) $(\vec{p} \times \vec{q}) \cdot \vec{q}$.

2. Use the cross product to help find the equations for the following planes:
   (a) the plane through the points $(2, 3, 1)$, $(0, 0, 0)$, and $(-1, 2, 9)$;
   (b) the plane through the points $(2, 3, 1)$, $(1, 0, 0)$, and $(-1, 2, 9)$;
   (c) the plane that contains the two lines parameterized by $\{(t - 1, 2t + 1, t - 1) \mid t \in \mathbf{R}\}$ and $\{(-1, t + 1, -1) \mid t \in \mathbf{R}\}$;
   (d) the plane that contains the two lines parameterized by $\{(t, 2t + 2, t - 1) \mid t \in \mathbf{R}\}$ and $\{(0, t + 2, -1) \mid t \in \mathbf{R}\}$.

3. (a) Find the area of the parallelogram in $\mathbf{R}^3$ with vertices $(0,0,0)$, $(1,2,2)$, $(2,3,5)$, and $(3,5,7)$.

   (b) Find the area of the triangle in $\mathbf{R}^3$ with vertices $(0,0,0)$, $(3,-1,2)$, and $(2,0,5)$.

4. (a) Find the area of the parallelogram in $\mathbf{R}^3$ with vertices $(1,0,1)$, $(1,2,3)$, $(2,0,5)$, and $(2,2,7)$.

   (b) Find the area of the triangle in $\mathbf{R}^3$ with vertices $(2,1,5)$, $(3,-1,2)$, and $(1,4,7)$.

5. (a) Find the distance between the point $R = (1,0,-2)$ in $\mathbf{R}^3$ and the plane $\mathcal{P}$ defined by the equation $x + y + 3z = 6$.

   (b) Find the distance between the point $O = (0,0,0)$ in $\mathbf{R}^3$ and the plane $\mathcal{Q}$ defined by the equation $2x + y + 2z = 3$.

6. (a) Find the distance between the point $Q = (0,0,2)$ in $\mathbf{R}^3$ and the plane $\mathcal{P}$ described parametrically by $\{(t+u, -t, u+5) \mid t, u \in \mathbf{R}\}$.

   (b) Find the distance between the point $O = (1,1,0)$ in $\mathbf{R}^3$ and the plane $\mathcal{Q}$ described parametrically by $\{(t+3u-1, -t+u, t-u-1) \mid t, u \in \mathbf{R}\}$.

7. (a) Find a parametric description of the line ℓ_1 in $\mathbf{R}^3$ containing $(3,2,1)$ which is perpendicular to the plane $\mathcal{P}$ passing through $(0,0,0)$, $(1,0,2)$, and $(-1,1,0)$.

   (b) Find a parametric description of the line ℓ_2 in $\mathbf{R}^3$ containing $(1,0,1)$ which is perpendicular to the plane $\mathcal{P}$ passing through $(1,0,0)$, $(0,1,0)$, and $(0,0,1)$.

8. (a) Find a parametric description of the line ℓ_1 in $\mathbf{R}^3$ containing $(1,2,1)$ which is perpendicular to the plane given by the equation $x-3y+4z=9$.

   (b) Find a parametric description of the line ℓ_2 in $\mathbf{R}^3$ containing $(-1,2,1)$ which is perpendicular to the plane given by the equation $y-2z=8$.

9. Find the equation for the plane $\mathcal{P}$ that is perpendicular to the plane $x-y+z=1$ and contains the line ℓ parameterized by $\{(t,t,1) \mid t \in \mathbf{R}\}$.

10. Find the equation for the plane that passes through $(1,0,0)$ and $(0,1,0)$ and is parallel to line ℓ parameterized by $\{(t,t,t) \mid t \in \mathbf{R}\}$.

11. (a) Give a sketch illustrating the meaning of the dihedral angle.

    (b) Find the dihedral angle between the planes $x+y=5$ and $x-z=2$.

    (c) Find the dihedral angle between the planes $x=3$ and $-x+z=2$.

12. Find the dihedral angle between the faces of a regular octahedron. (A regular octahedron is a polyhedron with eight equilateral triangles for faces and where four faces meet at each vertex.)

13. Suppose that $\vec{u} + \vec{v} + \vec{w} = \vec{0}$. Show that $\vec{u} \times \vec{v} = \vec{v} \times \vec{w} = \vec{w} \times \vec{u}$.

Group Project: The Cross Product and Volume

Suppose that $\vec{u}, \vec{v}, \vec{w} \in \mathbf{R}^3$. We denote by P the *parallelotope* with edges $\vec{u}, \vec{v}, \vec{w}$ starting from the origin. (A parallelotope is the three-dimensional analogue of the parallelogram; it is a polyhedron with eight vertices, twelve edges, and six faces. Each face of a parallelotope is a parallelogram that is congruent and parallel to the opposite face.)

(a) Show that the volume of P is $|\vec{u} \cdot (\vec{v} \times \vec{w})|$. (Hint: Rewrite $|\vec{u} \cdot (\vec{v} \times \vec{w})| = \|\vec{u}\| \cdot \|\vec{v} \times \vec{w}\| \cdot \cos(\theta)$ and use the geometric interpretation of $\vec{v} \times \vec{w}$.)

(b) Relate $|\vec{u} \cdot (\vec{v} \times \vec{w})|$ to a 3×3 determinant.

(c) Use Theorem 41 to show that the area of a parallelogram with edges given by two vectors in $\mathbf{R}^2$ is given by a determinant.

CHAPTER 7

EIGENVALUES AND EIGENVECTORS OF MATRICES

7.1 Eigenvalues and Eigenvectors

Throughout this chapter we will consider square matrices only. We shall see that many properties of an $n \times n$ matrix A can be understood by determining which (if any) vectors $\vec{v} \in \mathbf{R}^n$ satisfy $A\vec{v} = k\vec{v}$ for some real number k.

Long-Term Epidemic Effects

In Sec. 4.2 we considered a model that described changes caused by an epidemic in a rabbit population. At that time it was observed that the numbers of uninfected $U(n)$, sick $S(n)$, and immune $I(n)$ rabbits (in millions) after n months were given by multiplication by the nth power of the transition matrix

$$\begin{pmatrix} U(n) \\ S(n) \\ I(n) \end{pmatrix} = \begin{pmatrix} \frac{3}{4} & \frac{1}{12} & \frac{1}{12} \\ \frac{1}{3} & 0 & 0 \\ 0 & \frac{3}{4} & 1 \end{pmatrix}^n \begin{pmatrix} 12 \\ 0 \\ 0 \end{pmatrix}.$$

If we examine what happens after six months, and one, two, three, and four years we find (using a calculator to evaluate) that the population vectors are

$$\begin{matrix} \text{6 mo.} & \text{12 mo.} & \text{24 mo.} & \text{36 mo.} & \text{48 mo.} \\ \begin{pmatrix} 4.22 \\ 1.50 \\ 9.72 \end{pmatrix} & \begin{pmatrix} 4.76 \\ 1.51 \\ 16.1 \end{pmatrix} & \begin{pmatrix} 10.1 \\ 3.16 \\ 35.3 \end{pmatrix} & \begin{pmatrix} 22.0 \\ 6.88 \\ 76.9 \end{pmatrix} & \begin{pmatrix} 48.0 \\ 15.0 \\ 167 \end{pmatrix} \end{matrix}.$$

At first glance one may not notice any significant pattern in these numbers, but if we compute the relative percentage in each category, we find

$$\begin{matrix} \text{6 mo.} & \text{12 mo.} & \text{24 mo.} & \text{36 mo.} & \text{48 mo.} \\ \begin{pmatrix} 27.3\% \\ 8.8\% \\ 63.0\% \end{pmatrix} & \begin{pmatrix} 21.3\% \\ 6.75\% \\ 71.9\% \end{pmatrix} & \begin{pmatrix} 20.8\% \\ 6.51\% \\ 72.7\% \end{pmatrix} & \begin{pmatrix} 20.8\% \\ 6.50\% \\ 72.7\% \end{pmatrix} & \begin{pmatrix} 20.8\% \\ 6.50\% \\ 72.7\% \end{pmatrix} \end{matrix}.$$

This shows that the percentages in each category become stable after enough time (at least within the three decimal points of this calculation).

Why did this happen? Note that the difference between each successive 12-month interval was obtained by multiplying by the twelfth power of the transition matrix. But also note that the yearly population vectors for the second, third, and fourth years can be obtained by multiplying the previous year's vector by 2.18. In other words, if T^{12} is our 12-month transition matrix, and if $\vec{v}_n$ is our population vector after n months, we have

$$\vec{v}_{n+12} = T^{12}\vec{v}_n \approx 2.18\vec{v}_n.$$

This is an important observation. The vector $\vec{v}_n$ is called an *eigenvector* of *eigenvalue*[1] 2.18 for the matrix T^{12}. Eigenvectors often show up when phenomena exhibit stable behavior (as did our rabbit epidemic).

Definition of Eigenvalues and Eigenvectors

Definition. Suppose that A is an $n \times n$ matrix. If $\vec{v} \in \mathbf{R}^n$ is a nonzero vector and there exists a real number k such that $A\vec{v} = k\vec{v}$, we call $\vec{v}$ an *eigenvector* of A. The real number k is called the *eigenvalue* of A associated with the eigenvector $\vec{v}$.

Notice that an eigenvalue can be zero but that $\vec{0}$ *is never allowed to be an eigenvector.* The reason for this is that the zero vector satisfies the equation $A\vec{0} = k\vec{0}$ for all real numbers k. So if we allowed $\vec{0}$ to be an eigenvector, then every real number k would be an eigenvalue! It is important, however, to allow 0 to be an eigenvalue, for if $\vec{v}$ is an eigenvector with associated eigenvalue 0, then $A\vec{v} = 0\vec{v} = \vec{0}$, so in fact, $\vec{v} \in \ker(A)$.

[1] Strictly speaking, the eigenvalue is only approximately 2.18, and $\vec{v}_n$ is an approximate eigenvector.

For example, if we consider the 3×3 matrix

$$B = \begin{pmatrix} 3 & 2 & 2 \\ 1 & 2 & 2 \\ -1 & -1 & 0 \end{pmatrix} \quad \text{and} \quad \vec{v} = \begin{pmatrix} 2 \\ 0 \\ -1 \end{pmatrix},$$

then we can check that

$$B\vec{v} = \begin{pmatrix} 3 & 2 & 2 \\ 1 & 2 & 2 \\ -1 & -1 & 0 \end{pmatrix} \begin{pmatrix} 2 \\ 0 \\ -1 \end{pmatrix} = \begin{pmatrix} 4 \\ 0 \\ -2 \end{pmatrix} = 2\vec{v}.$$

Thus, $\vec{v}$ is an eigenvector of B with associated eigenvalue 2.

Column Buckling

In Fig. 7.1(a) we have indicated a vertical column that is subject to a compressive force (represented by $\vec{P}$). For example, this column could be supporting the roof of a building, and the force $\vec{P}$ could be the weight of the roof. An important problem for structural engineers is to determine when the column buckles under the force. When a column buckles, as indicated in Fig. 7.1(b), the situation becomes unstable and the roof could perhaps collapse.

One way to study the column buckling problem is represent the column by a function and assume this function is a solution to a specific differential equation. This method requires calculus and gives what is called a *continuous model* of the column. We will *not* do this. Instead we will use what is called a finite-difference approximation of the column. This is a sequence of points on the column equally spaced by height. We will then assume that this finite approximation of the column satisfies a *difference equation* (which is closely related to the differential equation in the continuous solution). As noted in Chap. 4, difference equations are best understood using matrix methods, and so we can use matrix techniques to study column buckling.

Here's how the finite-difference approximation of column buckling works. First, the nonbuckled column is represented in the xy plane by

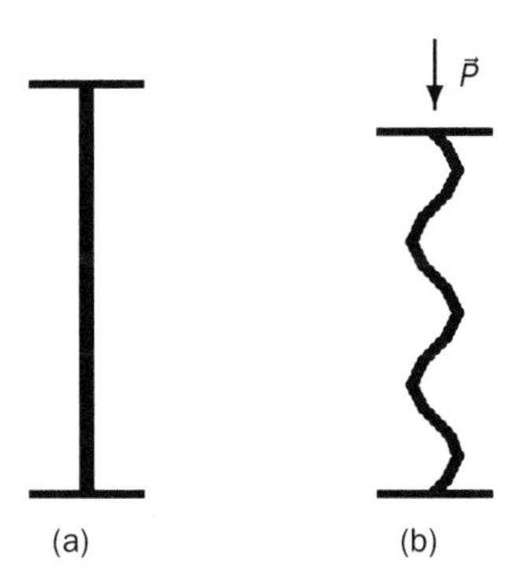

Fig. 7.1. A column (a) and the column buckling (b)

the points with x-coordinate 0 and y-coordinates between 0 and 6. We use the seven equally spaced points $(0,0)$, $(0,1)$, $(0,2)$, $(0,3)$, $(0,4)$, $(0,5)$, and $(0,6)$ on the column to give a discrete model of the column (see Fig. 7.2). Suppose the column buckles a little bit. Then we have the seven points $(0,0)$, $(x_1,1)$, $(x_2,2)$, $(x_3,3)$, $(x_4,4)$, $(x_5,5)$ and $(0,6)$, where some of the x_i are nonzero. We will assume that the top and bottom remain fixed in this model, ignoring the fact that the y-coordinates drop a little bit during buckling. Our problem is to determine the possibilities for the five values x_1, x_2, x_3, x_4, and x_5, and this information will give us a picture of what the buckling looks like.

Engineers have determined that a good model for buckling requires that the sequence of x-coordinates satisfy a difference equation of the form

$$(x_{i-1} - 2x_i + x_{i+1}) + \lambda x_i = 0,$$

where the values of λ depend on the force $\vec{P}$ and the bending stiffness of the column,[2] and where i ranges from 1 to 5. Since $x_0 = x_6 = 0$ in our model, finding solutions to this difference equation is the same as finding solutions to the matrix equation

$$\begin{pmatrix} 2 & -1 & 0 & 0 & 0 \\ -1 & 2 & -1 & 0 & 0 \\ 0 & -1 & 2 & -1 & 0 \\ 0 & 0 & -1 & 2 & -1 \\ 0 & 0 & 0 & -1 & 2 \end{pmatrix} \begin{pmatrix} x_1 \\ x_2 \\ x_3 \\ x_4 \\ x_5 \end{pmatrix} = \lambda \begin{pmatrix} x_1 \\ x_2 \\ x_3 \\ x_4 \\ x_5 \end{pmatrix}.$$

Of course, the solution $x_1 = 0$, $x_2 = 0$, $x_3 = 0$, $x_4 = 0$, and $x_5 = 0$ works for all λ. This corresponds to the column being straight. But what about other nonzero solutions?

It turns out that for most numbers λ, the zero solution for the x's is the only one possible. However, for those λ's that are *eigenvalues* of this 5×5 matrix, there are nonzero solutions for our x's, and these correspond to possible buckling positions of our column. For the above matrix, it turns out that $\lambda = 1$, 2, and 3 are eigenvalues. For example, if $\lambda = 1$, then it can be checked that $x_1 = 1$, $x_2 = 1$, $x_3 = 0$, $x_4 = -1$, and $x_5 = -1$ is a solution. If $\lambda = 2$, then $x_1 = 1$, $x_2 = 0$, $x_3 = -1$, $x_4 = 0$, and $x_5 = 1$ is a solution; and if $\lambda = 3$, we find that $x_1 = 1$, $x_2 = -1$, $x_3 = 0$, $x_4 = 1$, and $x_5 = -1$ is a solution. These three solutions for the buckling problem, together with the solution representing the straight column, are pictured in Fig. 7.2.

These solutions show that when this type of column buckles, it bends back and forth, assuming a wavelike appearance. If we wanted a more accurate picture of the column, we could use more than seven points, placing them closer together to help us visualize more points on the column. The computations would be similar—only the matrices would get bigger.

[2]The polynomial $x_{i-1} - 2x_i + x_{i+1}$ gives a discrete approximation to the second derivative d^2x/dy^2 of x as a function of y (near x_i), and therefore our difference equation is a discrete version of the differential equation $d^2x/dy^2 + \lambda x = 0$.

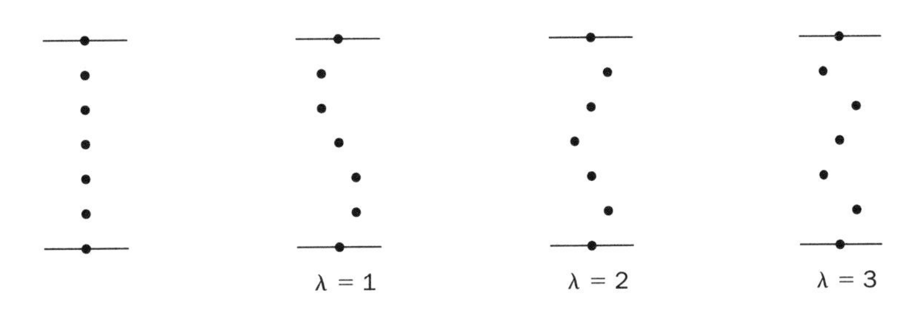

Fig. 7.2. Four discrete buckling solutions

In this particular problem, the crucial question is that of finding the possible eigenvalues of a tridiagonal matrix. The computation of eigenvalues of large tridiagonal matrices is important for applications such as the column buckling problem, and engineers and mathematicians have developed many methods for studying this question. As you might guess, when the matrices get large, computer techniques for computation are needed. Eigenvalues of some more tridiagonal matrices are studied in the next section.

Eigenspaces

We shall shortly describe a procedure for computing all the eigenvalues of an $n \times n$ matrix. In order to do this, we give an alternate characterization of the eigenvectors with a fixed eigenvalue. Recall for any matrix M that $\ker(M)$ denotes the null space of M, that is, $\ker(M) = \{\vec{v} \mid M\vec{v} = \vec{0}\}$.

Theorem 42. *Let A be an $n \times n$ matrix. Then $\vec{v}$ is an eigenvector of A with associated eigenvalue r if and only if $\vec{0} \neq \vec{v} \in \ker(rI_n - A)$. Consequently, the set of eigenvectors of A with associated eigenvalue r, together with the vector $\vec{0}$, is a subspace of $\mathbf{R}^n$.*

Proof. Note that $A\vec{v} = r\vec{v}$ is equivalent to $rI_n\vec{v} - A\vec{v} = \vec{0}$, and so $\vec{v}$ is an eigenvector with eigenvalue r if and only if $(rI_n - A)\vec{v} = \vec{0}$. This gives the first assertion. The second statement is a consequence of the first since Theorem 23 shows the kernel of any matrix is always a subspace. □

As a consequence of this theorem it makes sense to give the following definition.

Definition. Let r be an eigenvalue of an $n \times n$ matrix A. The subspace of $\mathbf{R}^n$ of all eigenvectors with associated eigenvalue r together with the vector $\vec{0}$ is called the *eigenspace* associated with r.

In this textbook the eigenspace associated with r will be denoted by E_r. For example, it is instructive to look at a diagonal matrix. Consider

$$D = \begin{pmatrix} 3 & 0 & 0 \\ 0 & 2 & 0 \\ 0 & 0 & 3 \end{pmatrix}.$$

Then

$$\vec{e}_1 = \begin{pmatrix} 1 \\ 0 \\ 0 \end{pmatrix}, \quad \vec{e}_3 = \begin{pmatrix} 0 \\ 0 \\ 1 \end{pmatrix}, \quad \text{and} \quad \vec{e}_2 = \begin{pmatrix} 0 \\ 1 \\ 0 \end{pmatrix}$$

are eigenvectors for D. $\vec{e}_1$ and $\vec{e}_3$ have associated eigenvalue 3, and $\vec{e}_2$ has associated eigenvalue 2. The theorem shows that the eigenspace E_3 contains the two-dimensional subspace span$\{\vec{e}_1, \vec{e}_3\}$ and that the eigenspace E_2 contains the one-dimensional subspace span$\{\vec{e}_2\}$. In fact, E_3 must be two-dimensional. If it were larger, it would be all of $\mathbf{R}^3$, but this is impossible since not every vector is an eigenvector for A. In addition, E_2 can only be one-dimensional. If it were two-dimensional, it would have to have a nonzero common vector with E_3. (The intersection of two nonparallel planes in $\mathbf{R}^3$ must contain a line.) This is impossible since an eigenvector cannot have two different eigenvalues. Thus $E_3 = \text{span}\{\vec{e}_1, \vec{e}_3\}$ and $E_2 = \text{span}\{\vec{e}_2\}$.

It must be emphasized that the sum of two eigenvectors with different eigenvalues is *not* an eigenvector. The sum of two eigenvectors is another eigenvector only if they have the same eigenvalue. For example, the vector $\vec{e}_1 + \vec{e}_2$ is *not* an eigenvector for D since

$$D(\vec{e}_1 + \vec{e}_2) = \begin{pmatrix} 3 & 0 & 0 \\ 0 & 2 & 0 \\ 0 & 0 & 3 \end{pmatrix} \begin{pmatrix} 1 \\ 1 \\ 0 \end{pmatrix} = \begin{pmatrix} 3 \\ 2 \\ 0 \end{pmatrix},$$

which is not a multiple of $\vec{e}_1 + \vec{e}_2$.

The Characteristic Polynomial

In order to study more examples, we must show how to compute the eigenvalues of a matrix. Theorem 42 shows that $\vec{v}$ is an eigenvector for A with eigenvalue r precisely if $(rI_n - A)\vec{v} = \vec{0}$. This observation motivates the definition of the characteristic polynomial.

Definition. Let A be an $n \times n$ matrix. The *characteristic polynomial* of A, $C_A(X)$, is the polynomial that is the determinant of the matrix $(XI_n - A)$. In other words, $C_A(X) = \det(XI_n - A)$.

For example, the characteristic polynomial of matrix D considered earlier is

$$\begin{aligned} C_D(X) &= \det\left(X \begin{pmatrix} 1 & 0 & 0 \\ 0 & 1 & 0 \\ 0 & 0 & 1 \end{pmatrix} - \begin{pmatrix} 3 & 0 & 0 \\ 0 & 2 & 0 \\ 0 & 0 & 3 \end{pmatrix} \right) \\ &= \det \begin{pmatrix} X-3 & 0 & 0 \\ 0 & X-2 & 0 \\ 0 & 0 & X-3 \end{pmatrix} \\ &= (X-3)(X-2)(X-3) = (X-3)^2(X-2). \end{aligned}$$

Note that in this case $C_D(X)$ is a polynomial of degree 3.

In general, if A is an $n \times n$ matrix, then the polynomial $C_A(X)$ is a polynomial of degree n. The reasons for this are explored in Prob. 8 at the end of this section. The next theorem shows that the roots of the characteristic polynomial of A are the eigenvalues of A. Recall for any matrix A that null(A) denotes the dimension of the null space ker(A) of A.

Theorem 43. *Suppose that A is an $n \times n$ matrix.*

(i) *A real number r is an eigenvalue of A if and only if $C_A(r) = 0$.*

(ii) *If r is an eigenvalue of A, the dimension of the eigenspace E_r of eigenvalue r is* $\text{null}(rI_n - A) = n - rk(rI_n - A)$.

Proof. (i) By Theorem 42, a real number r is an eigenvalue of A if and only if there exists some nonzero vector $\vec{v} \in \mathbf{R}^n$ such that $(A - rI_n)\vec{v} = \vec{0}$. Applying Theorem 15, we see that this is equivalent to $0 = \det(A - rI_n) = C_A(r)$. This gives (i). For (ii), since $E_r = \ker(rI_n - A)$, the dimension theorem shows that $\dim(E_r) = \dim(\ker(rI_n - A)) = \text{null}(rI_n - A) = n - \text{rk}(rI_n - A)$, as required. □

You may have been wondering how many different eigenvalues a matrix can have. The characteristic polynomial provides the answer.

Corollary. *Any $n \times n$ matrix A has at most n distinct eigenvalues.*

Proof. As observed, the degree of the polynomial $C_A(X)$ is n. The result now follows from Theorem 43, since $C_A(X)$ can have at most n distinct roots. □

For example, consider

$$B = \begin{pmatrix} 2 & 3 & 1 \\ 3 & 2 & 4 \\ 0 & 0 & -1 \end{pmatrix}.$$

We compute that

$$C_B(X) = \det \begin{pmatrix} X-2 & -3 & -1 \\ -3 & X-2 & -4 \\ 0 & 0 & X+1 \end{pmatrix} = [(X-2)(X-2) - 9](X+1)$$

$$= [X^2 - 4X - 5](X+1) = (X-5)(X+1)^2,$$

which has the two roots -1 and 5. Hence, B has the two eigenvalues -1 and 5. To find the associated eigenspaces, we must compute the null spaces of $-I - B$ and $5I - B$. The eigenspace associated with 5 is

$$E_5 = \ker(5I - B) = \ker\begin{pmatrix} 3 & -3 & -1 \\ -3 & 3 & -4 \\ 0 & 0 & 6 \end{pmatrix} = \operatorname{span}\left\{\begin{pmatrix} 1 \\ 1 \\ 0 \end{pmatrix}\right\}$$

and the eigenspace associated with -1 is

$$E_{-1} = \ker(-I - B) = \ker\begin{pmatrix} -3 & -3 & 1 \\ -3 & -3 & 4 \\ 0 & 0 & 0 \end{pmatrix} = \operatorname{span}\left\{\begin{pmatrix} -1 \\ 1 \\ 0 \end{pmatrix}\right\}.$$

The Epidemic Revisited

In our study of the rabbit epidemic at the beginning of this section, we used the transition matrix

$$T = \begin{pmatrix} \frac{3}{4} & \frac{1}{12} & \frac{1}{12} \\ \frac{1}{3} & 0 & 0 \\ 0 & \frac{3}{4} & 1 \end{pmatrix} = \frac{1}{12}\begin{pmatrix} 9 & 1 & 1 \\ 4 & 0 & 0 \\ 0 & 9 & 12 \end{pmatrix}.$$

We can compute its eigenvalues by determining the roots of the characteristic polynomial $C_{12T}(X)$. It is

$$\det\begin{pmatrix} X-9 & -1 & -1 \\ -4 & X & 0 \\ 0 & -9 & X-12 \end{pmatrix} = (X-9)X(X-12) + (-4)(X-12) - 36$$
$$= X^3 - 21X^2 + 104X + 12.$$

This cubic polynomial doesn't factor in any nice way, but we can use a graphing calculator to find close approximations to its roots. They are approximately $-.113$, 8.31, and 12.81. Now our transition matrix T is $\frac{1}{12}$ of the matrix $12T$, so its eigenvalues are $\frac{1}{12}$ of these. We find, therefore, that the eigenvalues of T are approximately $-.0094$, $.693$, and 1.067. Observe that the eigenvalue 1.067 raised to the twelfth power is $1.067^{12} \approx 2.18$. This is the eigenvalue (to two decimal places) of T^{12} found earlier by investigating the stable behavior of our rabbit population.

Fibonacci Numbers and the Golden Ratio

Recall that in Sec. 4.2 we noted that the Fibonacci numbers could be computed by iterative matrix multiplication. We saw there that the $(n+2)$nd Fibonacci number $P(n+2)$ was determined by $P(n+1)$, $P(n)$, and the matrix equation

$$\begin{pmatrix} P(n+2) \\ P(n+1) \end{pmatrix} = \begin{pmatrix} 1 & 1 \\ 1 & 0 \end{pmatrix}\begin{pmatrix} P(n+1) \\ P(n) \end{pmatrix}$$

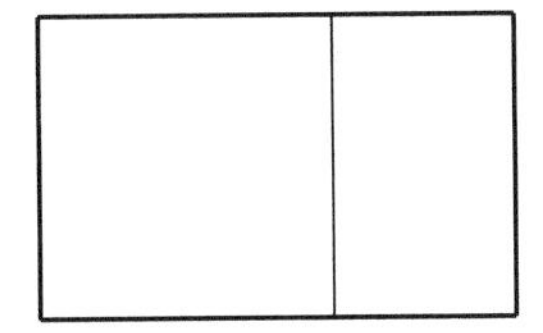

Fig. 7.3. A golden rectangle with square inside

for all numbers n. Just as our rabbit population ratios became stable above, ratios of successive Fibonacci numbers become stable, too. The reason is that the vectors of pairs of Fibonacci numbers are close to eigenvectors of the transition matrix. The eigenvalues of the matrix

$$P = \begin{pmatrix} 1 & 1 \\ 1 & 0 \end{pmatrix}$$

are the roots of its characteristic polynomial $C_P(X) = (X - 1)X - 1 = X^2 - X - 1$. Using the quadratic formula we find that these are $\frac{1 \pm \sqrt{5}}{2}$.

For large values of n the vector

$$\begin{pmatrix} P(n+2) \\ P(n+1) \end{pmatrix}$$

of pairs of Fibonacci numbers is very close to an eigenvector for the matrix T with eigenvalue $\frac{1+\sqrt{5}}{2}$. This means that for large n we should have

$$\begin{pmatrix} P(n+2) \\ P(n+1) \end{pmatrix} \approx \frac{1+\sqrt{5}}{2} \begin{pmatrix} P(n+1) \\ P(n) \end{pmatrix}, \quad \text{so} \quad \frac{P(n+1)}{P(n)} \approx \frac{1+\sqrt{5}}{2}.$$

Indeed, using a calculator we compute that $P(30) = 832,040$ and $P(29) = 514,229$. From this we finds $\frac{P(30)}{P(29)} \approx 1.61803398875$. You can check on your calculator that this agrees with $\frac{1+\sqrt{5}}{2}$ to all 12 decimal points given. The ratio $\frac{1+\sqrt{5}}{2}$ is known as the golden ratio since it is the ratio of the sides of what many people believe is an aesthetically perfect rectangle. Such a rectangle is characterized by the property that if a square is cut off one end the remaining piece is another (smaller) golden rectangle (see Fig. 7.3).

Problems

1. Find the characteristic polynomial, the eigenvalues, and the eigenspaces (if any) of each of the following matrices.

   (a) $\begin{pmatrix} 2 & 3 \\ 3 & 2 \end{pmatrix}$ (b) $\begin{pmatrix} 0 & 1 \\ 1 & 0 \end{pmatrix}$ (c) $\begin{pmatrix} 0 & 1 \\ -1 & 0 \end{pmatrix}$

2. Find the characteristic polynomial, the eigenvalues, and the eigenspaces (if any) of each of the following matrices.

(a) $\begin{pmatrix} 0 & 0 & 1 \\ 1 & 0 & 0 \\ 0 & 1 & 0 \end{pmatrix}$ (b) $\begin{pmatrix} 2 & 1 & 1 \\ 0 & 1 & 0 \\ 1 & 0 & 1 \end{pmatrix}$ (c) $\begin{pmatrix} 1 & 1 & 1 \\ 1 & 1 & 1 \\ 1 & 1 & 1 \end{pmatrix}$

3. Find a 2×2 matrix R with eigenvalues 1 and -1 and with

$$E_1 = \text{span}\left\{\begin{pmatrix} 1 \\ -2 \end{pmatrix}\right\} \quad \text{and} \quad E_{-1} = \text{span}\left\{\begin{pmatrix} 2 \\ 1 \end{pmatrix}\right\}.$$

If you multiply a vector on the left by R, you obtain what is called its reflection. Explain geometrically what this means.

4. Use a calculator or a computer to approximate the eigenvalues of the following matrices.

(a) $\begin{pmatrix} 0 & .3 & .4 \\ .1 & 0 & .2 \\ .9 & .7 & .4 \end{pmatrix}$ (b) $\begin{pmatrix} 2 & 3 & 1 \\ 3 & 4 & 3 \\ 1 & 3 & 4 \end{pmatrix}$ (c) $\begin{pmatrix} 1 & 3 & 9 \\ 1 & -3 & -1 \\ 2 & 4 & -2 \end{pmatrix}$

5. Show that the characteristic polynomial of

$$B = \begin{pmatrix} 3 & 2 & 2 \\ 1 & 2 & 2 \\ -1 & -1 & 0 \end{pmatrix}$$

is

$$C_B(X) = (X - 1)(X - 2)^2.$$

Find E_1 and E_2.

6. (a) Assume that the real number k is an eigenvalue of the matrix A. Show then that k^2 is an eigenvalue of the matrix A^2.

(b) If k is an eigenvalue of an invertible matrix A, show that k^{-1} is an eigenvalue of A^{-1}.

(c) Show that if N is nilpotent, then 0 is the only eigenvalue of N. (Recall that a square matrix N is called *nilpotent* if N^k is the zero matrix for some positive integer k.)

7. Show that the matrix

$$G = \begin{pmatrix} 2 & 0 & 1 \\ 1 & 1 & 0 \\ 1 & 0 & 1 \end{pmatrix}$$

has the three eigenvalues 1, $\frac{1}{2}(3 + \sqrt{5})$, and $\frac{1}{2}(3 - \sqrt{5})$.

8. Write a paragraph explaining why the characteristic polynomial of an $n \times n$ matrix has degree n. Use the fact that the determinant is a sum of (signed) products of entries of a matrix, where each summand consists of a product of entries representing exactly one row and exactly one column.

9. Suppose that A is a 3×3 matrix. Show that there is some real number c with $\text{rk}(A - cI) \leq 2$.

10. Assume that A is an $n \times n$ matrix with the special property that the sum of the entries of each row is the same number r. Show that r is an eigenvalue of A. (Hint: Find an eigenvector.)

Group Project: Eigenvalues and Determinants

For parts (a) and (b), suppose that A is an $n \times n$ matrix with n distinct eigenvalues $c_1, c_2, \ldots, c_n$.

(a) Show that $\det(A) = c_1 c_2 \cdots c_n$. (Hint: What must the characteristic polynomial be?)

(b) The *trace* of A, denoted by $\mathrm{tr}(A)$, is defined to be the sum of the diagonal entries of A. Show that $\mathrm{tr}(A) = c_1 + c_2 + \cdots + c_n$. (Note: For this problem you should first observe that the result is clear for diagonal matrices. Next try 2×2 and 3×3 matrices and find out which coefficient of the characteristic polynomial shows up as the trace.)

(c) Suppose that A is an $n \times n$ matrix with characteristic polynomial $X^n + a_{n-1}X^{n-1} + \cdots + a_1 X + a_0$. Show that $\det(A) = (-1)^n a_0$. Deduce that for any matrix A, A is invertible if and only if $C_A(0) \neq 0$.

Group Project: Estimating Eigenvalues with Matrix Powers

For this project you will need to know how to compute matrix powers on a calculator or computer. You will also need to find out how to calculate characteristic polynomials and approximate their roots on your machine. Consider the matrix

$$A = \begin{pmatrix} 0.2 & 0.3 & 0.1 & 0.4 \\ 0.3 & 0.0 & 0.6 & 0.2 \\ 0.1 & 0.6 & 0.2 & 0.3 \\ 0.4 & 0.2 & 0.3 & 0.1 \end{pmatrix}.$$

(a) Find A^{25}. The columns of A are eigenvectors of A (at least within a few decimal points). Check this. What eigenvalues do you obtain?

(b) Using a graphing calculator or a computer, find the roots of the characteristic polynomial of A to three decimal points.

(c) Which eigenvalues in part (b) did you find in part (a), and which did you miss? What happened to the eigenvalues that you missed? Why didn't they show up as columns of A^{25}?

7.2 Eigenspaces and Diagonalizability

Diagonalizable Matrices

If D is the diagonal matrix with diagonal entries $d_1, d_2, \ldots, d_n$, then the characteristic polynomial of D is $(X - d_1)(X - d_2)\cdots(X - d_n)$. Therefore, the eigenvalues of D are $d_1, d_2, \ldots, d_n$. Moreover, an eigenvector with eigenvalue d_i is the ith basis element $\vec{e}_i$. (Recall that $\vec{e}_i$ is our notation for the ith standard basis vector that has all zero entries except the ith entry, which is 1.) In particular, we see that D has n linearly independent eigenvectors.

Suppose that A is an $n \times n$ matrix and that the matrix $P^{-1}AP = D$ is diagonal for some invertible matrix P. Assume that the diagonal entries of D are $d_1, d_2, \ldots, d_n$. Consider the vector $P\vec{e}_i \in \mathbf{R}^n$. Multiplying the equation $P^{-1}AP = D$ by P shows $AP = PD$, and consequently

$$A(P\vec{e}_i) = (AP)\vec{e}_i = (PD)\vec{e}_i = P(D\vec{e}_i) = Pd_i\vec{e}_i = d_i(P\vec{e}_i).$$

This shows that $P\vec{e}_i$ is an eigenvector of A with eigenvalue d_i. In particular, we see that the matrix A has n linearly independent eigenvectors, namely, $P\vec{e}_1, P\vec{e}_2, \ldots, P\vec{e}_n$.

The problem of finding an invertible matrix P so that the matrix $P^{-1}AP$ is diagonal turns out to be an extremely important problem in matrix theory. We give the following definition.

Definition. A matrix A is called *diagonalizable* if there exists an invertible matrix P such that the matrix $P^{-1}AP = D$ is a diagonal matrix.

We first note that not every matrix is diagonalizable. For example, the matrix

$$J = \begin{pmatrix} 1 & 1 \\ 0 & 1 \end{pmatrix}$$

has characteristic polynomial $C_J(X) = (X - 1)^2$. Its only eigenspace, E_1, is one-dimensional, and therefore J cannot have two linearly independent eigenvectors. Thus, J is not diagonalizable.

Often, however, a matrix that looks nothing like a diagonal matrix can be diagonalizable. Consider the matrix

$$C = \begin{pmatrix} 1 & 3 & 3 \\ 20 & 5 & -10 \\ -2 & -1 & 14 \end{pmatrix}.$$

Let

$$P = \begin{pmatrix} 1 & 0 & 1 \\ 2 & -1 & -2 \\ 1 & 1 & 0 \end{pmatrix}.$$

Then by direct calculation,

$$\begin{aligned} P^{-1}CP &= \begin{pmatrix} \frac{2}{5} & \frac{1}{5} & \frac{1}{5} \\ -\frac{2}{5} & -\frac{1}{5} & \frac{4}{5} \\ \frac{3}{5} & -\frac{1}{5} & \frac{1}{5} \end{pmatrix} \begin{pmatrix} 1 & 3 & 3 \\ 20 & 5 & -10 \\ -2 & -1 & 14 \end{pmatrix} \begin{pmatrix} 1 & 0 & 1 \\ 2 & -1 & -2 \\ 1 & 1 & 0 \end{pmatrix} \\ &= \begin{pmatrix} 10 & 0 & 0 \\ 0 & 15 & 0 \\ 0 & 0 & -5 \end{pmatrix}, \end{aligned}$$

which shows that C is diagonalizable. The meaning of the matrix P in this example (as well as how to find it) is addressed next.

Theorem 44. *Suppose A is an $n \times n$ matrix. Then A is diagonalizable if and only if A has n linearly independent eigenvectors.*

Proof. We noted above that if $P^{-1}AP = D$ is diagonal, then the vectors $P\vec{e}_i$ are eigenvectors of A. Since the $P\vec{e}_i$ are the columns of the invertible matrix P, they are linearly independent. Conversely, suppose that $\vec{v}_1, \vec{v}_2, \ldots, \vec{v}_n$ are linearly independent eigenvectors for A with eigenvalues $k_1, k_2, \ldots, k_n$, respectively. Let $P = (\vec{v}_1\ \vec{v}_2\ \cdots\ \vec{v}_n)$ be the matrix whose columns are these vectors. By the definition of matrix multiplication, $P\vec{e}_i = \vec{v}_i$ since it is the ith column of P. Likewise, $(P^{-1}AP)\vec{e}_i$ is the ith column of $P^{-1}AP$. We find that $(P^{-1}AP)\vec{e}_i = P^{-1}A\vec{v}_i = P^{-1}k_i\vec{v}_i = k_iP^{-1}\vec{v}_i = k_iP^{-1}P\vec{e}_i = k_i\vec{e}_i$. This shows that $P^{-1}AP$ is the diagonal matrix whose ith diagonal entry is k_i. □

Note that the proof of Theorem 44 tells us how to find the diagonalizing matrix P once we know our matrix A is diagonalizable. P is the matrix whose columns are linearly independent eigenvectors for A. For example, recall the matrix C, which we diagonalized above. The diagonal entries of $P^{-1}CP$ were 10, 15, and -5. The columns of P, namely

$$\begin{pmatrix} 1 \\ 2 \\ 1 \end{pmatrix}, \quad \begin{pmatrix} 0 \\ -1 \\ 1 \end{pmatrix}, \quad \text{and} \quad \begin{pmatrix} 1 \\ -2 \\ 0 \end{pmatrix},$$

are eigenvectors of C with eigenvalues 10, 15, and -5 respectively. This can be checked by direct calculation.

Linear Independence of Eigenvectors

We have shown that a matrix is diagonalizable whenever it has enough linearly independent eigenvectors. How do we know if eigenvectors are linearly independent? This next result helps.

Theorem 45. *Suppose that $\vec{v}_1, \vec{v}_2, \ldots, \vec{v}_n$ are eigenvectors of a matrix A and that each corresponds to different eigenvalues. Then $\{\vec{v}_1, \vec{v}_2, \ldots, \vec{v}_n\}$ is linearly independent.*

Proof. Assume the contrary. Let k_i be the eigenvalue associated with $\vec{v}_i$, and suppose that

$$a_1\vec{v}_1 + a_2\vec{v}_2 + \cdots + a_n\vec{v}_n = \vec{0}, \tag{1}$$

where not all $a_i = 0$. Assume also that this expression has a *minimal* number of nonzero coefficients a_i among all possible such expressions.

Because each $\vec{v}_i$ is a nonzero vector, we see that at least two a_i must be nonzero. In particular, since the eigenvalues $k_1, k_2, \ldots, k_n$ are distinct, there is some i with $a_i \neq 0$ and $k_i \neq 0$. After reordering the vectors $\vec{v}_1, \vec{v}_2, \ldots, \vec{v}_n$ (if necessary), we can assume that $i = 1$. This means that both $a_1 \neq 0$ and $k_1 \neq 0$. Hence

$$\begin{aligned} \vec{0} = A\vec{0} &= A(a_1\vec{v}_1 + a_2\vec{v}_2 + \cdots + a_n\vec{v}_n) \\ &= a_1 A\vec{v}_1 + a_2 A\vec{v}_2 + \cdots + a_n A\vec{v}_n \\ &= a_1 k_1 \vec{v}_1 + a_2 k_2 \vec{v}_2 + \cdots + a_n k_n \vec{v}_n. \end{aligned}$$

Since $k_1 \neq 0$, this gives

$$a_1\vec{v}_1 + a_2\tfrac{k_2}{k_1}\vec{v}_2 + \cdots + a_n\tfrac{k_n}{k_1}\vec{v}_n = \vec{0}. \tag{2}$$

Subtracting (2) from our original expression (1), we find

$$0\vec{v}_1 + a_2\left(1 - \tfrac{k_2}{k_1}\right)\vec{v}_2 + \cdots + a_n\left(1 - \tfrac{k_n}{k_1}\right)\vec{v}_n = \vec{0}. \tag{3}$$

Note that since $k_1, k_2, \ldots, k_n$ are all distinct, each of $1 - \frac{k_2}{k_1}, 1 - \frac{k_3}{k_1}, \ldots, 1 - \frac{k_n}{k_1}$ is nonzero. Hence (3) is a nontrivial linear combination of $\vec{v}_1, \vec{v}_2, \ldots, \vec{v}_n$ that is $\vec{0}$. But (3) has fewer nonzero coefficients than (1), since $\vec{v}_1$ has 0 as a coefficient. This contradicts our original choice of expression (1). It follows that $\{\vec{v}_1, \vec{v}_2, \ldots, \vec{v}_n\}$ is linearly independent. □

As an immediate consequence of Theorems 44 and 45, we have the following corollary.

Corollary. *If an $n \times n$ matrix A has n distinct eigenvalues, then A is diagonalizable.*

Proof. According to Theorem 45, any set of n eigenvectors of A with distinct eigenvalues must be linearly independent. Theorem 44 shows that A is diagonalizable. □

Eigenvalues of the Tridiagonal Matrix Used in the Column Buckling Problem

Recall that the solutions found in the column buckling problem appeared to oscillate back and forth, like a trigonometric function.[3] This suggests we adopt a strategy to look for possible eigenvectors that are based on periodic

[3]Students familiar with the solutions to second-order linear differential equations won't be surprised, because the difference equation that generated the tridiagonal matrix corresponds to the differential equation $d^2x/dy^2 + \lambda x = 0$ whose solutions are generated by the trigonometric functions $x = \sin(\sqrt{\lambda}y)$ and $x = \cos(\sqrt{\lambda}y)$.

functions. The strategy will enable us to avoid computing the characteristic polynomial of tridiagonal matrices. We consider the $n \times n$ matrix

$$T_{n,\alpha} = \begin{pmatrix} \alpha & 1 & 0 & 0 & \cdots \\ 1 & \alpha & 1 & 0 & \cdots \\ 0 & 1 & \alpha & 1 & \cdots \\ \cdots & \cdots & \cdots & \cdots & \cdots \end{pmatrix},$$

which is symmetric and tridiagonal like the matrix considered in our buckling model. We desire to find for which values α this $n \times n$ matrix is singular. For this we must find a nonzero vector $\vec{x} = (x_1, x_2, \ldots, x_n)$ in the null space of $T_{n,\alpha}$. Because $\vec{x} \in \ker(T_{n,\alpha})$, we have the system of equations

$$\begin{aligned} \alpha x_1 + x_2 &= 0 \\ x_1 + \alpha x_2 + x_3 &= 0 \\ x_2 + \alpha x_3 + x_4 &= 0 \\ &\vdots \\ x_{n-1} + \alpha x_n &= 0. \end{aligned}$$

If we take $x_0 = 0$ and $x_{n+1} = 0$, then all of these equations can be represented by the single equation

$$x_{j-1} + \alpha x_j + x_{j+1} = 0,$$

where $j = 1, 2, \ldots, n$.

Our strategy of looking for periodic solutions to our buckling problem suggests that we try $x_0 = 0 = \sin(0\,\theta)$, $x_1 = \sin(\theta)$, $x_2 = \sin(2\,\theta)$, $x_3 = \sin(3\,\theta)$, and so on for various values of θ, with the final assumption that $x_{n+1} = 0 = \sin((n+1)\theta)$. This last assumption means that θ is a multiple of $\frac{\pi}{n+1}$. The trigonometric identity

$$\sin(\theta + \psi) = \cos(\theta)\sin(\psi) + \sin(\theta)\cos(\psi),$$

together with the facts $\sin(-\theta) = -\sin(\theta)$ and $\cos(-\theta) = \cos(\theta)$, applied to our proposed solution shows,

$$\begin{aligned} 0 &= x_{j-1} + \alpha x_j + x_{j+1} \\ &= \sin((j-1)\theta) + \alpha \sin(j\,\theta) + \sin((j+1)\theta) \\ &= \cos(j\,\theta)\sin(-\theta) + \sin(j\,\theta)\cos(-\theta) + \alpha\sin(j\,\theta) \\ &\quad + \cos(j\,\theta)\sin(\theta) + \sin(j\theta)\cos(\theta) \\ &= -\cos(j\theta)\sin(\theta) + \sin(j\theta)\cos(\theta) + \alpha\sin(j\,\theta) \\ &\quad + \cos(j\,\theta)\sin(\theta) + \sin(j\,\theta)\cos(\theta) \\ &= \alpha\sin(j\,\theta) + 2\sin(j\,\theta)\cos(\theta). \end{aligned}$$

The last expression is zero when $\alpha = -2\cos(\theta)$. This shows that whenever $0 = \sin((n+1)\theta)$ and $\alpha = -2\cos(\theta)$, the $n \times n$ symmetric tridiagonal matrix $T_{n,\alpha}$ is singular. For each n, we can choose $\theta = \frac{k\pi}{n+1}$, where $k = 1, 2, \ldots, n$, and our calculation will give n distinct values of α for which $T_{n,\alpha}$ is singular.

Note that $T_{n,b} - \lambda I_n = T_{n,b-\lambda}$. If $T_{n,\alpha}$ is singular, then setting $b - \lambda = \alpha$ gives $\lambda = b - \alpha$ as an eigenvalue of $T_{n,b}$. Since we have found n distinct possible values of α, this gives the n eigenvalues of the matrix $T_{n,b}$. Since $T_{n,b}$ has at most n distinct eigenvalues, we have found them all. This shows that $T_{n,b}$ is diagonalizable.

For example, if $n = 5$, the possible α values with $T_{5,\alpha}$ singular are $\alpha = 2\cos\frac{\pi}{6} = \sqrt{3}$, $\alpha = 2\cos\frac{\pi}{3} = 1$, $\alpha = 2\cos\frac{\pi}{2} = 0$, $\alpha = 2\cos\frac{4\pi}{3} = -1$, and $\alpha = 2\cos\frac{5\pi}{6} = -\sqrt{3}$. In our buckling problem earlier we needed to know the eigenvalues of the matrix

$$T_{5,2} = \begin{pmatrix} 2 & -1 & 0 & 0 & 0 \\ -1 & 2 & -1 & 0 & 0 \\ 0 & -1 & 2 & -1 & 0 \\ 0 & 0 & -1 & 2 & -1 \\ 0 & 0 & 0 & -1 & 2 \end{pmatrix}.$$

These eigenvalues are precisely the values λ for which $\lambda - 2 = \alpha$. We find that they are $\lambda = 2 + \sqrt{3}$, 3, 2, 1, and $2 - \sqrt{3}$. Recall that the eigenvalues 1, 2, and 3 were used in constructing the discrete solutions to our buckling problem pictured in Fig. 7.1.

Further Criteria for Diagonalizability

We conclude this section by discussing further criteria that determine when a matrix is diagonalizable. Recall that in the corollary to Theorem 45 we showed that if an $n \times n$ matrix has n distinct eigenvalues then it is diagonalizable. This shows that if the characteristic polynomial of a matrix A factors as $C_A(X) = (X - d_1)(X - d_2)\cdots(X - d_n)$, where $d_1, d_2, \ldots, d_n$ are *n distinct* real numbers, then A is diagonalizable. A partial converse to this is true.

Suppose that an $n \times n$ matrix A is diagonalizable. Then we can find P so that $P^{-1}AP = D$ is a diagonal matrix, with diagonal entries $d_1, d_2, \ldots, d_n$. Using $A = PDP^{-1}$, we can compute the characteristic polynomial of A using this expression:

$$\begin{aligned} C_A(X) &= \det(XI_n - A) = \det(XPI_nP^{-1} - PDP^{-1}) \\ &= \det(PXI_nP^{-1} - PDP^{-1}) = \det(P(XI_nP^{-1} - DP^{-1})) \\ &= \det(P(XI_n - D)P^{-1}) = \det(P)\det(XI_n - D)\det(P^{-1}) \\ &= \det(XI_n - D) = C_D(X) = (X - d_1)(X - d_2)\cdots(X - d_n). \end{aligned}$$

In particular, $C_A(X)$ can be expressed as a product of linear factors. We record this in the following result.

Theorem 46. *If A is an $n \times n$ diagonalizable matrix, then its characteristic polynomial can be expressed as a product of linear factors, that is, $C_A(X) = (X - d_1)(X - d_2)\cdots(X - d_n)$, where $d_1, d_2, \ldots, d_n$ are real numbers.*

Unfortunately, Theorem 46 does not handle all possibilities. Many polynomials cannot be factored into a product of linear polynomials $(X - d_i)$ where the d_i are real numbers. Even when they do, they may have multiple roots (which means the d_i in a factorization may not be distinct). What then?

For example, consider the following three matrices:

$$\begin{pmatrix} 2 & 0 & 0 \\ 0 & 2 & 0 \\ 0 & 0 & 2 \end{pmatrix}, \quad \begin{pmatrix} 2 & 1 & 0 \\ 0 & 2 & 0 \\ 0 & 0 & 2 \end{pmatrix}, \quad \begin{pmatrix} 2 & 1 & 0 \\ 0 & 2 & 1 \\ 0 & 0 & 2 \end{pmatrix}.$$

Each has the same characteristic polynomial, $(X - 2)^3$, and thus each has the single eigenvalue 3. If we subtract $2I_3$ from each, we obtain

$$\begin{pmatrix} 0 & 0 & 0 \\ 0 & 0 & 0 \\ 0 & 0 & 0 \end{pmatrix}, \quad \begin{pmatrix} 0 & 1 & 0 \\ 0 & 0 & 0 \\ 0 & 0 & 0 \end{pmatrix}, \quad \begin{pmatrix} 0 & 1 & 0 \\ 0 & 0 & 1 \\ 0 & 0 & 0 \end{pmatrix},$$

which have ranks 0, 1, and 2, respectively. This means the dimensions of the eigenspaces of these three matrices are 3, 2, and 1. Since the first matrix is diagonalizable, we see that it is possible for a matrix to be diagonalizable when it has a "repeated" eigenvalue. However, the later two matrices are not diagonalizable since it is impossible for them to have three linearly independent eigenvectors. These observations motivate the next definition.

Definition. Let A be an $n \times n$ matrix, and suppose k is an eigenvalue of A. The highest power of $(X - k)$ that divides $C_A(X)$ is called the *algebraic multiplicity* of the eigenvalue k. The dimension of the eigenspace E_k of A associated with k is called the *geometric multiplicity* of the eigenvalue k.

The three matrices just considered each have eigenvalue 2 with algebraic multiplicity 3. The geometric multiplicities of these matrices are 3, 2, and 1. Only the first is diagonalizable.

Note that if the characteristic polynomial of an $n \times n$ matrix is a product of linear factors, the sum of the algebraic multiplicities is n. It turns out (although we won't give a proof) that the geometric multiplicity of an eigenvalue never exceeds its algebraic multiplicity. Using this we see that there exist n linearly independent eigenvectors for a matrix A only when the geometric multiplicity equals the algebraic multiplicity for each distinct eigenvalue of A. Conversely, if the geometric multiplicity equals the algebraic multiplicity for each eigenvalue, then bases for each eigenspace, taken together, will give n linearly independent eigenvalues. This gives the following theorem.

Theorem 47. *Let A be an $n \times n$ matrix, and suppose that $C_A(X)$ factors into linear factors. Then A is diagonalizable if and only if the geometric multiplicity of each eigenvalue of A is the same as its algebraic multiplicity.*

For example, the matrices

$$\begin{pmatrix} 6 & 6 & 3 \\ 0 & 7 & 1 \\ 0 & 0 & 2 \end{pmatrix} \quad \text{and} \quad \begin{pmatrix} 1 & 0 & 0 \\ 2 & 9 & 0 \\ 5 & 0 & 3 \end{pmatrix}$$

are diagonalizable. In each case the characteristic polynomial has distinct linear factors. However, the matrix

$$B = \begin{pmatrix} 2 & 3 & 1 \\ 3 & 2 & 4 \\ 0 & 0 & -1 \end{pmatrix}$$

from Sec. 7.1 has characteristic polynomial $C_B(X) = (X - 5)(X + 1)^2$. The two eigenvalues 5 and -1 have algebraic multiplicity 1 and 2, respectively. The associated eigenspaces each have dimension 1. Theorem 47 shows that B is not diagonalizable.

Problems

1. For each of the following matrices, determine all (real) eigenvalues and eigenspaces. If the matrix is diagonalizable, show how to express its diagonalization in the form $P^{-1}AP = D$.

(a) $\begin{pmatrix} 0 & 2 & 2 \\ 2 & 0 & 2 \\ 2 & 2 & 0 \end{pmatrix}$ (b) $\begin{pmatrix} 0 & 1 & 0 \\ 0 & 0 & 1 \\ 1 & -3 & 3 \end{pmatrix}$ (c) $\begin{pmatrix} 0 & -1 \\ 1 & -1 \end{pmatrix}$

2. For each of the following matrices, determine all (real) eigenvalues and eigenspaces. If the matrix is diagonalizable, show how to express its diagonalization in the form $P^{-1}AP = D$.

(a) $\begin{pmatrix} 1 & 1 & 0 \\ 2 & 1 & 0 \\ 2 & 2 & 0 \end{pmatrix}$ (b) $\begin{pmatrix} 2 & 1 & -1 \\ 0 & 1 & 0 \\ 2 & -1 & 2 \end{pmatrix}$ (c) $\begin{pmatrix} 0 & 1 & 0 & 0 \\ 0 & 0 & 1 & 0 \\ 0 & 0 & 0 & 1 \\ 1 & 0 & 0 & 0 \end{pmatrix}$

3. Suppose that A is a 3×3 matrix with eigenvalues 0, 1, and 2 and eigenvectors $\vec{u}$, $\vec{v}$, and $\vec{w}$, respectively.
   (a) Find a basis for col(A). Justify your assertions.
   (b) Find a basis for ker(A). Justify your assertions.
   (c) Can you solve the equation $A\vec{X} = \vec{v}$ for $\vec{X}$? How about $A\vec{X} = \vec{v} + \vec{w}$? or $A\vec{X} = \vec{u}$?
4. Write five 5×5 matrices, each with the single eigenvalue 7, with all possible geometric multiplicities.
5. Suppose that A is an $n \times n$ matrix and ker(A) is $(n-1)$-dimensional. If A also has a nonzero eigenvalue, show that A is diagonalizable.
6. Find the eigenvalues of the tridiagonal matrices

$$\begin{pmatrix} 3 & -2 & 0 & 0 & 0 \\ -2 & 3 & -2 & 0 & 0 \\ 0 & -2 & 3 & -2 & 0 \\ 0 & 0 & -2 & 3 & -2 \\ 0 & 0 & 0 & -2 & 3 \end{pmatrix} \quad \text{and} \quad \begin{pmatrix} 1 & 1 & 0 & 0 & 0 & 0 \\ 1 & 1 & 1 & 0 & 0 & 0 \\ 0 & 1 & 1 & 1 & 0 & 0 \\ 0 & 0 & 1 & 1 & 1 & 0 \\ 0 & 0 & 0 & 1 & 1 & 1 \\ 0 & 0 & 0 & 0 & 1 & 1 \end{pmatrix}.$$

7. Suppose that $A^2 = A$. What are the possible eigenvalues of A?
8. (a) Let

$$A = \begin{pmatrix} 1 & 0 & 0 \\ 0 & 2 & 0 \\ 0 & 0 & 3 \end{pmatrix}.$$

   Suppose that B is a 3×3 matrix and that $AB = BA$. Show that B is also a diagonal 3×3 matrix.
   (b) Suppose that A is an 3×3 real matrix with three distinct eigenvalues. Let B be another 3×3 matrix for which $AB = BA$. Show that B is diagonalizable. (Hint: Think about part (a), and use a diagonalizing equation $P^{-1}AP = D$.)
   (c) Suppose that A and B are diagonalizable matrices with precisely the same eigenspaces (but not necessarily the same eigenvalues). Prove that $AB = BA$.
9. Two matrices A and B are called *similar* if there exists an invertible matrix P such that $A = P^{-1}BP$.
   (a) Show that if A is similar to B, and if B is similar to C, then A is similar to C.
   (b) Determine if the following pairs of matrices are similar.
      (i) $\begin{pmatrix} 2 & 1 \\ 0 & 2 \end{pmatrix}$ and $\begin{pmatrix} 2 & 0 \\ 0 & 2 \end{pmatrix}$
      (ii) $\begin{pmatrix} 0 & 1 \\ 0 & 2 \end{pmatrix}$ and $\begin{pmatrix} 0 & 0 \\ 0 & 1 \end{pmatrix}$

(iii) $\begin{pmatrix} 2 & 1 \\ 0 & 1 \end{pmatrix}$ and $\begin{pmatrix} 2 & 0 \\ 0 & 1 \end{pmatrix}$

10. Diagonalize the following real symmetric matrices. In each case find the angle between the eigenspaces.

    (a) $\begin{pmatrix} 2 & 4 \\ 4 & 3 \end{pmatrix}$ (b) $\begin{pmatrix} 3 & -2 \\ -2 & 3 \end{pmatrix}$ (c) $\begin{pmatrix} 1 & 1 \\ 1 & 1 \end{pmatrix}$

11. Suppose that A is an $n \times n$ diagonalizable matrix. Show that for any m, A^m is also diagonalizable.

Group Project: Triangular Matrices

Show that every matrix whose characteristic polynomial factors into linear factors is similar to an upper triangular matrix using the following outline as a guide: Proceed by induction on n where A is an $n \times n$ matrix. If A has the eigenvalue c, show that A is similar (see Prob. 9 for the definition) to the partitioned matrix

$$\left(\begin{array}{c|ccc} c & b_2 & \cdots & b_n \\ \hline 0 & & & \\ \vdots & & B & \\ 0 & & & \end{array}\right),$$

where B is an $(n-1)\times(n-1)$ matrix. Observe that $C_A(X) = (X - c)C_B(X)$. Apply the induction hypothesis to B and derive the result.

Group Project: Diagonalization and Matrix Powers

The matrix

$$A = \begin{pmatrix} 17 & -30 \\ 9 & -16 \end{pmatrix}$$

is diagonalizable. A has eigenvectors

$$\begin{pmatrix} 2 \\ 1 \end{pmatrix} \text{ and } \begin{pmatrix} 5 \\ 3 \end{pmatrix}.$$

(a) Find their eigenvalues, and find a matrix P so that $P^{-1}AP = D$ is diagonal.

(b) For any square matrix D and invertible matrix P, show that $(PDP^{-1})^2 = PD^2P^{-1}$. What is $(PDP^{-1})^{10}$?

(c) Use parts (a) and (b) to calculate A^{10} by hand. Check your answer using a calculator.

(d) Using the idea from part (c), find a formula for A^n.

7.3 Symmetric Matrices and Probability Matrices

Symmetric matrices arise in many applications of linear algebra. The most important fact is that any real symmetric matrix can be diagonalized.

Symmetric Matrices Have Real Eigenvalues

In our first result we shall need to use the set of complex numbers **C**. Recall that if $\alpha = a + bi \in \mathbf{C}$, where $a, b \in \mathbf{R}$, then the complex conjugate α is $\overline{\alpha} = a - bi$. We shall need to use the facts that whenever $\alpha, \beta \in \mathbf{C}$, $\overline{\alpha} + \overline{\beta} = \overline{\alpha + \beta}$ and $\overline{\alpha}\overline{\beta} = \overline{\alpha\beta}$. We shall also require the theorem that every nonconstant polynomial over **C** has a root in **C** (the fundamental theorem of algebra). First we show that symmetric matrices have real eigenvalues.

Theorem 48. *Every real symmetric matrix has a real eigenvalue. Moreover, all eigenvalues of a symmetric matrix are real.*

Proof. Let S be a real symmetric $n \times n$ matrix. We view S momentarily as a complex matrix. Since the characteristic polynomial of S has a root in **C**, we see that S has a complex eigenvalue α. (The reasoning given in Sec. 7.1 where we showed that the roots of the characteristic polynomial are eigenvalues, applies to complex roots as well, as long as we allow our eigenvectors to have complex coordinates.) To prove the theorem we must show that $\alpha \in \mathbf{R}$.

Let $\vec{v} \in \mathbf{C}^n$ be an eigenvector of eigenvalue α, which we view as a column vector. We denote by $\overline{\vec{v}}$ the vector obtained by replacing each entry of $\vec{v}$ by its conjugate, and similarly for S. Since S is a real symmetric matrix, we have $\overline{S} = S = S^t$. Note that since we use $\vec{v}$ as a column vector, it makes sense to write $\vec{v}^{\,t}$, which is a row vector. Using $S\vec{v} = \alpha\vec{v}$, we have the following sequence of equalities:

$$\begin{aligned}\alpha(\vec{v}^{\,t}\overline{\vec{v}}) &= (\alpha\vec{v})^t\overline{\vec{v}} = (S\vec{v})^t\overline{\vec{v}} = \vec{v}^{\,t}S^t\overline{\vec{v}} \\ &= \vec{v}^{\,t}(S\overline{\vec{v}}) = \vec{v}^{\,t}(\overline{S}\,\overline{\vec{v}}) = \vec{v}^{\,t}(\overline{S\vec{v}}) = \vec{v}^{\,t}(\overline{\alpha\vec{v}}) \\ &= \vec{v}^{\,t}(\overline{\alpha}\overline{\vec{v}}) = \overline{\alpha}\vec{v}^{\,t}\overline{\vec{v}}.\end{aligned}$$

Suppose that $\vec{v}$ is the vector whose jth entry is the complex number $a_j + b_j i$. Then multiplying out we find that $\vec{v}^{\,t}\overline{\vec{v}} = a_1^2 + b_1^2 + a_2^2 + b_2^2 + \cdots + a_n^2 + b_n^2$. Since $\vec{v} \neq \vec{0}$, this shows that $\vec{v}^{\,t}\overline{\vec{v}} \neq 0$. Consequently, our computation shows that $\alpha = \overline{\alpha}$. This shows that α is real and proves the theorem. □

Diagonalization of Symmetric Matrices

Using Theorem 48 one can show that symmetric matrices are diagonalizable. In fact, it turns out that the eigenspaces of a symmetric matrix are perpendicular. The proof of this result will be given in Theorem 66, which appears in Chap. 9.

Theorem 49. *Every real symmetric matrix S is diagonalizable. Moreover, the eigenspaces of S are orthogonal.*

For example, consider the symmetric matrix

$$T = \begin{pmatrix} 1 & 3 \\ 3 & 1 \end{pmatrix}.$$

The characteristic polynomial of the matrix T is $C_T(X) = (X - 1)^2 - 9 = (X + 2)(X - 4)$. The eigenspaces of T are checked to be

$$E_{-2} = \text{span}\left\{\begin{pmatrix} 1 \\ -1 \end{pmatrix}\right\} \quad \text{and} \quad E_4 = \text{span}\left\{\begin{pmatrix} 1 \\ 1 \end{pmatrix}\right\}.$$

Note that E_{-2} and E_4 are orthogonal in $\mathbf{R}^2$. If we choose an eigenbasis for T with length 1 eigenvectors and form the matrix

$$P = \begin{pmatrix} \frac{1}{\sqrt{2}} & \frac{1}{\sqrt{2}} \\ -\frac{1}{\sqrt{2}} & \frac{1}{\sqrt{2}} \end{pmatrix},$$

then

$$P^{-1}TP = \begin{pmatrix} \frac{1}{\sqrt{2}} & -\frac{1}{\sqrt{2}} \\ \frac{1}{\sqrt{2}} & \frac{1}{\sqrt{2}} \end{pmatrix} \begin{pmatrix} 1 & 3 \\ 3 & 1 \end{pmatrix} \begin{pmatrix} \frac{1}{\sqrt{2}} & \frac{1}{\sqrt{2}} \\ -\frac{1}{\sqrt{2}} & \frac{1}{\sqrt{2}} \end{pmatrix} = \begin{pmatrix} -2 & 0 \\ 0 & 4 \end{pmatrix}.$$

Observe that $P^{-1} = P^t$.

Quadratic Forms Associated with Symmetric Matrices

Symmetric matrices are closely related to homogeneous quadratic polynomials. A *homogeneous quadratic polynomial*, or *quadratic form*, in the variables $X_1, X_2, \ldots, X_n$ is a polynomial of the form

$$F(X_1, X_2, \ldots, X_n) = \sum_{ij} a_{ij} X_i X_j,$$

where every summand has degree two. For example, $X^2 + YX$ is a homogeneous quadratic polynomial, while $X^2 + Y$ is not.

Observe that if $A = (a_{ij})$ is an $n \times n$ matrix, then the polynomial obtained by multiplication,

$$Q_A(X_1, X_2, \ldots, X_n) = (X_1 \quad X_2 \quad \cdots \quad X_n)\, A \begin{pmatrix} X_1 \\ X_2 \\ \vdots \\ X_n \end{pmatrix} = \sum_{ij} a_{ij} X_i X_j,$$

is a homogeneous quadratic polynomial. Every homogeneous quadratic polynomial can be obtained by matrix multiplication in this way. In fact, we can require the matrix used be symmetric. For example, suppose the polynomial $F(X_1, X_2, \ldots, X_n) = \sum_{ij} a_{ij}X_iX_j$. If we define $S = (s_{ij})$ where $s_{ij} = \frac{1}{2}(a_i + a_j)$, we find that S is a symmetric matrix and that $F(X_1, X_2, \ldots, X_n) = \sum_{ij} a_{ij}X_iX_j = \sum_{ij} s_{ij}X_iX_j = Q_S(X_1, X_2, \ldots, X_n)$.

As an example, the polynomial $F(X, Y) = 3X^2 - 2XY$ can expressed as $F(X, Y) = Q_S(X, Y)$, where S is the symmetric matrix

$$S = \begin{pmatrix} 3 & -1 \\ -1 & 0 \end{pmatrix},$$

since

$$\begin{pmatrix} X & Y \end{pmatrix} \begin{pmatrix} 3 & -1 \\ -1 & 0 \end{pmatrix} \begin{pmatrix} X \\ Y \end{pmatrix} = 3X^2 - XY - YX = 3X^2 - 2XY.$$

Note that when D is the diagonal matrix with diagonal entries $d_1, d_2, \ldots, d_n$ then $Q_D(X_1, X_2, \ldots, X_n) = d_1X_1^2 + d_2X_2^2 + \cdots + d_nX_n^2$, which is called a *diagonal* polynomial.

The positivity or negativity of functions defined by diagonal polynomials can be understood by looking at their coefficients. For example, if each $d_i > 0$, then the function $Q_D(X_1, X_2, \ldots, X_n) = d_1X_1^2 + d_2X_2^2 + \cdots + d_nX_n^2$ increases whenever the variables do. Furthermore, the minimum value of this function is zero. On the other hand, if there are both positive and negative values for the d_i's, then there is no maximum or minimum value for the function. If all the d_i are positive, then Q_D is called *positive definite*; if all the d_i are negative; then Q_D is called *negative definite*; and if both signs occur, Q_D is called *indefinite*.

The fact that symmetric matrices can be diagonalized is extremely useful in this connection. It shows that every homogeneous quadratic polynomial can be diagonalized after a linear change of variables. Suppose that S is symmetric and that $Q_S(\vec{X}) = \vec{X}^tS\vec{X}$ is the associated quadratic polynomial. Assume that P is invertible with $P^t = P^{-1}$ and $D = P^tSP$ is diagonal. (The fact that P can be chosen with $P^t = P^{-1}$ is a special property of symmetric matrices that we prove in Theorem 66.) We define a new column of variables $\vec{Y}$ by $\vec{Y} = P^t\vec{X}$, and so $\vec{X} = P\vec{Y}$. Then we can express $Q_S(\vec{X}) = \vec{X}^tS\vec{X} = (P\vec{Y})^tSP\vec{Y} = \vec{Y}^t(P^tSP)\vec{Y} = Q_D(\vec{Y})$ as a diagonal polynomial in the new variables $Y_1, Y_2, \ldots, Y_n$.

For example, consider $F(X, Y) = X^2 + 6XY + Y^2$. Then $F(X, Y) = Q_T(X, Y)$, where T is the symmetric matrix

$$T = \begin{pmatrix} 1 & 3 \\ 3 & 1 \end{pmatrix}$$

considered at the start of this subsection. We saw there that if

$$P = \begin{pmatrix} \frac{1}{\sqrt{2}} & \frac{1}{\sqrt{2}} \\ -\frac{1}{\sqrt{2}} & \frac{1}{\sqrt{2}} \end{pmatrix}, \quad \text{then} \quad P^t TP = \begin{pmatrix} -2 & 0 \\ 0 & 4 \end{pmatrix}$$

where $P^t = P^{-1}$. This means that if we set

$$\begin{pmatrix} A \\ B \end{pmatrix} = P^t \begin{pmatrix} X \\ Y \end{pmatrix} = \begin{pmatrix} \frac{1}{\sqrt{2}} & -\frac{1}{\sqrt{2}} \\ \frac{1}{\sqrt{2}} & \frac{1}{\sqrt{2}} \end{pmatrix} \begin{pmatrix} X \\ Y \end{pmatrix} = \frac{1}{\sqrt{2}} \begin{pmatrix} X - Y \\ X + Y \end{pmatrix},$$

then $F(X, Y) = -2A^2 + 4B^2$. We check this:

$$\begin{aligned} -2A^2 + 4B^2 &= -2\tfrac{1}{2}(X - Y)^2 + 4\tfrac{1}{2}(X + Y)^2 \\ &= -X^2 + 2XY - Y^2 + 2X^2 + 4XY + 2Y^2 \\ &= X^2 + 6XY + Y^2 \end{aligned}$$

as desired.

Remark. The two-variable case just illustrated is a familiar result from high school mathematics. It is nothing other than completing the square to eliminate the mixed term of a homogeneous quadratic. Explicitly, consider $q_1 = aX^2 + bXY + cY^2$. Then $q_1 = a[X + \frac{b}{2a}Y]^2 + (c - \frac{b^2}{4a^2})Y^2$, which shows how to diagonalize the original polynomial as $q_2 = aR^2 + (c - \frac{b^2}{4a^2})S^2$. The matrix P giving this change of variables is

$$P = \begin{pmatrix} 1 & \frac{b}{2a} \\ 0 & 1 \end{pmatrix}$$

since

$$q_2\left(P\begin{pmatrix} X \\ Y \end{pmatrix}\right) = q_1(X, Y).$$

It is possible to diagonalize any quadratic polynomial by completing the square. You need to eliminate successively all the nondiagonal terms by appropriate replacements of the variables in the same way that the XY term was removed above.

Markov Chains and Random Events

In our study of the effect of an epidemic on a rabbit population we saw that the population growth could be estimated using iterative matrix multiplication. These same ideas can be used to compute the probability of the outcome of a large number of sequential random events. Suppose that a new gambling game is introduced. In this game a mouse is placed in room

Fig. 7.4. A mouse maze

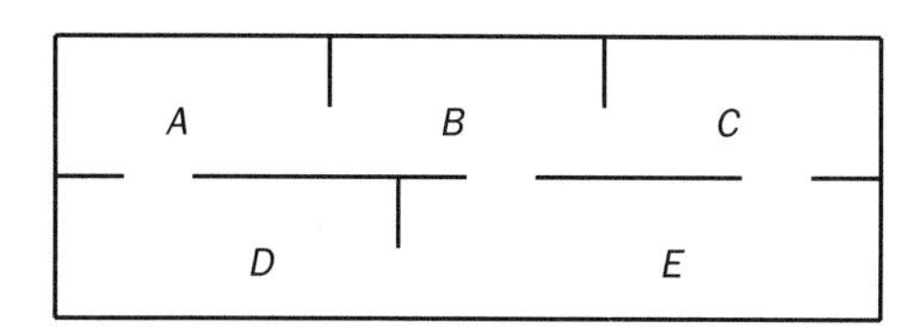

A in the maze depicted in Fig. 7.4. All players bet on where the mouse will be exactly one minute later. Different rooms have different payoff rates. You want to determine the probability that the mouse will be in any given room so that you can decide how to bet.

After observing the mouse that runs this maze, you decide that on the average, every two seconds the mouse will either move through a door or do nothing and that each of these events has equal probability. This means, for example, that after two seconds of its start in room A, the mouse has a one-in-three probability of being in room A, B, and D, and a zero probability of being in room C or E. Of course, after four seconds, the mouse could be anywhere. To find out what happens after 60 seconds, we construct a 5×5 transition matrix representing each two-second movement of the mouse, and we raise the matrix to the 30th power. In its first column this matrix, M, has the probabilities of going from room A to each of the rooms A, B, C, D, and E. The second column of M contains the probabilities of going from room B to each of the rooms A, B, C, D, and E, and so forth. We have

$$M = \begin{pmatrix} \frac{1}{3} & \frac{1}{4} & 0 & \frac{1}{3} & 0 \\ \frac{1}{3} & \frac{1}{4} & \frac{1}{3} & 0 & \frac{1}{4} \\ 0 & \frac{1}{4} & \frac{1}{3} & 0 & \frac{1}{4} \\ \frac{1}{3} & 0 & 0 & \frac{1}{3} & \frac{1}{4} \\ 0 & \frac{1}{4} & \frac{1}{3} & \frac{1}{3} & \frac{1}{4} \end{pmatrix}.$$

Using a calculator with matrix capabilities, we find that

$$M^{30} \approx \begin{pmatrix} .1762 & .1762 & .1762 & .1762 & .1762 \\ .2349 & .2349 & .2349 & .2349 & .2349 \\ .1762 & .1762 & .1762 & .1762 & .1762 \\ .1762 & .1762 & .1762 & .1762 & .1762 \\ .2349 & .2349 & .2349 & .2349 & .2349 \end{pmatrix}.$$

This matrix tells us the probabilities of the mouse position after 30 seconds. (Of course, this is all based on our theory of how the mouse runs. If we add food, or get a lazy mouse, the outcome probabilities will change.) Note that our initial mouse location is the column vector e_1, and consequently,

$$\begin{pmatrix} .1762 & .1762 & .1762 & .1762 & .1762 \\ .2349 & .2349 & .2349 & .2349 & .2349 \\ .1762 & .1762 & .1762 & .1762 & .1762 \\ .1762 & .1762 & .1762 & .1762 & .1762 \\ .2349 & .2349 & .2349 & .2349 & .2349 \end{pmatrix} \begin{pmatrix} 1 \\ 0 \\ 0 \\ 0 \\ 0 \end{pmatrix} = \begin{pmatrix} .1762 \\ .2349 \\ .1762 \\ .1762 \\ .2349 \end{pmatrix}$$

is the column representing the outcome probabilities. We see that on average, the mouse should end up in rooms A, C, and D 17.6% of the time and in rooms B and E 23.5% of the time. This should help us place our bets. Observe that if the mouse started in a different room the outcome probabilities would be the same since the columns of M^{30} are identical.

Probability Matrices

In our mouse example, the matrix M had some special features. The first noticeable fact is that its high powers, such as M^{30}, have identical columns. In fact, M^{31} would have these same columns (to the four decimal points listed). The reason for this is that M is a *probability matrix.* Observe that the columns of M represent the probabilities of a certain outcome, and therefore all sum to 1. More generally, we have the next definition.

Definition. A vector $(a_1, a_2, \ldots, a_n) \in \mathbf{R}^n$ is called a *probability vector* (also called a *distribution vector*) if each $a_i \geq 0$ and $a_1 + a_2 + \cdots + a_n = 1$. An $n \times n$ matrix T is called a *probability matrix* (also called a *stochastic matrix*) if each column is a probability vector.

A model that uses iterative matrix multiplication by a probability matrix is called a *Markov chain.* For example, our mouse probability determination used a Markov chain. In a Markov chain, the probabilities of the states of the system are represented by a probability vector. Then each change in these probabilities—that is, each transition—will be given by multiplying the probability vector by a probability matrix. Of course, we have to make sure that the result of this transition is another probability vector. This next lemma says exactly this.

Lemma. *A matrix M is a probability matrix if and only if all entries are nonnegative and if whenever $\vec{p}$ is a probability vector, $M\vec{p}$ is also a probability vector.*

What we noticed in our mouse example is true in general. After many iterations the outcome is a matrix whose columns are the same and in fact are eigenvectors with eigenvalue 1. We say that a probability matrix M is *regular* if, for some n, M^n has no zero entries. If a probability matrix is regular, then eventually (after enough iterations) there is a positive probability that an object can change from its original state into any other state. A probability vector is called a *stable* vector for M if it is an eigenvector for M with eigenvalue 1. Using this terminology, one has the following theorem.

Theorem 50. *Let M be a regular probability matrix. Then*

(i) 1 *is an eigenvalue of M and M has a unique stable vector $\vec{v}$,*

(ii) *For any probability vector $\vec{w}$ the limit $\lim_{n\to\infty} M^n \vec{w}$ makes sense and is the unique stable vector $\vec{v}$.*

In the previous subsection each column of M^{30} is a stable vector for M^{30} (they are all the same to the decimal points listed). Although we won't define the limit in part (b) of the theorem, it can be understood as saying that for large n the $M^n \vec{w}$ all are extremely close.

Example. Suppose that there are three taxi zones in a small town which are called Zone 1, Zone 2, and Zone 3. By studying the passenger logs over a one year period, the taxi company has determined the probabilities of passengers needing to travel from one zone to another. In the probability matrix

$$T = \begin{pmatrix} .15 & .30 & .25 \\ .40 & .25 & .55 \\ .45 & .45 & .20 \end{pmatrix}$$

the ijth entry represents the probability of a passenger starting in Zone i requiring a ride to Zone j.

If the vector, $\vec{v}$, represents the distribution of taxis in each zone, then the product $T\vec{v}$ represents the likely distribution of these taxis after each makes one trip. We make the assumption that each trip takes the same amount of time (which of course isn't quite true, but is needed make the model work). Then $T^n\vec{v}$ represents the distribution of taxis after n trips. The second part of Theorem 50 says that for large values of n, the matrices $T^n\vec{v}$ will all be (essentially) the same. This value represents the distribution of taxis that will be stable, and therefore provide a service to taxi passengers that (on average) will minimize their waiting time. Computation shows that

$$T^{50} \approx \begin{pmatrix} .25 & .25 & .25 \\ .39 & .39 & .39 \\ .36 & .36 & .36 \end{pmatrix}.$$

Therefore, if there are 100 taxis in the fleet, a stable distribution would be 25 taxis in Zone 1, 39 taxis in Zone 2, and 36 taxis in Zone 3.

The Closed Leontief Input-Output Model

An important problem in economics is to determine how goods and services are exchanged. Markov chains can provide a model for understanding what may happen in a closed economy (one where the goods and services exchanged remain confined to a single location under study). These models are known as closed Leontief input-output models.

As an illustration we consider an economy with n "industries" that buy and sell products from one another. These industries could be manufacturers, stores, consulting firms, or similar entities. We will analyze this economy using an $n \times n$ matrix $L = (\ell_{ij})$ where the entry ℓ_{ij} represents the *fraction of the total output of the jth industry purchased by the ith industry*. If you add the entries of the jth column of L, you are adding the various fractions of the total output of this industry and consequently obtain 1. This shows that L is a probability matrix. L will be used as the transition matrix in this model and in this setting is sometimes referred to as an *exchange matrix*.

Next we consider a vector $\vec{p} = (p_j)$ whose jth component is the *total price (say in dollars) charged for the entire output of the jth industry during one month*. We want to study the vector arising as the product $L\vec{p}$. What do its entries mean? Note that $\ell_{ij}p_j$ is the product of the fraction of the jth industry output purchased by i and the total price of all of the jth industry goods, the total dollars spent by the ith industry in the jth industry goods. The ith entry of $L\vec{p}$ is the sum

$$\ell_{in}p_1 + \ell_{in}p_2 + \cdots + \ell_{in}p_n,$$

and so it represents *the total income by the jth industry*.

An economy is most healthy when the amount each industry spends is close to its income; otherwise some industry will risk bankruptcy. So for a healthy economy one would like prices to be set so that $\vec{p}$ is close to an eigenvector for L with eigenvalue 1. Such eigenvectors exist by Theorem 50. An alternate interpretation would be that in a closed economy prices tend to migrate to those eigenvectors since they give stability to the economy. This means that when new industries are added to an economy, one might try to predict their influence on prices by finding the new transition matrix and looking for its eigenvectors. A specific Leontief model is developed in a group project at the end of this section.

Problems

1. For each symmetric matrix S below, find a matrix P for which $P^{-1}SP$ is diagonal.

   (a) $\begin{pmatrix} 1 & 2 \\ 2 & 3 \end{pmatrix}$ (b) $\begin{pmatrix} 0 & 2 & 0 \\ 2 & 0 & 2 \\ 0 & 2 & 0 \end{pmatrix}$

(c) $\begin{pmatrix} 1 & 1 & 0 \\ 1 & 0 & 1 \\ 0 & 1 & 1 \end{pmatrix}$ (d) $\begin{pmatrix} 0 & 1 & 0 & 0 \\ 1 & 0 & 0 & 0 \\ 0 & 0 & 0 & 1 \\ 0 & 0 & 1 & 0 \end{pmatrix}$

2. For each matrix S in Prob. 1 find the homogeneous quadratic polynomial $Q_S(\vec{X})$. Then find a diagonalization of each of these polynomials.
3. Find the symmetric matrix associated with each quadratic form below. Find diagonalizations of these quadratic forms by completing the square or by diagonalizing the associated symmetric matrix.
   (a) $2X^2 - 2XY + 2Y^2 + XZ - Z^2$
   (b) $X^2 + YZ$
4. Consider

$$S = \begin{pmatrix} 2 & -2 & -4 \\ -2 & 5 & -2 \\ -4 & -2 & 2 \end{pmatrix}.$$

   Then $C_S(X) = (X-6)^2(X+3)$. Using this, find three orthogonal eigenvectors for S. Use them to obtain a diagonalizing matrix P whose columns are orthogonal unit vectors. Check that $P^{-1} = P^t$ for your matrix.
5. Show that the quadratic polynomial $q(X, Y) = aX^2 + 2bXY + cY^2$ gives a positive-definite quadratic form on $\mathbf{R}^2$ if and only if $ac - b^2 > 0$ and $a > 0$.
6. Which of the following probability matrices are regular?

$$\begin{pmatrix} \frac{1}{2} & 1 \\ \frac{1}{2} & 0 \end{pmatrix}, \quad \begin{pmatrix} 0 & 0 & \frac{1}{3} \\ 1 & 0 & \frac{1}{3} \\ \frac{1}{2} & \frac{1}{2} & \frac{1}{3} \end{pmatrix}, \quad \begin{pmatrix} \frac{2}{3} & \frac{1}{3} & \frac{1}{3} \\ \frac{1}{3} & \frac{2}{3} & \frac{1}{3} \\ 0 & 0 & \frac{1}{3} \end{pmatrix}$$

7. What are the following matrix limits to two decimal places? Use a calculator or a computer to help if you like.
   (a) $\lim_{n\to\infty} P^n$ if

$$P = \begin{pmatrix} \frac{3}{4} & \frac{1}{4} \\ \frac{1}{4} & \frac{3}{4} \end{pmatrix}$$

   (b) $\lim_{n\to\infty} A^n$ if

$$A = \frac{1}{5}\begin{pmatrix} 4 & 1 & 0 \\ 1 & 3 & 1 \\ 1 & 2 & 2 \end{pmatrix}$$

8. If r is an eigenvalue of a probability matrix, show that $|r| \leq 1$.
9. Suppose that the Fibonacci sequence started with $a_1 = 1$ and $a_2 = 3$. What is the limit of the ratios $\frac{a_n}{a_{n+1}}$ as $n \to \infty$ for this sequence?
10. Prove the lemma in the subsection on probability matrices.

Group Project: The Signature of Quadratic Forms

The results of this section show that every quadratic form (with real numbers for coefficients) can be "diagonalized" in the form

$$X_1^2 + X_2^2 + \cdots + X_r^2 - X_{r+1}^2 - X_{r+2}^2 - \cdots - X_{r+s}^2.$$

(a) Explain why this follows from the results in this section.

(a′) Alternatively, use high school algebra to explain how, generalizing the idea of completing the square, this result can be understood.

(b) The number $r - s$ in the above representation is called the *signature* of the quadratic form. Find the signatures of each quadratic form studied in Prob. 5.

(c) It is possible to diagonalize a quadratic form according to various variable changes. Explain why the signature of a quadratic form does not depend on its diagonalization. (Hint: Think about the eigenspaces of the associated symmetric matrix.)

(d) Two quadratic forms are called *equivalent* if one can be transformed into another by a linear change of variables. The classification problem for quadratic forms is to determine all possible equivalences of quadratic forms. Show that the signature, together with the dimension (or number of variables), enables one to "classify" quadratic forms with real coefficients. How many possible inequivalent quadratic forms with three variables are there?

Group Project: A Leontief Model

There is a small economy on an isolated island. It's sunny and warm all year long, and food is plentiful and can picked from trees whenever anyone is hungry. The only problem is that the soil is full of sharp volcanic rocks, so the inhabitants must wear shoes to avoid cutting their feet. The island economy is thus based almost entirely on shoes.

There are three industries in this economy. There are 25 leather hunters, 10 twine gatherers, and 15 shoemakers. The shoemakers have to buy 60% of the leather and 80% of the twine from those industries for shoemaking. The remainder of the leather and twine is purchased equally by the entire population for personal use. The shoemakers keep enough of the shoes they make for their own use, and they sell the rest to the leather hunters and twine gatherers. The leather hunters need three times as many shoes as the rest of the island's inhabitants, since they wear out their shoes while chasing animals for leather.

The island leaders have calculated how much leather, twine, and shoes need to be made monthly in order to keep everyone happy. However, they would like to set a price for each item and have it result in a stable exchange of currency. Your task is to write a short report explaining how a Leontief model can help them. You will have to make up more to this story to complete your report, perhaps making additional assumptions not given here. Be sure to find an appropriate probability matrix that describes a Leontief model for the exchanges of goods in this economy.

Group Project: The Second-Derivative Test for Functions of Several Variables

This project is accessible to students who have seen the second-derivative test from multivariable calculus.

(a) First recall the second-derivative test from differential calculus. What is the geometric reason that $f''(a) > 0$ at a critical point a shows that a is a local minimum of f?

(b) Now suppose that $\mathcal{U} \subseteq \mathbf{R}^n$ is an open subset and that $f : \mathcal{U} \to \mathbf{R}$ is a function with continuous first and second partial derivatives. Suppose that $(a_1, a_2, \ldots, a_n)$ is a critical point of f. Consider the symmetric matrix

$$S = \left(\frac{\partial^2 f}{\partial x_i \, \partial x_j}(a_1, a_2, \ldots, a_n) \right)$$

(that is, the matrix whose ijth entry is the ijth second-order partial derivative). Why is S symmetric? What is S if the function $f(x_1, x_2, \ldots, x_n)$ is given by a quadratic form?

(c) Suppose now that S is nonsingular, and let Q be its associated quadratic form. If Q is negative definite, explain why $(a_1, a_2, \ldots, a_n)$ is a local maximum, and if Q is positive definite, explain $(a_1, a_2, \ldots, a_n)$ is a local minimum. In case the quadratic form Q is indefinite and nonsingular, then $(a_1, a_2, \ldots, a_n)$ is called a *saddle* point. What does that mean? Use the second-degree Taylor expansion of the function f,

$$f(a_1, a_2, \ldots, a_n) + \frac{1}{2} \sum_{i,j} \frac{\partial^2 f}{\partial x_i \, \partial x_j}(a_1, a_2, \ldots, a_n)(X_i - a_i)(X_j - a_j),$$

near your critical point in your explanation.

MATRICES AS LINEAR TRANSFORMATIONS

8.1 Linear Transformations

Reflections in the Plane

We all learned at an early age that in spite of the similarities, there is a significant difference between a left shoe and a right shoe. How does a mathematician recognize and describe this difference? To a mathematician, the right shoe is the reflection of the left shoe, and vice versa. Look in the mirror at a left shoe's reflection next to its companion right shoe to see why. It turns out that the difference between some objects and their reflections can be substantial, even more so than for shoes!

Consider the two pairs of triangles pictured in Fig. 8.1. They are reflections of one another across the y-axis. In spite of the fact that the reflections of the triangles are congruent, there is a fundamental difference between the two cases considered. The upper triangles, being isosceles, are also congruent through a rotation. In other words, the 180° rotation centered at the point $(0, 3)$ moves one triangle onto the other. However, there is no way to

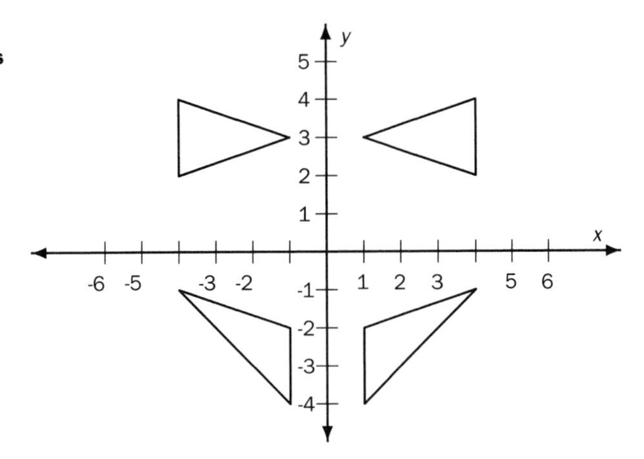

Fig. 8.1. Reflections of triangles across the *y*-axis

rotate the lower triangles onto one another. In fact, there is no combination of rotations or slides that can move one of the lower triangles onto the other. A reflection is needed to see their congruence (aside from cutting the triangles out of the page and flipping them over).

The difference between our lower triangles is the same as the difference between right and left shoes. A left shoe looks like a right shoe in the mirror, but unfortunately it will never fit well on a right foot. The reflection relationship is also important in organic chemistry, where it is known that reflections of chemical compounds can sometimes behave differently. If you imagine a molecule bouncing around in a gas or liquid you realize it can move in many directions and it can rotate in various ways, but it may not be able to transform into its reflection. In this chapter we study *linear transformations*, which are helpful in analyzing motions and symmetry. It is also important to understand how reflections and rotations are related, and for this we will need to use their matrix descriptions.

Matrix Descriptions of Reflections and Rotations

The reflection across the y-axis used in Fig. 8.1 is obtained algebraically by multiplying the x-coordinate of a point by -1. Hence it is given by left multiplication by the matrix

$$M_y = \begin{pmatrix} -1 & 0 \\ 0 & 1 \end{pmatrix}.$$

In other words, if $\vec{v} \in \mathbf{R}^2$ is a column vector, then its reflection across the y-axis is the product $M_y\vec{v}$. We will denote this reflection function from $\mathbf{R}^2$ to $\mathbf{R}^2$ by the symbol T_{M_y}, and we write $T_{M_y} : \mathbf{R}^2 \to \mathbf{R}^2$ to demonstrate that the domain and range of the reflection function are $\mathbf{R}^2$. Our matrix representation shows for all $\vec{v} \in \mathbf{R}^2$ that $T_{M_y}(\vec{v}) = M_y\vec{v}$.

Observe that we are making a distinction between the reflection function and the matrix that describes it. We will shortly find that all linear transformations are represented by matrix multiplication in this fashion, and it is important conceptually to make this distinction. In later sections this distinction may at times become a bit blurred, and you should always be careful to keep the difference in mind.

In Sec. 6.2 we studied the rotation matrix

$$R_\theta = \begin{pmatrix} \cos(\theta) & -\sin(\theta) \\ \sin(\theta) & \cos(\theta) \end{pmatrix},$$

where we showed that the counterclockwise rotation by the angle θ is given by left multiplication by R_θ. If we denote the rotation function by $T_{R_\theta} : \mathbf{R}^2 \to \mathbf{R}^2$, then we have for all $\vec{v} \in \mathbf{R}^2$ that $T_{R_\theta}(\vec{v}) = R_\theta \vec{v}$. Observe that $\det(M_y) = -1$ and $\det(R_\theta) = \cos^2(\theta) + \sin^2(\theta) = 1$. This shows that a reflection can never be a rotation, and vice versa.

In $\mathbf{R}^3$ we can consider the three matrices

$$M_{xy} = \begin{pmatrix} 1 & 0 & 0 \\ 0 & 1 & 0 \\ 0 & 0 & -1 \end{pmatrix}, \quad M_{xz} = \begin{pmatrix} 1 & 0 & 0 \\ 0 & -1 & 0 \\ 0 & 0 & 1 \end{pmatrix}, \quad M_{yz} = \begin{pmatrix} -1 & 0 & 0 \\ 0 & 1 & 0 \\ 0 & 0 & 1 \end{pmatrix}.$$

The matrix M_{xy} gives reflection through the xy-plane, M_{xz} gives reflection through the xz-plane, and M_{yz} gives reflection through the yz-plane. For example, if we left-multiply a vector by M_{xy}, then its x- and y-coordinates are unchanged, while the sign of its z-coordinate is reversed.

We also can express some rotations in $\mathbf{R}^3$ using matrices

$$R_{\theta,z} = \begin{pmatrix} \cos(\theta) & -\sin(\theta) & 0 \\ \sin(\theta) & \cos(\theta) & 0 \\ 0 & 0 & 1 \end{pmatrix}, \quad R_{\theta,y} = \begin{pmatrix} \cos(\theta) & 0 & -\sin(\theta) \\ 0 & 1 & 0 \\ \sin(\theta) & 0 & \cos(\theta) \end{pmatrix},$$

and

$$R_{\theta,x} = \begin{pmatrix} 1 & 0 & 0 \\ 0 & \cos(\theta) & -\sin(\theta) \\ 0 & \sin(\theta) & \cos(\theta) \end{pmatrix}.$$

The matrix $R_{\theta,z}$ describes a rotation around (and fixing) the z-axis by θ, $R_{\theta,y}$ describes a rotation around the y-axis by θ, and $R_{\theta,x}$ describes a rotation around the x-axis by θ. More examples can be found in the second group project in the problems at the end of Sec. 8.3.

Linear Transformations

In our study of rotations and reflections we used matrix multiplication to define functions. This can be done more generally. Suppose that A is an $m \times n$ matrix. Multiplication on the left by A defines a function from $\mathbf{R}^n$

to $\mathbf{R}^m$. We denote this function by $T_A : \mathbf{R}^n \to \mathbf{R}^m$. In other words,

$$T_A(\vec{v}) = A\vec{v}$$

for all $n \times 1$ matrices $\vec{v}$ in $\mathbf{R}^n$. *This change in viewpoint, from a matrix to the function it defines, is the key to the applications in this section.* Next we give the definition of a linear transformation from $\mathbf{R}^n$ to $\mathbf{R}^m$. Note that we are discussing functions, not matrices, in this definition.

Definition. A *linear transformation* from $\mathbf{R}^n$ to $\mathbf{R}^m$ is a function $T : \mathbf{R}^n \to \mathbf{R}^m$ that satisfies these conditions:

(i) If $\vec{u}, \vec{v} \in \mathbf{R}^n$, then $T(\vec{u} + \vec{v}) = T(\vec{u}) + T(\vec{v})$.

(ii) If $\vec{v} \in \mathbf{R}^n$ and $k \in \mathbf{R}$, then $T(k\vec{v}) = kT(\vec{v})$.

If $T : \mathbf{R}^n \to \mathbf{R}^m$ is a linear transformation, then $\mathbf{R}^n$ is called the *domain* of the linear transformation T. $\mathbf{R}^m$ is called the *range space* (or range) of T. The term "image" is sometimes used for the range, but we will reserve it for a slightly different meaning in the next section.

Observe that a linear function *need not* be a linear transformation. Note that if $T : \mathbf{R}^m \to \mathbf{R}^n$ is a linear transformation, then $T(\vec{0}) = T(0\vec{v}) = 0 \cdot T(\vec{v}) = \vec{0}$ for all $\vec{v} \in \mathbf{R}^m$. In other words, *for every linear transformation T we have*

$$T(\vec{0}) = \vec{0}.$$

However, the linear function $L : \mathbf{R} \to \mathbf{R}$ defined by $L(x) = x + 1$ has $L(0) = 1$. So this linear function L is *not* a linear transformation. In fact, a linear function $F : \mathbf{R}^m \to \mathbf{R}$ is a linear transformation only if $F(\vec{0}) = 0$. This distinction causes a bit of confusion at first, but it is necessary.

Usually, instead of calling a function a linear transformation, we will say that it is "linear" or we will call it a "transformation" for short. A linear transformation $T : \mathbf{R}^n \to \mathbf{R}^n$ from $\mathbf{R}^n$ to itself is called a *linear operator* on $\mathbf{R}^n$. The properties of matrix multiplication guarantee that for every $m \times n$ matrix the function $T_A : \mathbf{R}^n \to \mathbf{R}^m$ is a linear transformation. The notation is quite convenient, too. For example, suppose that

$$A = \begin{pmatrix} 2 & 1 \\ 1 & 1 \\ 2 & 4 \end{pmatrix}.$$

Then $T_A : \mathbf{R}^2 \to \mathbf{R}^3$ is the linear transformation defined by

$$T_A\begin{pmatrix} x \\ y \end{pmatrix} = \begin{pmatrix} 2 & 1 \\ 1 & 1 \\ 2 & 4 \end{pmatrix}\begin{pmatrix} x \\ y \end{pmatrix} = \begin{pmatrix} 2x + y \\ x + y \\ 2x + 4y \end{pmatrix}.$$

Observe that the entries of the matrix A appear as the coefficients of the variables in the formula for T_A.

For any n, the function $I : \mathbf{R}^n \to \mathbf{R}^n$ defined by $I(\vec{v}) = \vec{v}$ and the function $O : \mathbf{R}^n \to \mathbf{R}^n$ defined by $O(\vec{v}) = \vec{0}$ are both linear transformations. For obvious reasons, I is called the *identity transformation*, and O is called the *zero transformation*.

The next theorem is used so often that we do not refer to it by name, but rather we shall simply say "by linearity."

Theorem 51 (General linearity). *Suppose that $T : \mathbf{R}^n \to \mathbf{R}^m$ is a linear transformation. Let $v_1, v_2, \ldots, v_s \in \mathbf{R}^n$. Then for all real numbers $a_1, a_2, \ldots, a_s$,*

$$T(a_1 v_1 + a_2 v_2 + \cdots + a_s v_s) = a_1 T(v_1) + a_2 T(v_2) + \cdots + a_s T(v_s).$$

Proof. We repeatedly apply both (i) and (ii) of the definition:

$$\begin{aligned} T(a_1 v_1 + \cdots + a_s v_s) &= T(a_1 v_1) + T(a_2 v_2 + \cdots + a_s v_s) \\ &= \cdots = T(a_1 v_1) + T(a_2 v_2) + \cdots + T(a_s v_s), \\ &= a_1 T(v_1) + a_2 T(v_2) + \cdots + a_s T(v_s) \end{aligned}$$

as required. □

Linear Transformations and Bases

The next result says that a linear transformation from $\mathbf{R}^n$ to $\mathbf{R}^m$ is uniquely determined by how it "acts" on a basis.

Theorem 52. *Let $\{\vec{v}_1, \vec{v}_2, \ldots, \vec{v}_n\}$ be a basis of $\mathbf{R}^n$ and $\vec{w}_1, \vec{w}_2, \ldots, \vec{w}_n \in \mathbf{R}^m$ be arbitrary vectors (not necessarily distinct). Then there is a unique linear transformation $T : \mathbf{R}^n \to \mathbf{R}^m$ such that $T(\vec{v}_1) = \vec{w}_1$, $T(\vec{v}_2) = \vec{w}_2, \ldots, T(\vec{v}_n) = \vec{w}_n$.*

Proof. Since $\{\vec{v}_1, \vec{v}_2, \ldots, \vec{v}_n\}$ is a basis of $\mathbf{R}^n$, every element of $\mathbf{R}^n$ can be written *uniquely* as a linear combination $a_1\vec{v}_1 + a_2\vec{v}_2 + \cdots + a_n\vec{v}_n$. This means we can define the value of a function T on such a linear combination by

$$\begin{aligned} T(a_1\vec{v}_1 + a_2\vec{v}_2 + \cdots + a_n\vec{v}_n) &= a_1 T(\vec{v}_1) + a_2 T(\vec{v}_2) + \cdots + a_n T(\vec{v}_n) \\ &= a_1\vec{w}_1 + a_2\vec{w}_2 + \cdots + a_n\vec{w}_n. \end{aligned}$$

We must check that T defined in this manner is linear. Let $\vec{u} = a_1\vec{v}_1 + a_2\vec{v}_2 + \cdots + a_n\vec{v}_n$ and $\vec{u}\,' = b_1\vec{v}_1 + b_2\vec{v}_2 + \cdots + b_n\vec{v}_n$ be two vectors in $\mathbf{R}^n$. Then $\vec{u} + \vec{u}\,' = (a_1 + b_1)\vec{v}_1 + (a_2 + b_2)\vec{v}_2 + \cdots + (a_n + b_n)\vec{v}_n$, and so by the definition of T,

$$\begin{aligned} T(\vec{u} + \vec{u}\,') &= (a_1 + b_1)\vec{w}_1 + (a_2 + b_2)\vec{w}_2 + \cdots + (a_n + b_n)\vec{w}_n \\ &= (a_1\vec{w}_1 + a_2\vec{w}_2 + \cdots + a_n\vec{w}_n) \\ &\quad + (b_1\vec{w}_1 + b_2\vec{w}_2 + \cdots + b_n\vec{w}_n) \\ &= T(\vec{u}) + T(\vec{u}\,'). \end{aligned}$$

Next, if k is a real number, then

$$\begin{aligned} T(k\vec{u}) &= T(ka_1\vec{v}_1 + ka_2\vec{v}_2 + \cdots + ka_n\vec{v}_n) \\ &= ka_1\vec{w}_1 + ka_2\vec{w}_2 + \cdots + ka_n\vec{w}_n \\ &= k(a_1\vec{w}_1 + a_2\vec{w}_2 + \cdots + a_n\vec{w}_n) = kT(\vec{u}). \end{aligned}$$

This shows that T is a linear transformation.

Now observe that according to Theorem 51, we defined T in the only way possible. The uniqueness assertion of the theorem follows. □

Matrix Representations of Linear Transformations

Using Theorem 52 we next show that every linear transformation from $\mathbf{R}^n$ to $\mathbf{R}^m$ arises from matrix multiplication. This result will enable us to translate questions about linear transformations into questions about matrices, and vice versa.

Theorem 53. *Suppose that $S : \mathbf{R}^n \to \mathbf{R}^m$ is a linear transformation. Then there exists a unique $m \times n$ matrix A such that $S = T_A$. The matrix A is characterized by the property that $S(\vec{v}) = A\vec{v}$ for all $\vec{v} \in \mathbf{R}^n$.*

Proof. We let $\{\vec{e}_1, \vec{e}_2, \ldots, \vec{e}_n\}$ denote the standard basis of $\mathbf{R}^n$, and let $\{\vec{\mathrm{e}}_1, \vec{\mathrm{e}}_2, \ldots, \vec{\mathrm{e}}_m\}$ denote the standard basis of $\mathbf{R}^m$. (Note the different notation between $\vec{e}_i$ and $\vec{\mathrm{e}}_i$.) We define real numbers a_{ij} for $1 \le i \le m$, $1 \le j \le n$ using the expression $S(\vec{e}_i) = a_{1i}\vec{\mathrm{e}}_1 + a_{2i}\vec{\mathrm{e}}_2 + \cdots + a_{mi}\vec{\mathrm{e}}_m$. Such a_{ij} are uniquely determined since $\{\vec{\mathrm{e}}_1, \vec{\mathrm{e}}_2, \ldots, \vec{\mathrm{e}}_m\}$ is a basis for $\mathbf{R}^m$. Then $A = (a_{ij})$ is an $m \times n$ matrix. We claim that $S = T_A$. To see this, by the uniqueness assertion of Theorem 52 it suffices to show that $S(\vec{e}_1) = T_A(\vec{e}_1)$, $S(\vec{e}_2) = T_A(\vec{e}_2)$, ..., $S(\vec{e}_n) = T_A(\vec{e}_n)$.

According to the definition of A, the ith column of A consists of (the coordinates of) the vector $G(\vec{e}_i)$. As $A\vec{e}_i$ is precisely the ith column of A (according to the definition of matrix multiplication), we see that $T_A(\vec{e}_i) = A\vec{e}_i = G(\vec{e}_i)$. The theorem is proved. □

Definition. If $S : \mathbf{R}^n \to \mathbf{R}^m$ and if $S = T_A$ as guaranteed by Theorem 53, then matrix A is called the *standard matrix of* S. We often denote this matrix A by $[S]$, in other words, $S = T_{[S]}$ and $[T_A] = A$.

The proof of Theorem 53 gives the procedure for finding the matrix $A = [S]$ corresponding to the linear transformation $S : \mathbf{R}^n \to \mathbf{R}^m$. First you compute the values $S(\vec{e}_1)$, $S(\vec{e}_2)$, $\ldots$, $S(\vec{e}_n)$ of S applied to the standard basis vectors $\vec{e}_1, \vec{e}_2, \ldots, \vec{e}_n$ of $\mathbf{R}^n$. Then the ith column of the desired matrix A is precisely the (coordinates of the) vector $S(\vec{e}_i)$.

For example, suppose that $S : \mathbf{R}^3 \to \mathbf{R}^4$ is defined by

$$S\begin{pmatrix} x \\ y \\ z \end{pmatrix} = \begin{pmatrix} x + 2y \\ x + 3z \\ x + y \\ x + y \end{pmatrix}.$$

First we compute the images of the standard basis under S:

$$S\begin{pmatrix} 1 \\ 0 \\ 0 \end{pmatrix} = \begin{pmatrix} 1 \\ 1 \\ 1 \\ 1 \end{pmatrix}, \quad S\begin{pmatrix} 0 \\ 1 \\ 0 \end{pmatrix} = \begin{pmatrix} 2 \\ 0 \\ 1 \\ 1 \end{pmatrix}, \quad S\begin{pmatrix} 0 \\ 0 \\ 1 \end{pmatrix} = \begin{pmatrix} 0 \\ 3 \\ 0 \\ 0 \end{pmatrix}.$$

Then it follows that $S = T_A$, where

$$A = \begin{pmatrix} 1 & 2 & 0 \\ 1 & 0 & 3 \\ 1 & 1 & 0 \\ 1 & 1 & 0 \end{pmatrix}.$$

Observe that the entries of the matrix A are exactly the coefficients that appeared in front of the variables in the definition of S.

Composition of Linear Transformations

In many mathematical problems it is important to study the composition of functions. Consider an example where we compose two linear transformations: Suppose $T_1 : \mathbf{R}^3 \to \mathbf{R}^2$ is defined by $T_1(x, y, z) = (x + y, 2x + z)$ and $T_2 : \mathbf{R}^2 \to \mathbf{R}^4$ is defined by $T_2(r, s) = (r + s, 2s, r, 2r + s)$. Then the composition $T_2 \circ T_1(x, y) = T_2(T_1(x, y)) = ((x + y) + (2x + z), 2(2x + z), x + y, 2(x + y) + (2x + z))$. This row is the same as column in the matrix product

$$\begin{pmatrix} 1 & 1 \\ 0 & 2 \\ 1 & 0 \\ 2 & 1 \end{pmatrix} \begin{pmatrix} x + y \\ 2x + z \end{pmatrix}.$$

Note that the left matrix is the standard matrix of T_2, and the right matrix is the product

$$\begin{pmatrix} x+y \\ 2x+z \end{pmatrix} = \begin{pmatrix} 1 & 1 & 0 \\ 2 & 0 & 1 \end{pmatrix} \begin{pmatrix} x \\ y \\ z \end{pmatrix}$$

of the standard matrix of T_1 and the column of variables x, y, and z. Altogether this shows that

$$T_2 \circ T_1 \begin{pmatrix} x \\ y \end{pmatrix} = \begin{pmatrix} 1 & 1 \\ 0 & 2 \\ 1 & 0 \\ 2 & 1 \end{pmatrix} \begin{pmatrix} 1 & 1 & 0 \\ 2 & 0 & 1 \end{pmatrix} \begin{pmatrix} x \\ y \\ z \end{pmatrix}.$$

In other words, $T_2 \circ T_1 = T_A$ where A is the product $[T_2][T_1]$ of the standard matrices of the transformations T_2 and T_1.

This observation is true more generally.

Theorem 54. *Suppose that A is an $n \times m$ matrix and B is an $s \times n$ matrix. Then the composition of linear transformations $T_B \circ T_A : \mathbf{R}^m \to \mathbf{R}^s$ is a linear transformation and $T_B \circ T_A = T_{BA}$.*

Proof. Consider any $\vec{v} \in \mathbf{R}^m$, expressed as a column vector. Then $T_B \circ T_A(\vec{v}) = T_B(T_A(\vec{v})) = T_B(A\vec{v}) = B(A\vec{v}) = (BA)\vec{v} = T_{BA}(\vec{v})$, where the second to last equality is the associativity of matrix multiplication. This shows that the functions $T_B \circ T_A$ and T_{BA} must be the same. □

Products of Reflections and Rotations

For some problems that utilize geometric symmetry it is important to understand the composition of reflection and rotation transformations. According to Theorem 54 we need to determine products of reflection and rotation matrices to do this. The reflections across the x-axis and the y-axis are given by the matrices

$$T_x = \begin{pmatrix} 1 & 0 \\ 0 & -1 \end{pmatrix} \quad \text{and} \quad T_y = \begin{pmatrix} -1 & 0 \\ 0 & 1 \end{pmatrix}.$$

Their product

$$T_y T_x = \begin{pmatrix} -1 & 0 \\ 0 & 1 \end{pmatrix} \begin{pmatrix} 1 & 0 \\ 0 & -1 \end{pmatrix} = \begin{pmatrix} -1 & 0 \\ 0 & -1 \end{pmatrix} = \begin{pmatrix} \cos(180°) & \sin(180°) \\ -\sin(180°) & \cos(180°) \end{pmatrix}$$

is the matrix describing 180° rotation.

In general, the product of two reflections in $\mathbf{R}^2$ is also a rotation. This can be understood geometrically. Consider the two lines ℓ_1 and ℓ_2 in Fig. 8.2.

Fig. 8.2. The reflection through ℓ_1 followed by the reflection through ℓ_2

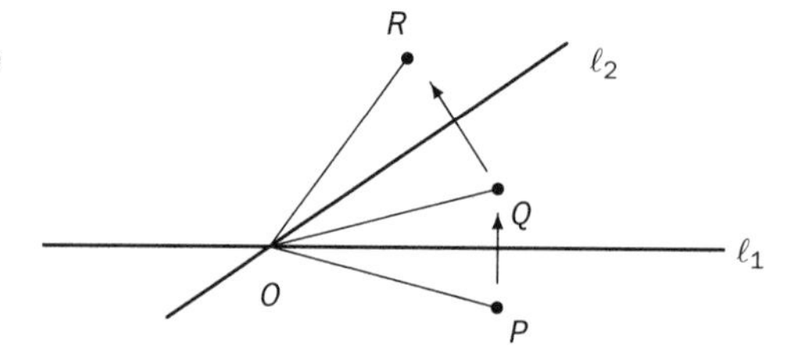

First the point P is reflected across the line ℓ_1 to obtain Q, and then Q is reflected across the line ℓ_2 to obtain the point R.

Observe that the line ℓ_1 bisects the angle $\angle POQ$ and the line ℓ_2 bisects the angle $\angle QOR$. This shows that the measure of $\angle POR$ is exactly twice the measure of the angle between the lines ℓ_1 and ℓ_2. Note also, by triangle congruences, that the distances PO, QO, and RO are all equal. This shows that the point R is obtained from the point P by rotating around O counterclockwise by the angle 2θ, where θ is the angle between ℓ_1 and ℓ_2. Similar reasons show this composite rotates all points in the plane counterclockwise by the angle 2θ (see Prob. 8 at the end of this section). Therefore, the product of two reflections is a rotation.

The product of a rotation and a reflection,

$$P = T_x R_\theta = \begin{pmatrix} 1 & 0 \\ 0 & -1 \end{pmatrix} \begin{pmatrix} \cos(\theta) & -\sin(\theta) \\ \sin(\theta) & \cos(\theta) \end{pmatrix} = \begin{pmatrix} \cos(\theta) & -\sin(\theta) \\ -\sin(\theta) & -\cos(\theta) \end{pmatrix},$$

cannot be another rotation since its matrix has determinant -1. What is it? Observe that the characteristic polynomial of this matrix is

$$C_P(X) = (X - \cos(\theta))(X + \cos(\theta)) - \sin^2(\theta) = X^2 - 1.$$

Therefore, the linear transformation T_P has an eigenvector $\vec{v}_1$ of eigenvalue 1 and an eigenvector $\vec{v}_{-1}$ of eigenvalue -1. Theorem 49 shows that $\vec{v}_1$ and $\vec{v}_{-1}$ are perpendicular. This means that T_P must fix a line ℓ, which is the line containing $\vec{v}_1$, and reflect the vector $\vec{v}_{-1}$ across the line ℓ. It follows from this that T_P is the reflection through the line ℓ. The geometric determination of the line ℓ and details why T_P is the reflection through ℓ are set up in the second group project at the end of this section.

Problems

1. Find the standard matrix of the linear transformation $F : \mathbf{R}^2 \to \mathbf{R}^2$, where F is geometrically described by
   (a) the reflection across the line $X = Y$;
   (b) the projection onto the line $X = 0$.
2. Determine whether each of the following functions is linear. If it is, find its standard matrix.

(a) $T_1 : \mathbf{R}^2 \to \mathbf{R}^2$ defined by $T_1(x, y) = (x, y^2)$
(b) $T_2 : \mathbf{R}^2 \to \mathbf{R}^2$ defined by $T_2(x, y) = (x - y, y - x)$
(c) $T_3 : \mathbf{R}^2 \to \mathbf{R}^2$ defined by $T_3(x, y) = (0, -xy)$
(d) $T_4 : \mathbf{R}^2 \to \mathbf{R}^4$ defined by $T_4(x, y) = (0, 0, 0, x)$
(e) $T_5 : \mathbf{R}^3 \to \mathbf{R}^2$ defined by $T_5(x, y, z) = (7x - 3, 2y)$
(f) $T_6 : \mathbf{R}^3 \to \mathbf{R}^2$ defined by $T_6(x, y, z) = (\sin(x), \cos(y))$
(g) $T_7 : \mathbf{R}^3 \to \mathbf{R}^4$ defined by $T_7(x, y, z) = (x, x, x, x)$

3. (a) Find two linear transformations $T_1, T_2 : \mathbf{R}^2 \to \mathbf{R}^2$ such that $T_1 \circ T_2 = T_2 \circ T_1$.
   (b) Find two linear transformations $T_1, T_2 : \mathbf{R}^2 \to \mathbf{R}^2$ such that $T_1 \circ T_2 \neq T_2 \circ T_1$.
4. For each part below, answer the question "Is there a linear transformation $T : \mathbf{R}^3 \to \mathbf{R}^3$ such that ...?" and justify your answer.
   (a) $T(1, 0, 0) = (0, 1, 0)$, $T(0, 1, 0) = (0, 1, 0)$, and $T(0, 0, 1) = (1, 0, 0)$
   (b) $T(0, 0, 0) = (1, 0, 0)$, $T(1, 0, 0) = (0, 1, 0)$, $T(0, 1, 0) = (0, 1, 0)$, and $T(0, 0, 1) = (0, 0, 0)$
5. Assume that $T_1 : \mathbf{R}^3 \to \mathbf{R}^2$ and $T_2 : \mathbf{R}^2 \to \mathbf{R}^3$ are the linear transformations described below. Show how Theorem 54 can be used to find the standard matrices for $T_2 \circ T_1$, and $T_1 \circ T_2$.
   (a) $T_1(x, y, z) = (x + y, 0)$ and $T_2(x, y) = (x, y, x)$
   (b) $T_1(x, y, z) = (0, 0)$ and $T_2(x, y)$ is unknown.
   (c) $T_1(x, y, z) = (x, x)$ and $T_2(x, y) = (y, x, x + y)$
6. Suppose that $T_1 : \mathbf{R}^s \to \mathbf{R}^n$ and $T_2 : \mathbf{R}^s \to \mathbf{R}^m$ are both linear transformations. Show that the function $T : \mathbf{R}^s \to \mathbf{R}^{n+m}$ defined by $T(\vec{v}) = (T_1(\vec{v}), T_2(\vec{v}))$ is a linear transformation.
7. (a) Assume that $T : \mathbf{R}^n \to \mathbf{R}^m$ is a linear transformation and that the collection of image vectors $\{T(\vec{v}_1), T(\vec{v}_2), \ldots, T(\vec{v}_n)\}$ is linearly independent in $\mathbf{R}^m$. Show that $\{\vec{v}_1, \vec{v}_2, \ldots, \vec{v}_n\}$ is linearly independent in $\mathbf{R}^n$.
   (b) Is the converse to part (a) true? In other words, if we know that $\{\vec{v}_1, \vec{v}_2, \ldots, \vec{v}_n\}$ is linearly independent, does it follow that $\{T(\vec{v}_1), T(\vec{v}_2), \ldots, T(\vec{v}_n)\}$ is linearly independent?
8. Study Fig. 8.2 and give geometric arguments extending the discussion there to all points of the plane.

Group Project: Transformations of the Plane

In this project you will investigate more of the geometric properties of linear transformations from $\mathbf{R}^2$ to itself.

(a) Using the angle addition formulas, show that the product of R_θ and R_α is $R_{\theta+\alpha}$.

(b) Which 2×2 matrix describes the reflection of the plane through the line $y = kx$, where k is a real number?

(c) Both rotations and reflections "preserve area" in the sense that if $\mathcal{C}$ is a region in $\mathbf{R}^2$ then the area of $\mathcal{C}$ is the same as the transformed area. Give a short explanation of why this is true.

(d) The linear transformation $T_A : \mathbf{R}^2 \to \mathbf{R}^2$ where A is the matrix kI_2 and k is a nonzero real number is called a *similarity transformation*. This means that although it does not preserve area, it does transform triangles into similar triangles. Why is this?

(e) Give some more examples of similarity transformations that do not preserve area. (Hint: Consider composite transformations.)

(f) Translations of the plane are defined by fixing a vector $\vec{v}_0$ and then defining $T(\vec{v}) = \vec{v} + \vec{v}_0$ for all $\vec{v} \in \mathbf{R}^2$. Show that translations are *not* linear transformations, but that they do preserve area.

Group Project: The Product of a Reflection and a Rotation

In this section we considered the composite $P = T_{M_x} \circ R_\theta$ of the reflection across the x-axis (denoted T_{M_x}), and a rotation by angle θ (denoted R_θ). We noted there that P has characteristic polynomial $C_P(X) = X^2 - 1$. Your task here is to study the picture below and write out a geometric proof of this fact. You will be able to deduce from your work that P is a reflection.

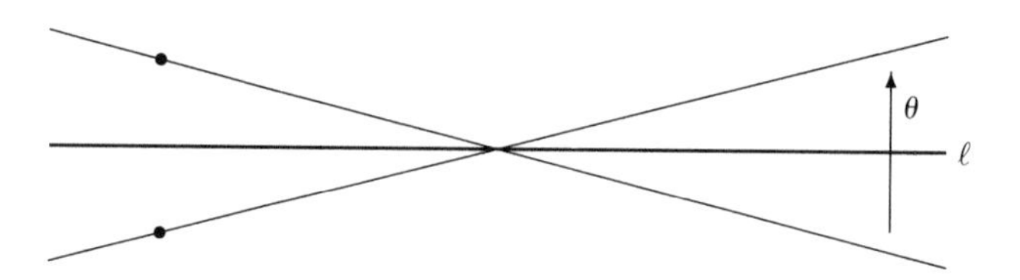

In this figure, the bold line is the line ℓ of reflection, and the rotation by the angle θ is indicated between the thin lines.

(a) The dots in the figure are reflections of one another across the line ℓ. They are also related by the rotation. Which of the smaller lines is fixed by the composite $P = T_x \circ R_\theta$? Give a full, careful explanation why. Include in your explanation any assumptions needed for this picture to work.

(b) Show geometrically that the points on the line perpendicular to the fixed line you found in (a) are transformed to their opposites by the composite $P = T_x \circ R_\theta$.

(c) Now you can show that the linear transformation P is the reflection through the fixed line you found in (a). For this, show that any vector on the plane can be expressed as a linear combination of a vector on this line and a vector perpendicular to this line. Use the linearity property of linear transformations, together with geometry, to show that P is the desired reflection.

(d) What happens when the order of the composition is reversed, that is, can you use the geometry just developed to describe $R_\theta \circ T_{M_x}$?

8.2 Using Linear Transformations

Linear Transformations and Two-Port Electrical Networks

Next we consider electrical networks built from networks similar to those considered earlier in Sec. 3.1. In Fig. 8.3 we have modified the representation of such a network by removing the batteries and replacing them with circles. Our problem is to understand what happens when we connect networks like this together at the circles. These are called *two-port networks* because there are two places at which power sources or other networks can be attached.

In Fig. 8.3 we assume that the voltage between the left-hand ports is V_1 and the current between these ports is I_1. Similarly, we assume that the voltage between the right-hand ports is V_2 and the current is I_2. This means that the current across the upper left resistor is I_1, the current across the upper right resistor is I_2, and the current across the center resistor is $I_1 + I_2$.

Recall that the voltage across the left port will be the sum of voltages across the upper left resistor and the center resistor. With the resistances as indicated, the voltage across the left port must be given by $V_1 = r_1 I_1 + r_3(I_1 + I_2)$. Similarly, the voltage across the right port must be given by $V_2 = r_2 I_2 + r_3(I_1 + I_2)$. This means that we can write a matrix equation

$$\begin{pmatrix} V_1 \\ V_2 \end{pmatrix} = \begin{pmatrix} r_1 + r_3 & r_3 \\ r_3 & r_2 + r_3 \end{pmatrix} \begin{pmatrix} I_1 \\ I_2 \end{pmatrix}$$

that relates the quantities V_1, V_2, I_1, and I_2.

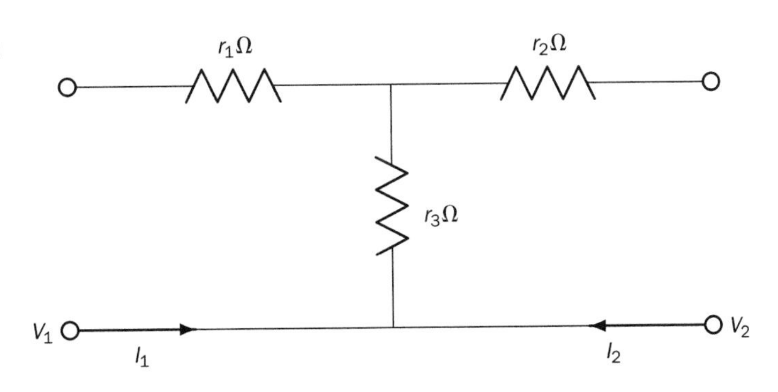

Fig. 8.3. The circuit from Fig. 3.1 viewed as a two-port network

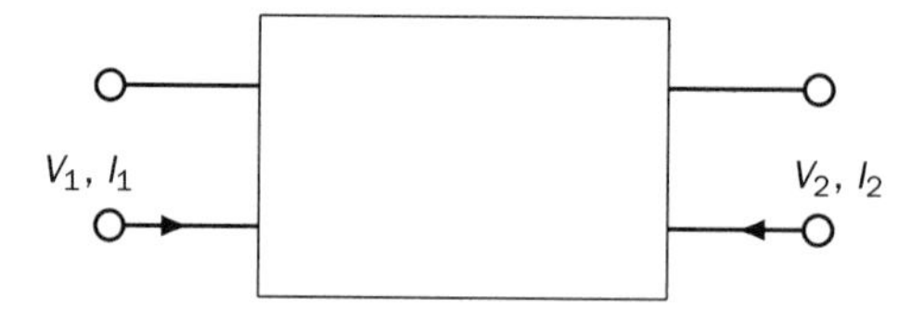

Fig. 8.4. A black box two-port network

More generally, suppose that you have a two-port "black box" network, where you don't know exactly what the contents are. This situation is indicated in Fig. 8.4.

Then you could still measure the voltages and currents V_1, V_2, I_1, and I_2 and see what happens. Based on what we have learned, we will assume that the relationship is given by some linear transformation $T_Z(I_1, I_2) = (V_1, V_2)$, where

$$Z = \begin{pmatrix} z_{11} & z_{12} \\ z_{21} & z_{22} \end{pmatrix}$$

is some 2×2 matrix. We will call the matrix $Z = (z_{ij})$ the *characteristic matrix* for the two-port network. In a group project at the end of this section you will explore ways to determine the entries z_{ij} of the matrix Z.

It turns out that there are other useful matrix descriptions of a two-port network. Using the characteristic equation $V_2 = z_{21}I_1 + z_{22}I_2$, we can solve for I_1 as $I_1 = z_{21}^{-1}V_2 - z_{22}z_{21}^{-1}I_2$. Then, substituting this in the other characteristic equation $V_1 = z_{11}I_1 + z_{12}I_2$ gives

$$V_1 = z_{11}[z_{21}^{-1}V_2 - z_{22}z_{21}^{-1}I_2] + z_{12}I_2 = z_{11}z_{21}^{-1}V_2 - (z_{11}z_{22}z_{21}^{-1} - z_{12})I_2.$$

Together these new expressions for I_1 and V_1 give the matrix equation

$$\begin{pmatrix} V_1 \\ I_1 \end{pmatrix} = \begin{pmatrix} z_{11}z_{21}^{-1} & z_{11}z_{22}z_{21}^{-1} - z_{12} \\ z_{21}^{-1} & z_{22}z_{21}^{-1} \end{pmatrix} \begin{pmatrix} V_2 \\ -I_2 \end{pmatrix}.$$

Note that we have used $-I_2$ instead of I_2 in this equation. This is not a typographical error. The reason is that we need to reverse the orientation of the current I_2 when we apply this matrix to cascade networks, which we do ahead. The matrix

$$A = \begin{pmatrix} z_{11}z_{21}^{-1} & z_{11}z_{22}z_{21}^{-1} - z_{12} \\ z_{21}^{-1} & z_{22}z_{21}^{-1} \end{pmatrix}$$

is often referred to as the *cascade parameter matrix* or sometimes is called the *transfer matrix.*

Cascading Two-Port Networks

In Fig. 8.5 we consider two two-port black boxes, which we will subsequently connect at the adjacent open circles. Our goal is to understand how

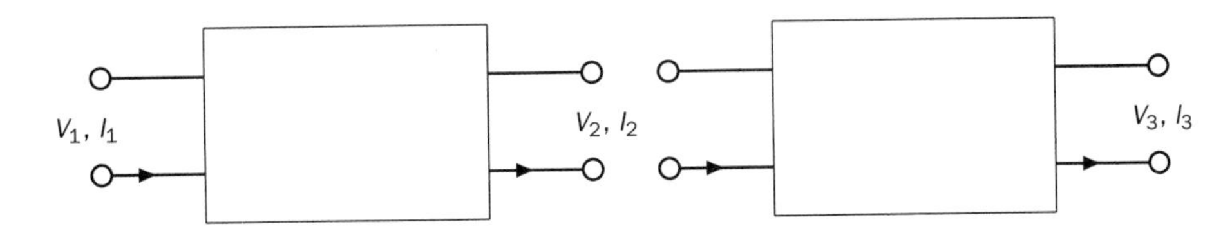

Fig. 8.5. Two black box two-port networks

the behavior of the two individual black boxes can be used to understand the new network. Note the orientations of the currents on the right-hand side of each black box. This is necessary so that when they are connected, both variables I_2 (from either box) will have the same orientation. This also means that when we use the cascade parameter matrix just constructed, we will not need the minus sign.

We assume that the relationships for these two black box circuits are given by the cascade parameter matrices

$$\begin{pmatrix} V_1 \\ I_1 \end{pmatrix} = A_1 \begin{pmatrix} V_2 \\ I_2 \end{pmatrix} \quad \text{and} \quad \begin{pmatrix} V_2 \\ I_2 \end{pmatrix} = B_2 \begin{pmatrix} V_3 \\ I_3 \end{pmatrix}.$$

Then, substituting the first matrix equation into the second shows that

$$\begin{pmatrix} V_1 \\ I_1 \end{pmatrix} = B_1 \begin{pmatrix} V_2 \\ I_2 \end{pmatrix} = B_1 \left[B_2 \begin{pmatrix} V_3 \\ I_3 \end{pmatrix} \right] = [B_1 B_2] \begin{pmatrix} V_3 \\ I_3 \end{pmatrix}.$$

We have found that the product matrix $B_1 B_2$ is the cascade parameter matrix describing the relationship between the quantities V_1, I_1, V_3, and I_3 that results if the adjacent ports are connected and a new two-port network is formed. Note we have shown that the linear transformation representing a cascaded network is the composite of the linear transformations representing the component networks.

We illustrate this technique by considering the network in Fig. 8.6. Our problem is to determine the currents I_1 and I_3 discharged by the two bat-

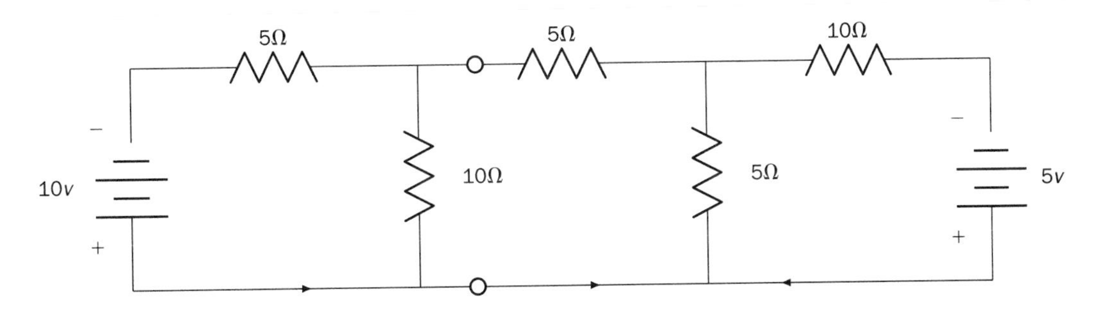

Fig. 8.6. A cascaded network

teries, where the voltages and resistances are as indicated. We will have to be careful with the sign of I_3 at the end of our calculation.

In order to analyze this network, we must find the cascade parameter matrices for the left- and right-hand networks (on opposite sides of the circles). We will denote by V_2 and I_2 the voltage and current flow, respectively, between the two open circles. Applying the previous computations for each of the left and right networks, we find

$$\begin{pmatrix} V_1 \\ V_2 \end{pmatrix} = \begin{pmatrix} 15 & 10 \\ 10 & 10 \end{pmatrix} \begin{pmatrix} I_1 \\ -I_2 \end{pmatrix} \quad \text{and} \quad \begin{pmatrix} V_2 \\ V_3 \end{pmatrix} = \begin{pmatrix} 10 & 5 \\ 5 & 15 \end{pmatrix} \begin{pmatrix} I_2 \\ I_3 \end{pmatrix}.$$

(Note the orientation of I_2 in the diagram.) Our formula for the cascade parameter matrices gives

$$\begin{pmatrix} V_1 \\ I_1 \end{pmatrix} = \begin{pmatrix} \frac{3}{2} & 5 \\ \frac{1}{10} & 1 \end{pmatrix} \begin{pmatrix} V_2 \\ I_2 \end{pmatrix} \quad \text{and} \quad \begin{pmatrix} V_2 \\ I_2 \end{pmatrix} = \begin{pmatrix} 2 & 25 \\ \frac{1}{5} & 3 \end{pmatrix} \begin{pmatrix} V_3 \\ -I_3 \end{pmatrix}.$$

Multiplying our cascade matrices shows

$$\begin{aligned} \begin{pmatrix} V_1 \\ I_1 \end{pmatrix} &= \begin{pmatrix} \frac{3}{2} & 5 \\ \frac{1}{10} & 1 \end{pmatrix} \begin{pmatrix} 2 & 25 \\ \frac{1}{5} & 3 \end{pmatrix} \begin{pmatrix} V_3 \\ -I_3 \end{pmatrix} \\ &= \begin{pmatrix} 4 & \frac{105}{2} \\ \frac{2}{5} & \frac{11}{2} \end{pmatrix} \begin{pmatrix} V_3 \\ -I_3 \end{pmatrix}. \end{aligned}$$

We have found the cascade matrix for the total network. If we desire the matrix representation for our network in the standard form, we must solve for standard representation matrix A using the matrix equation

$$\begin{pmatrix} z_{11} z_{21}^{-1} & z_{11} z_{22} z_{21}^{-1} - z_{12} \\ z_{21}^{-1} & z_{22} z_{21}^{-1} \end{pmatrix} = \begin{pmatrix} 4 & \frac{105}{2} \\ \frac{2}{5} & \frac{11}{2} \end{pmatrix}.$$

This equation is readily solved, and we find $z_{21} = \frac{5}{2}$, $z_{22} = \frac{5}{2} \cdot \frac{11}{2} = \frac{55}{4}$, $z_{11} = \frac{5}{2} \cdot 4 = 10$, and $z_{12} = 10 \cdot \frac{55}{4} / \frac{5}{2} - \frac{105}{2} = \frac{5}{2}$. Altogether we have found that the standard representation of our total network is

$$\begin{pmatrix} V_1 \\ V_3 \end{pmatrix} = \begin{pmatrix} 10 & \frac{5}{2} \\ \frac{5}{2} & \frac{55}{4} \end{pmatrix} \begin{pmatrix} I_1 \\ I_3 \end{pmatrix}.$$

If we like, we can invert our standard representation matrix and obtain

$$\begin{pmatrix} I_1 \\ I_3 \end{pmatrix} = \frac{1}{105} \begin{pmatrix} 11 & -2 \\ -2 & 8 \end{pmatrix} \begin{pmatrix} V_1 \\ V_3 \end{pmatrix}.$$

Substituting $V_1 = 10$ and $V_3 = 5$ into this equation gives $I_1 = 1$ and $I_3 = \frac{4}{21}$. This is the answer to our original question about how the current flows.

The network we just analyzed could have been studied by considering larger systems of equations. However, the technique of using matrix multiplication to study the cascading of basic electric circuits (including those that involve components other than resistors) is extremely important in circuit analysis. More information is contained in the group projects at the end of this section.

The Image and Kernel of a Linear Transformation

In our analysis of matrices, the column space and the null space were extremely important. These subspaces can be studied using linear transformations where they are called the image and the kernel, respectively. We begin with these two basic definitions.

Definition. Suppose $T : \mathbf{R}^n \to \mathbf{R}^m$ is a linear transformation.

(i) The *image* of T, denoted $\mathrm{im}(T)$, is defined by

$$\mathrm{im}(T) = \{T(\vec{v}) \mid \vec{v} \in \mathbf{R}^n\}.$$

(ii) The *kernel* of T, denoted $\ker(T)$, is defined by

$$\ker(T) = \{\vec{v} \in \mathbf{R}^n \mid T(\vec{v}) = \vec{0}\}.$$

It is important to keep track of where these subsets are located. Note that $\mathrm{im}(T) \subseteq \mathbf{R}^m$ and $\ker(T) \subseteq \mathbf{R}^n$. Often, $\ker(T)$ is called the *null space* of T.

The kernel of $T : \mathbf{R}^n \to \mathbf{R}^m$ is the set of solutions to the equation $T(\vec{v}) = \vec{0}$. Thus, the kernel of $T_A : \mathbf{R}^n \to \mathbf{R}^m$ is precisely the null space of the matrix A. The image of T_A also has a familiar interpretation. By definition, $\mathrm{im}(T_A) = \{\vec{u} \in \mathbf{R}^m \mid \text{there exists } \vec{v} \in \mathbf{R}^n \text{ with } T_A(\vec{v}) = \vec{u}\}$. Later we shall see that the image of T_A is precisely the column space of the matrix A.

The next lemma shows that both the kernel and image of a linear transformation are subspaces.

Lemma. *Suppose $T : \mathbf{R}^n \to \mathbf{R}^m$ is a linear transformation. Then $\ker(T)$ is a subspace of $\mathbf{R}^n$ and $\mathrm{im}(T)$ is a subspace of $\mathbf{R}^m$.*

Proof. Since $T = T_A$, where A is the standard matrix of T, $\ker(T) = \ker(A)$ is a subspace of $\mathbf{R}^n$. Now choose $\vec{v}_1, \vec{v}_2 \in \mathrm{im}(T)$. By the definition of $\mathrm{im}(T)$, there exist $\vec{u}_1, \vec{u}_2 \in \mathbf{R}^n$ with $T(\vec{u}_1) = \vec{v}_1$ and $T(\vec{u}_2) = \vec{v}_2$. Then, since T is linear, $T(\vec{u}_1 + \vec{u}_2) = \vec{v}_1 + \vec{v}_2$ and $T(k\vec{u}_1) = k\vec{v}_1$. Hence, $\vec{v}_1 + \vec{v}_2, k\vec{v}_1 \in \mathrm{im}(T)$, and so $\mathrm{im}(T)$ is a subspace of $\mathbf{R}^m$. □

The Rank and Nullity

Now that we know that $\text{im}(T)$ and $\ker(T)$ are subspaces, we give names to their dimensions.

Definition. Suppose that $T : \mathbf{R}^n \to \mathbf{R}^m$ is a linear transformation. The dimension of $\text{im}(T)$ is called the *rank* of T and is denoted $\text{rk}(T)$. The dimension of $\ker(T)$ is called the *nullity* of T and is denoted by $\text{null}(T)$.

In the next theorem we prove the dimension theorem using linear transformations. The proof illustrates the value of studying subspaces and linear transformations as opposed to always computing with matrices.

Theorem 55. *Suppose that A is an $m \times n$ matrix and consider the linear transformation $T_A : \mathbf{R}^n \to \mathbf{R}^m$. Then*

(i) $\ker(T)$ *is the set of solutions to* $AX = \vec{0}$, $\text{im}(T)$ *is the column space of* A, *and* $\text{rk}(T) = \text{rk}(A)$.

(ii) *(Dimension theorem)* $\text{rk}(T) + \text{null}(T) = n$.

Proof. (i) We have already noted that $T_A(\vec{v}) = \vec{0}$ if and only if $A\vec{v} = \vec{0}$. Now note that a vector $\vec{u} \in \mathbf{R}^m$ lies in $\text{im}(T)$ if and only if there exists some $\vec{v} \in \mathbf{R}^n$ such that $\vec{u} = A\vec{v}$. According to Sec. 5.1, this occurs if and only if $\vec{u}$ is a linear combination of the columns of A, that is, if and only if $\vec{u}$ lies in the column space of A. According to Theorem 31, $\text{rk}(A)$ is the dimension of the column space of A. This proves (i).

(ii) Let $\{\vec{w}_1, \vec{w}_2, \ldots, \vec{w}_s\}$ be a basis for $\ker(T_A)$. Suppose that $\vec{w}_{s+1}, \vec{w}_{s+2}, \ldots, \vec{w}_n$ extend these vectors to a basis of $\mathbf{R}^n$, that is, $\{\vec{w}_1, \vec{w}_2, \ldots, \vec{w}_n\}$ is now a basis for $\mathbf{R}^n$ (see the group project at the end of Sec. 5.3 for discussion of this). We claim that $\{T_A(\vec{w}_{s+1}), T_A(\vec{w}_{s+2}), \ldots, T_A(\vec{w}_n)\}$ is a basis for $\text{im}(T_A)$.

Note that if $\vec{w} = a_1\vec{w}_1 + a_2\vec{w}_2 + \cdots + a_n\vec{w}_n \in \mathbf{R}^n$, then by linearity $T_A(\vec{w}) = a_1T_A(\vec{w}_1) + a_2T_A(\vec{w}_2) + \cdots + a_nT_A(\vec{w}_n) = a_{s+1}T_A(\vec{w}_{s+1}) + a_{s+2}T_A(\vec{w}_{s+2}) + \cdots + a_nT_A(\vec{w}_n)$, which shows that the vectors $T_A(\vec{w}_{s+1})$, $T_A(\vec{w}_{s+2}), \ldots, T_A(\vec{w}_n)$ span $\text{im}(T_A)$.

Next suppose that $a_{s+1}T_A(\vec{w}_{s+1}) + a_{s+2}T_A(\vec{w}_{s+2}) + \cdots + a_nT_A(\vec{w}_n) = \vec{0}$. Then $T_A(a_{s+1}\vec{w}_{s+1} + a_{s+2}\vec{w}_{s+2} + \cdots + a_n\vec{w}_n) = \vec{0}$, so we see that $a_{s+1}\vec{w}_{s+1} + a_{s+2}\vec{w}_{s+2} + \cdots + a_n\vec{w}_n \in \ker(T_A)$. Therefore, there exist real numbers $a_1, a_2, \ldots, a_s$ such that $a_1\vec{w}_1 + a_2\vec{w}_2 + \cdots + a_s\vec{w}_s = a_{s+1}\vec{w}_{s+1} + a_{s+2}\vec{w}_{s+2} + \cdots + a_n\vec{w}_n$. The linear independence of $\{\vec{w}_1, \vec{w}_2, \ldots, \vec{w}_n\}$ guarantees that $a_1 = 0, a_2 = 0, \ldots, a_n = 0$. This shows that $\{T_A(\vec{w_{s+1}}), T_A(\vec{w}_{s+2}), \ldots, T_A(\vec{w}_n)\}$ is linearly independent.

We have shown that if $\dim(\ker(T_A)) = s$, then $\dim(\text{im}(T_A)) = n - s$. The dimension theorem follows. □

For example, consider the linear transformation $T_A : \mathbf{R}^3 \to \mathbf{R}^5$ where

$$A = \begin{pmatrix} 5 & 0 & 3 \\ 4 & 2 & 8 \\ -1 & -2 & -3 \\ 0 & 3 & 2 \\ 2 & 10 & 10 \end{pmatrix}$$

has the following reduced row-echelon form:

$$R = \begin{pmatrix} 1 & 0 & \frac{5}{3} \\ 0 & 1 & \frac{2}{3} \\ 0 & 0 & 0 \\ 0 & 0 & 0 \\ 0 & 0 & 0 \end{pmatrix}.$$

Applying back-substitution, we find that

$$\ker(T_A) = \operatorname{span}\left\{ \begin{pmatrix} -\frac{5}{3} \\ -\frac{2}{3} \\ 1 \end{pmatrix} \right\}.$$

Using the fact that $\operatorname{im}(T_A) = \operatorname{col}(A)$, we find

$$\operatorname{im}(T_A) = \operatorname{span}\left\{ \begin{pmatrix} 5 \\ 4 \\ -1 \\ 0 \\ 1 \end{pmatrix}, \begin{pmatrix} 0 \\ 2 \\ -2 \\ 3 \\ 5 \end{pmatrix} \right\}.$$

So, $\operatorname{rk}(T_A) + \operatorname{null}(T_A) = 2 + 1 = 3$, as it should.

Linear Transformations and Systems of Equations

We have shown that the dimension theorem can be proved using properties of linear transformations. We next consider the connections between linear transformations and the nature of solutions to systems of equations.

Definition. A linear transformation $T : \mathbf{R}^n \to \mathbf{R}^m$ is called *one-one* if whenever $\vec{v}_1, \vec{v}_2 \in \mathbf{R}^n$ and $T(\vec{v}_1) = T(\vec{v}_2)$, then necessarily $\vec{v}_1 = \vec{v}_2$.

For example, consider the linear transformation $T_A : \mathbf{R}^2 \to \mathbf{R}^3$, where A is the 3×2 matrix

$$A = \begin{pmatrix} 1 & 2 \\ 1 & 1 \\ 0 & 1 \end{pmatrix}.$$

In order to determine if T_A is one-one, we must check to see that whenever $T_A(\vec{v}_1) = T_A(\vec{v}_2)$, necessarily $\vec{v}_1 = \vec{v}_2$. So we suppose that $A\vec{v}_1 = A\vec{v}_2$ for $\vec{v}_1, \vec{v}_2 \in \mathbf{R}^2$. Explicitly, we have

$$A\vec{v}_1 = \begin{pmatrix} 1 & 2 \\ 1 & 1 \\ 0 & 1 \end{pmatrix} \begin{pmatrix} r \\ s \end{pmatrix} = \begin{pmatrix} 1 & 2 \\ 1 & 1 \\ 0 & 1 \end{pmatrix} \begin{pmatrix} t \\ u \end{pmatrix} = A\vec{v}_2.$$

Multiplying this expression out shows that

$$\begin{pmatrix} r + 2s \\ r + s \\ s \end{pmatrix} = \begin{pmatrix} t + 2u \\ t + u \\ u \end{pmatrix}.$$

Inspecting the bottom entry gives $s = u$, and then the remaining entries show that $r = t$. This shows $\vec{v}_1 = \vec{v}_2$ and consequently that T_A is one-one.

Theorem 56. *Suppose that $T : \mathbf{R}^n \to \mathbf{R}^m$ is a linear transformation. Then T is one-one if and only if $\ker(T) = \{\vec{0}\}$. In particular, if A is an $m \times n$ matrix, then T_A is one-one if and only if the homogeneous system $A\vec{X} = \vec{0}$ has only the zero solution.*

Proof. Suppose that T is one-one. If $T(\vec{v}) = \vec{0}$, then $T(\vec{v}) = T(\vec{0}) = \vec{0}$. Since T is one-one, we must have $\vec{v} = \vec{0}$. Hence $\ker(T) = \{\vec{0}\}$. Conversely, assume $\ker(T) = \{\vec{0}\}$ and $T(\vec{v}_1) = T(\vec{v}_2)$. Then $T(\vec{v}_1 - \vec{v}_2) = \vec{0}$; that is, $\vec{v}_1 - \vec{v}_2 \in \ker(T)$. Thus $\vec{v}_1 - \vec{v}_2 = \vec{0}$, and $\vec{v}_1 = \vec{v}_2$ follows. This shows that T is one-one. The second assertion follows since $\ker(T_A)$ is the set of solutions to the homogeneous system $A\vec{X} = \vec{0}$. □

Our next definition concerns the image of a linear transformation.

Definition. A linear transformation $T : \mathbf{R}^n \to \mathbf{R}^m$ is called *onto* if $\text{im}(T) = \mathbf{R}^m$.

In our preceding example we saw that $T_A : \mathbf{R}^2 \to \mathbf{R}^3$ had nullity 0. Applying Theorem 55, we find $\dim(\text{im}(T_A)) = \text{rk}(A) = 2$. Since $\mathbf{R}^3$ is a three-dimensional vector space, it is impossible for $\text{im}(T_A) = \mathbf{R}^3$. We conclude that T_A is not onto. The relationship between T_A being onto and systems of equations is spelled out next.

Theorem 57. *Suppose that $T : \mathbf{R}^n \to \mathbf{R}^m$ is a linear transformation. Then T is onto if and only if $\text{rk}(T) = m$. In particular, if A is an $m \times n$ matrix, then T_A is onto if and only if the system $A\vec{X} = \vec{v}$ can be solved for all $\vec{v} \in \mathbf{R}^m$.*

Proof. By definition, T is onto if and only if $\text{im}(T) = \mathbf{R}^m$. However, by Theorem 55, $\dim(\text{im}(T)) = \text{rk}(T)$. Thus, $\text{im}(T) = \mathbf{R}^m$ if and only if $\text{rk}(T) = \dim(\mathbf{R}^m) = m$. The second assertion follows since we know that $A\vec{X} = \vec{v}$ can be solved for $\vec{v}$ if and only if $\vec{v}$ lies in the column space of A. □

Finally, we consider the conditions of being one-one and onto simultaneously.

Definition. A linear transformation T is called an *isomorphism* if T is both one-one and onto. An isomorphism $T : \mathbf{R}^n \to \mathbf{R}^n$ is often called a *nonsingular linear operator* on $\mathbf{R}^n$.

In this situation, Theorems 56 and 57 combine to give the next result.

Theorem 58. *Suppose that $T : \mathbf{R}^n \to \mathbf{R}^m$ is a linear transformation. If T is an isomorphism, then $n = m$. In this case T^{-1} exists as a function and $T^{-1} : \mathbf{R}^n \to \mathbf{R}^n$ is also an isomorphism. In particular, if A is an $n \times n$ matrix, then T_A is invertible if and only if the system $A\vec{X} = \vec{v}$ always has a unique solution.*

Proof. If T is an isomorphism, we know by Theorems 56 and 57 that $\text{null}(T) = 0$ and $\text{rk}(T) = m$. By the dimension theorem we find that $0 + m = n$, giving the first statement. Suppose now that the standard matrix for T is A. Then A is invertible since it is an $n \times n$ rank-n matrix. Since $A\vec{v} = \vec{u}$ is equivalent to $\vec{v} = A^{-1}\vec{u}$, we see that the inverse function of T_A must be $T_{A^{-1}}$. Since $rk(A^{-1}) = n$, we see that $T_{A^{-1}}$ is also an isomorphism. The final assertion of the theorem follows from the invertibility of the matrix A. □

Note that Theorem 58 shows that a linear transformation T_A is invertible as a function if and only if the matrix A is invertible.

Problems

1. Show that the determinant of any cascade parameter matrix is always 1.
2. Find the standard matrix for the two-port network obtained by removing the batteries and connecting the two black boxes in Fig. 8.6 in reverse order. Put the batteries back on and determine the current flows.
3. Find the standard matrix, the rank, and the nullity of each of the following linear transformations.
   (a) $T_1 : \mathbf{R}^3 \to \mathbf{R}^4$ defined by $T_1(x, y, z) = (x + z, 3x + 3z, 2x + 2z, -x - z)$
   (b) $T_2 : \mathbf{R}^4 \to \mathbf{R}$ defined by $T_2(x, y, z, w) = y - z - w$

(c) $T_3 : \mathbf{R}^4 \to \mathbf{R}^4$ given by $T_3(\vec{v}) = \vec{v}$ for all $\vec{v} \in \mathbf{R}^4$

(d) $T_4 : \mathbf{R} \to \mathbf{R}$ where $T_4(r) = 7r$

(e) $T_5 : \mathbf{R}^5 \to \mathbf{R}^5$ given by $T_5(\vec{v}) = \vec{0}$ for all $\vec{v} \in \mathbf{R}^5$

4. Find the rank and nullity and bases for the range and kernel of T_A whenever A is

(a) $\begin{pmatrix} 1 & 0 & 0 & 0 \\ 0 & 1 & 0 & 0 \\ 1 & 1 & 0 & 0 \end{pmatrix}$ (b) $\begin{pmatrix} -1 & -1 \\ 2 & 2 \\ -3 & 3 \end{pmatrix}$ (c) $\begin{pmatrix} 1 & 1 & 1 & 1 & 1 \end{pmatrix}$.

5. Find a linear transformation $T : \mathbf{R}^3 \to \mathbf{R}^3$ such that

$$\operatorname{im}(T) = \operatorname{span}\left\{\begin{pmatrix} 1 \\ 1 \\ 1 \end{pmatrix}\right\}.$$

What is a basis for $\ker(T)$?

6. Suppose that $T : \mathbf{R}^4 \to \mathbf{R}^3$ is defined by

$$T\begin{pmatrix} x \\ y \\ z \\ w \end{pmatrix} = \begin{pmatrix} y - x + w \\ z - w \\ x + 2y \end{pmatrix}.$$

(a) Find the standard matrix of T.

(b) Find $\operatorname{rk}(T)$ and $\operatorname{null}(T)$.

(c) Is T one-one? Is T onto?

(d) Find a basis for $\ker(T)$ and $\operatorname{im}(T)$.

7. Suppose that $T : \mathbf{R}^3 \to \mathbf{R}^3$ satisfies $T(1, 0, 0) = (0, 1, 0)$, $T(0, 1, 0) = (0, 0, 1)$. What conditions must $T(0, 0, 1)$ satisfy so that T is invertible?

8. (a) Suppose that $T, U : \mathbf{R}^n \to \mathbf{R}^n$ are both one-one linear transformations. Show that $T \circ U$ is also one-one.

(b) Suppose that $T, U : \mathbf{R}^n \to \mathbf{R}^n$ are both onto linear transformations. Show that $T \circ U$ is also onto.

9. (a) Show that a linear transformation $T : \mathbf{R}^3 \to \mathbf{R}^4$ can never be onto.

(b) Show that a linear transformation $T : \mathbf{R}^4 \to \mathbf{R}^3$ can never be one-one.

(c) Suppose that $T : \mathbf{R}^4 \to \mathbf{R}^3$ is linear and

$$\operatorname{im}(T) = \{(x, y, z) \mid x - y - 2z = 0\}.$$

What is $\operatorname{null}(T)$?

10. (a) Give an example of a linear operator $T : \mathbf{R}^2 \to \mathbf{R}^2$ for which $\ker(T) = \operatorname{im}(T)$.

(b) Does there exist a linear operator $T : \mathbf{R}^3 \to \mathbf{R}^3$ for which $\ker(T) = \operatorname{im}(T)$?

11. Suppose a linear transformation $T : \mathbf{R}^3 \to \mathbf{R}^3$ satisfies $T \circ T = O$ (the zero operator). Is it true that $T = O$? Show why or give a counterexample.

12. Suppose that A and B are $m \times n$ matrices. Show that $\mathrm{rk}(A) + \mathrm{rk}(B) \geq \mathrm{rk}(A + B)$ by studying the ranks of T_A, T_B, and T_{A+B}.

13. Suppose that $\{\vec{v}_1, \vec{v}_2, \ldots, \vec{v}_n\}$ is a basis for $\mathbf{R}^n$. Let $T : \mathbf{R}^n \to \mathbf{R}^n$ be the linear transformation defined by $T(\vec{v}_1) = \vec{v}_2$, $T(\vec{v}_2) = \vec{v}_3$, $\ldots$, $T(\vec{v}_{n-1}) = \vec{v}_n$, and $T(\vec{v}_n) = \vec{0}$. Give a basis for $\mathrm{im}(T)$, $\mathrm{im}(T \circ T)$, $\mathrm{im}(T \circ T \circ T)$, $\ldots$. Find the rank and nullity of T^i for all i.

14. Assume that $T : \mathbf{R}^n \to \mathbf{R}^n$ is a linear transformation and $T = T \circ T$. Show that $\ker(T) + \mathrm{im}(T) = \mathbf{R}^n$.

Group Project: Network Analysis with Larger Systems of Equations

Analyze the network studied in Fig. 8.6 by letting five variables represent the current across each of the five resistors. You can obtain five equations between these variables: two from the voltage drops across the batteries and three from the junctions where resistors come together. (Recall both of Kirchoff's laws here.) Your answer will agree with those found in this section, but which two of the five you computed correspond to the I_1 and I_3 computed in the text?

Group Project: Two-Port Networks Connected in Series

Figure 8.7 shows two two-port black boxes connected in *series*. The matrix of the new network can be computed by adding the matrices corresponding to each box. But which matrices do you add, the standard or the cascade parameter matrices? Find out, and give a careful written explanation of your findings.

Fig. 8.7. Two black box two-port networks connected in series

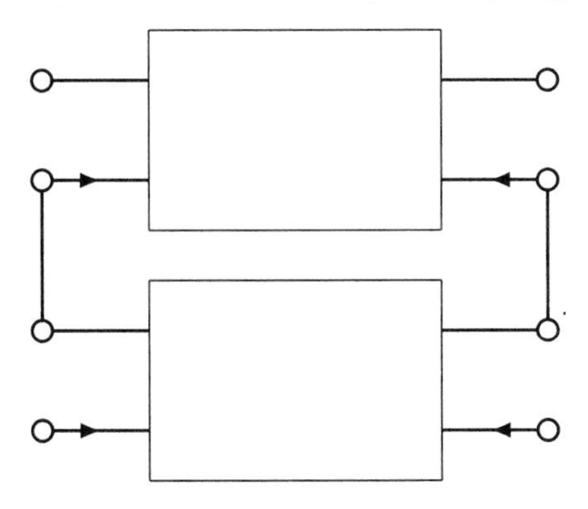

Group Project: Determining the Characterisitic Matrix of a Black Box

Suppose you have a two-port network as illustrated in Fig. 8.4 but you don't know what the resistances are and the resistors are hidden inside the black box. Assume you have a 10-volt battery and electronic equipment that enables you to measure the voltage and current flows at various ports. Explain how to use this equipment to determine the standard matrix for the black box. You may connect your battery across any two ports, measure current or voltage across the other two, and you can short across any two ports with a wire. However, you can't open the box!

Group Project: Matrix Multiplication and Rank

The following questions relate matrix ranks and products. The key in each case is to think about the linear transformations involved, not just matrices.

1. Suppose that A is an $n \times n$ matrix such that $A^2 = 0$.
   (a) Show that $\mathrm{im}(T_A) \subseteq \ker(T_A)$.
   (b) Show $\mathrm{rk}(A) \leq \frac{n}{2}$.
2. (a) Suppose that C is a $k \times p$ matrix and D is a $p \times r$ matrix. Show that $\mathrm{im}(T_{CD})$ is a subspace of $\mathrm{im}(T_C)$.
   (b) Using (a), conclude that whenever C is a $k \times p$ matrix and D is a $p \times r$ matrix, $\mathrm{rk}(CD) \leq \mathrm{rk}(C)$.
3. Suppose that A and B are matrices and the product AB makes sense. Show that $\mathrm{null}(AB) \geq \mathrm{null}(B)$ by showing that $\ker(T_B) \subseteq \ker(T_{AB})$.
4. If A is invertible, show that AB and B have the same nullity and rank.
5. The concept of a linear transformation helped you solve parts 1–4 much more easily than if you tried to use matrices and row operations. Why is this?

8.3 Change of Basis

In the previous sections we saw how to relate information about an $m \times n$ matrix A to the linear transformation $T_A : \mathbf{R}^n \to \mathbf{R}^m$. In this section we shall study linear transformations $T : \mathbf{R}^n \to \mathbf{R}^n$ and *different* $n \times n$ matrices that can represent the same operator T. To obtain these different representations, we will use different bases for $\mathbf{R}^n$.

Coordinates with Respect to a Basis

We recall that if $\{\vec{v}_1, \vec{v}_2, \ldots, \vec{v}_n\}$ is a basis for $\mathbf{R}^n$, then for every vector $\vec{w} \in \mathbf{R}^n$ there is a unique sequence of real numbers $a_1, a_2, \ldots, a_n$ such that

$\vec{w} = a_1\vec{v}_1 + a_2\vec{v}_2 + \cdots + a_n\vec{v}_n$. Because of this important property, it makes sense to give the following definition.

Definition. An *ordered basis* $\mathcal{B} = \{\vec{v}_1, \vec{v}_2, \ldots, \vec{v}_n\}$ for $\mathbf{R}^n$ is a basis of $\mathbf{R}^n$ whose elements are listed in a specific order. If $\mathcal{B}$ is an ordered basis for $\mathbf{R}^n$ and $\vec{u} \in \mathbf{R}^n$, we say that $(a_1, a_2, \ldots, a_n)$ are the *$\mathcal{B}$-coordinates of $\vec{u}$* if $\vec{u} = a_1\vec{v}_1 + a_2\vec{v}_2 + \cdots + a_n\vec{v}_n$. We denote this by the column

$$(\vec{u})_{\mathcal{B}} = \begin{pmatrix} a_1 \\ a_2 \\ \vdots \\ a_n \end{pmatrix}.$$

We emphasize that the $\mathcal{B}$-coordinates of a vector $\vec{u}$ depend not only on the choice of the basis but also on the order in which the basis elements are listed.

First we note that if $\mathcal{S} = \{\vec{e}_1, \vec{e}_2, \ldots, \vec{e}_n\}$ is the standard basis for $\mathbf{R}^n$, then $\mathcal{S}$-coordinates are the usual coordinates. In our notation, this says that if $P = (a_1, a_2, \ldots, a_n) \in \mathbf{R}^n$ and $\vec{u} = \overrightarrow{OP}$, then $\vec{u} = a_1\vec{e}_1 + a_2\vec{e}_2 + \cdots + a_n\vec{e}_n$ and

$$(\vec{u})_{\mathcal{S}} = \begin{pmatrix} a_1 \\ a_2 \\ \vdots \\ a_n \end{pmatrix}.$$

Now consider the ordered basis

$$\mathcal{B} = \left\{ \begin{pmatrix} 1 \\ 1 \\ 0 \end{pmatrix}, \begin{pmatrix} 1 \\ 0 \\ 1 \end{pmatrix}, \begin{pmatrix} 0 \\ 1 \\ 1 \end{pmatrix} \right\}$$

of $\mathbf{R}^3$. To find the coordinates of the vector $(1, 0, 0)$ with respect to $\mathcal{B}$, we must express

$$\begin{pmatrix} 1 \\ 0 \\ 0 \end{pmatrix} = a_1 \begin{pmatrix} 1 \\ 1 \\ 0 \end{pmatrix} + a_2 \begin{pmatrix} 1 \\ 0 \\ 1 \end{pmatrix} + a_3 \begin{pmatrix} 0 \\ 1 \\ 1 \end{pmatrix}$$

for appropriate real numbers a_1, a_2, a_3. Using the definition of matrix multiplication this is the same as

$$\begin{pmatrix} 1 & 1 & 0 \\ 1 & 0 & 1 \\ 0 & 1 & 1 \end{pmatrix} \begin{pmatrix} a_1 \\ a_2 \\ a_3 \end{pmatrix} = \begin{pmatrix} 1 \\ 0 \\ 0 \end{pmatrix}.$$

A calculation shows

$$\begin{pmatrix} 1 & 1 & 0 \\ 1 & 0 & 1 \\ 0 & 1 & 1 \end{pmatrix}^{-1} = \frac{1}{2}\begin{pmatrix} 1 & 1 & -1 \\ 1 & -1 & 1 \\ -1 & 1 & 1 \end{pmatrix},$$

and therefore we can find the a_i as

$$\begin{pmatrix} 1 \\ 0 \\ 0 \end{pmatrix}_{\mathcal{B}} = \begin{pmatrix} a_1 \\ a_2 \\ a_3 \end{pmatrix} = \frac{1}{2}\begin{pmatrix} 1 & 1 & -1 \\ 1 & -1 & 1 \\ -1 & 1 & 1 \end{pmatrix}\begin{pmatrix} 1 \\ 0 \\ 0 \end{pmatrix} = \begin{pmatrix} \frac{1}{2} \\ \frac{1}{2} \\ -\frac{1}{2} \end{pmatrix}.$$

We obtain

$$\begin{pmatrix} 1 \\ 0 \\ 0 \end{pmatrix}_{\mathcal{B}} = \begin{pmatrix} \frac{1}{2} \\ \frac{1}{2} \\ -\frac{1}{2} \end{pmatrix}.$$

More generally, this same calculation shows that whenever $(x, y, z) \in \mathbf{R}^3$,

$$\begin{pmatrix} x \\ y \\ z \end{pmatrix}_{\mathcal{B}} = \frac{1}{2}\begin{pmatrix} 1 & 1 & -1 \\ 1 & -1 & 1 \\ -1 & 1 & 1 \end{pmatrix}\begin{pmatrix} x \\ y \\ z \end{pmatrix} = \frac{1}{2}\begin{pmatrix} x + y - z \\ x - y + z \\ -x + y + z \end{pmatrix}.$$

The Coordinate-Change Matrix

The previous example shows that the coordinates of a vector in $\mathbf{R}^n$ with respect to a nonstandard basis can be determined by matrix multiplication. This is spelled out in the next theorem.

Theorem 59. *Let $\mathcal{B} = \{\vec{v}_1, \vec{v}_2, \ldots, \vec{v}_n\}$ be an ordered basis of $\mathbf{R}^n$. Set P to be the matrix $(\vec{v}_1 \quad \vec{v}_2 \quad \cdots \quad \vec{v}_n)$; that is, P is the matrix whose columns are the standard coordinates of the vectors in $\mathcal{B}$. Then P is invertible, and for any $\vec{v} \in \mathbf{R}^n$, $(\vec{v})_{\mathcal{B}} = P^{-1}\vec{v}$ and $P(\vec{v})_{\mathcal{B}} = \vec{v}$.*

The theorem motivates the following definition.

Definition. The matrix P^{-1} characterized by $(\vec{v})_{\mathcal{B}} = P^{-1}\vec{v}$ is called the *coordinate-change matrix* or *transition* from the standard basis to the basis $\mathcal{B}$. The matrix P (without the inverse) characterized by $P(\vec{v})_{\mathcal{B}} = \vec{v}$ is called the coordinate-change matrix from the basis $\mathcal{B}$ to the standard basis.

The existence of coordinate-change matrices shows immediately (using matrix algebra) that if $\mathcal{B} = \{\vec{v}_1, \vec{v}_2, \ldots, \vec{v}_n\}$ is an ordered basis of $\mathbf{R}^n$ and if $\vec{w}_1, \vec{w}_2 \in \mathbf{R}^n$, then

$$(\vec{w}_1 + \vec{w}_2)_{\mathcal{B}} = (\vec{w}_1)_{\mathcal{B}} + (\vec{w}_2)_{\mathcal{B}} \quad \text{and} \quad (k\vec{w}_1)_{\mathcal{B}} = k(\vec{w}_1)_{\mathcal{B}}.$$

As an example, consider the ordered basis

$$\mathcal{B} = \left\{ \begin{pmatrix} 1 \\ 2 \end{pmatrix}, \begin{pmatrix} 2 \\ 5 \end{pmatrix} \right\}$$

of $\mathbf{R}^2$. We set

$$P = \begin{pmatrix} 1 & 2 \\ 2 & 5 \end{pmatrix} \quad \text{and note} \quad P^{-1} = \begin{pmatrix} 5 & -2 \\ -2 & 1 \end{pmatrix}.$$

Then P^{-1} is the coordinate-change matrix from the standard basis to $\mathcal{B}$. This shows that

$$\left(\begin{pmatrix} 1 \\ 1 \end{pmatrix} \right)_{\mathcal{B}} = P^{-1} \begin{pmatrix} 1 \\ 1 \end{pmatrix} = \begin{pmatrix} 3 \\ -1 \end{pmatrix}.$$

We can check this by expressing the vector as a linear combination of the elements of the basis $\mathcal{B}$:

$$\begin{pmatrix} 1 \\ 1 \end{pmatrix} = 3 \begin{pmatrix} 1 \\ 2 \end{pmatrix} + (-1) \begin{pmatrix} 2 \\ 5 \end{pmatrix}.$$

The Matrix of a Linear Operator with Respect to a Basis

In Sec. 7.2 we discussed criteria for a matrix to be diagonalizable. Let's find out what this means from the geometric point of view.

Definition. Suppose that $T : \mathbf{R}^n \to \mathbf{R}^n$ is a linear operator and $\mathcal{B} = \{\vec{v}_1, \vec{v}_2, \ldots, \vec{v}_n\}$ is an ordered basis for $\mathbf{R}^n$. We define the $n \times n$ matrix $[T]_{\mathcal{B}}$ by $[T]_{\mathcal{B}} = (a_{ij})$, where for each j the real numbers a_{ij} are uniquely determined by the equations

$$T(\vec{v}_j) = a_{1j}\vec{v}_1 + a_{2j}\vec{v}_2 + \cdots + a_{nj}\vec{v}_n.$$

In other words, $[T]_{\mathcal{B}}$ is the matrix whose jth column is the coordinates of $T(\vec{v}_j)$ with respect to the basis $\mathcal{B}$.

Note that before the matrix representation $[T]_{\mathcal{B}}$ could be defined, we had to *fix* the basis $\mathcal{B}$. This is crucial, for as we shall see later a change in the choice of bases can wildly change the matrix $[T]_{\mathcal{B}}$.

The definition gives a recipe for computing the matrix $[T]_{\mathcal{B}}$. For example, suppose that $T : \mathbf{R}^3 \to \mathbf{R}^3$ is the linear transformation defined by $T(x, y, z) = (x - y, z + y, 0)$. Let $\mathcal{B} = \{(1, 1, 0), (0, 1, 1), (0, 0, 1)\}$ be an ordered basis for $\mathbf{R}^3$. To compute $[T]_{\mathcal{B}}$, we must compute the $\mathcal{B}$-coordinates of the images of the basis elements in $\mathcal{B}$ and list these coordinates as the columns of the matrix $[T]_{\mathcal{B}}$. Direct calculation shows

$$T(1, 1, 0) = (0, 1, 0) = (0, 1, 1) - (0, 0, 1),$$

$$T(0, 1, 1) = (-1, 2, 0) = -(1, 1, 0) + 3(0, 1, 1) - 3(0, 0, 1),$$

$$T(0, 0, 1) = (0, 1, 0) = (0, 1, 1) - (0, 0, 1).$$

We find that $(T(1,1,0))_{\mathcal{B}} = (0,1,-1)$, $(T(0,1,1))_{\mathcal{B}} = (-1,3,-3)$, and $(T(0,0,1))_{\mathcal{B}} = (0,1,-1)$. (Do not forget this second step of computing the $\mathcal{B}$-coordinates!) From this we obtain

$$[T]_{\mathcal{B}} = \begin{pmatrix} 0 & -1 & 0 \\ 1 & 3 & 1 \\ -1 & -3 & -1 \end{pmatrix}.$$

Consider the special case where $T : \mathbf{R}^n \to \mathbf{R}^n$ and $\mathcal{S}$ is the standard basis for $\mathbf{R}^n$. Then the matrix $[T]_{\mathcal{S}}$ just defined is precisely the standard matrix of T; that is, $[T]_{\mathcal{S}} = A$, where $T = T_A$. This occurs because the recipe just described gives the standard matrix for T whenever the standard bases are used.

The Defining Property of $[T]_{\mathcal{B}}$

This next result shows that the matrix $[T]_{\mathcal{B}}$ describes the linear operator $T : \mathbf{R}^n \to \mathbf{R}^n$ by matrix multiplication, except that all vector representations must be made by their $\mathcal{B}$-coordinates. The proof is analogous to the case considered in Theorem 53 and is thus omitted.

Theorem 60. *Let $T : \mathbf{R}^n \to \mathbf{R}^n$ be a linear transformation, and suppose $\mathcal{B}$ is an ordered basis for $\mathbf{R}^n$. Then for all $\vec{v} \in \mathbf{R}^n$,*

$$[T]_{\mathcal{B}}(\vec{v})_{\mathcal{B}} = (T(\vec{v}))_{\mathcal{B}}.$$

Moreover, this condition uniquely determines the matrix $[T]_{\mathcal{B}}$.

To illustrate the theorem we consider the ordered basis $\mathcal{B} = \{(3,1), (-1,1)\}$ of $\mathbf{R}^2$. Suppose that $T : \mathbf{R}^2 \to \mathbf{R}^2$ is defined by $T(x,y) = (x+3y, x-y)$. We compute that $T(3,1) = (6,2) = 2(3,1)$ and $T(-1,1) = (2,-2) = -2(-1,1)$. Therefore, the definition shows that

$$[T]_{\mathcal{B}} = \begin{pmatrix} 2 & 0 \\ 0 & -2 \end{pmatrix}.$$

This representation for T is quite nice since it is a diagonal matrix.

The theorem shows that the value of T can be determined from the $\mathcal{B}$-coordinates of any vector in $\mathbf{R}^2$. For example,

$$\begin{pmatrix} 4 \\ 4 \end{pmatrix}_{\mathcal{B}} = \begin{pmatrix} 2 \\ 2 \end{pmatrix},$$

and therefore the $\mathcal{B}$-coordinates of $T(4,4)$ are given by

$$[T]_{\mathcal{B}} \begin{pmatrix} 4 \\ 4 \end{pmatrix}_{\mathcal{B}} = \begin{pmatrix} 2 & 0 \\ 0 & -2 \end{pmatrix} \begin{pmatrix} 2 \\ 2 \end{pmatrix} = \begin{pmatrix} 4 \\ -4 \end{pmatrix} = \left(T \begin{pmatrix} 4 \\ 4 \end{pmatrix} \right)_{\mathcal{B}}.$$

Unravelling all of this we can conclude that $T(4,4) = 4(3,1) - 4(-1,1) = (16,0)$, which can be checked by direct computation using the definition of T.

Basis Change for Operators

It is important to learn to work with different bases when studying $n \times n$ matrices (which we view as linear operators on $\mathbf{R}^n$). Suppose that a linear operator $T : \mathbf{R}^n \to \mathbf{R}^n$ is specified by a matrix $[T]$ in terms of the standard basis. Let $\mathcal{B}$ be an ordered basis for $\mathbf{R}^n$. An important problem is finding the matrix $[T]_{\mathcal{B}}$ for this operator in terms of the matrix $[T]$. The next theorem will solve this problem. This result is also the key tool needed to understand the geometry of linear operators.

Theorem 61. *Suppose that $T : \mathbf{R}^n \to \mathbf{R}^n$ is a linear operator and $\mathcal{B}$ is an ordered basis of $\mathbf{R}^n$. Let P^{-1} denote the transition matrix from the standard basis to the ordered basis $\mathcal{B}$. Then*

$$[T] = P[T]_{\mathcal{B}}P^{-1} \quad \text{or} \quad P^{-1}[T]P = [T]_{\mathcal{B}}.$$

Proof. Since P^{-1} is the transition matrix from the standard basis to the ordered basis $\mathcal{B}$, we know for all $\vec{v} \in \mathbf{R}^n$ that

$$P^{-1}\vec{v} = (\vec{v})_{\mathcal{B}} \quad \text{and} \quad \vec{v} = P(\vec{v})_{\mathcal{B}}.$$

Applying the associativity of matrix multiplication together with the last theorem, we find

$$P[T]_{\mathcal{B}_1}P^{-1}\vec{v} = P[T]_{\mathcal{B}}(\vec{v})_{\mathcal{B}} = P(T(\vec{v}))_{\mathcal{B}} = T(\vec{v}).$$

This shows that the matrix $P[T]_{\mathcal{B}}P^{-1}$ satisfies the defining property of $[T]$ given in Theorem 53, and consequently they are equal. □

The theorem motivates the following definition.

Definition. Suppose that A and B are $n \times n$ matrices. If there exists an invertible matrix P such that $A = PBP^{-1}$, we say that A and B are *similar*.

Example. Consider $T : \mathbf{R}^3 \to \mathbf{R}^3$ defined by $T(a,b,c) = (a+b+c, 2(b+c), 3c)$. With respect to the standard basis $\mathcal{S}$ we have

$$[T] = \begin{pmatrix} 1 & 1 & 1 \\ 0 & 2 & 2 \\ 0 & 0 & 3 \end{pmatrix}.$$

Now consider the basis $\mathcal{C} = \{(1,0,0),(1,1,0),(3,4,2)\}$ for $\mathbf{R}^3$. The transition matrix from $\mathcal{C}$ to $\mathcal{S}$ is the matrix whose columns are the standard coordinates of the elements of $\mathcal{C}$; that is, it is

$$P = \begin{pmatrix} 1 & 1 & 3 \\ 0 & 1 & 4 \\ 0 & 0 & 2 \end{pmatrix}.$$

The transition matrix from $\mathcal{S}$ to $\mathcal{C}$ is given by

$$P^{-1} = \begin{pmatrix} 1 & -1 & \frac{1}{2} \\ 0 & 1 & -2 \\ 0 & 0 & \frac{1}{2} \end{pmatrix}.$$

The theorem now gives

$$\begin{aligned} [T]_{\mathcal{C}} &= P^{-1}[T]P \\ &= \begin{pmatrix} 1 & -1 & \frac{1}{2} \\ 0 & 1 & -2 \\ 0 & 0 & \frac{1}{2} \end{pmatrix} \begin{pmatrix} 1 & 1 & 1 \\ 0 & 2 & 2 \\ 0 & 0 & 3 \end{pmatrix} \begin{pmatrix} 1 & 1 & 3 \\ 0 & 1 & 4 \\ 0 & 0 & 2 \end{pmatrix} \\ &= \begin{pmatrix} 1 & 0 & 0 \\ 0 & 2 & 0 \\ 0 & 0 & 3 \end{pmatrix}. \end{aligned}$$

The change of basis enables us to better understand the geometry of the linear operator just considered. It shows that the operator T stretches by factors of 1, 2, and 3 in the three directions specified by the first, second, and third basis vectors of $\mathcal{C}$, respectively. In other words, $\mathcal{C}$ is an eigenbasis for T with eigenvalues 1, 2, and 3. The reader should compare this calculation with those presented in Sec. 7.2. When we diagonalized matrices there, we were really constructing the matrix of the operator in an eigenbasis. The only difference is that in Sec. 7.2 we didn't use the terminology of linear operators and their matrices with respect to a basis.

Change of Basis and Rotations

In Sec. 8.1 we noted that the rotation $T_{\theta,x}$ by θ degrees around the x-axis in $\mathbf{R}^3$ had as standard matrix

$$R_{\theta,x} = \begin{pmatrix} 1 & 0 & 0 \\ 0 & \cos(\theta) & -\sin(\theta) \\ 0 & \sin(\theta) & \cos(\theta) \end{pmatrix}.$$

Suppose we needed to find the standard matrix of the rotation by θ degrees around the line ℓ through the origin and $(1, 1, 0)$ in $\mathbf{R}^3$. How could we find this matrix? One method is to consider the basis

$$\mathcal{B} = \left\{ \begin{pmatrix} \frac{1}{\sqrt{2}} \\ \frac{1}{\sqrt{2}} \\ 0 \end{pmatrix}, \begin{pmatrix} \frac{1}{\sqrt{2}} \\ -\frac{1}{\sqrt{2}} \\ 0 \end{pmatrix}, \begin{pmatrix} 0 \\ 0 \\ 1 \end{pmatrix} \right\}$$

and compute the matrix $[T_{\theta,x}]_{\mathcal{B}}$. This is the matrix

$$\begin{pmatrix} \frac{1}{\sqrt{2}} & \frac{1}{\sqrt{2}} & 0 \\ \frac{1}{\sqrt{2}} & -\frac{1}{\sqrt{2}} & 0 \\ 0 & 0 & 1 \end{pmatrix} \begin{pmatrix} 1 & 0 & 0 \\ 0 & \cos(\theta) & -\sin(\theta) \\ 0 & \sin(\theta) & \cos(\theta) \end{pmatrix} \begin{pmatrix} \frac{1}{\sqrt{2}} & \frac{1}{\sqrt{2}} & 0 \\ \frac{1}{\sqrt{2}} & -\frac{1}{\sqrt{2}} & 0 \\ 0 & 0 & 1 \end{pmatrix}$$

$$= \begin{pmatrix} \frac{1+\cos(\theta)}{2} & \frac{1-\cos(\theta)}{2} & \frac{-\sin(\theta)}{\sqrt{2}} \\ \frac{1-\cos(\theta)}{2} & \frac{1+\cos(\theta)}{2} & \frac{\sin(\theta)}{\sqrt{2}} \\ \frac{\sin(\theta)}{\sqrt{2}} & \frac{-\sin(\theta)}{\sqrt{2}} & \cos(\theta) \end{pmatrix}.$$

In order to understand why the process works, we view the transition matrix P as giving the linear transformation T_P. The transformation T_P moves the line ℓ to the x-axis and takes the plane perpendicular to ℓ to the plane perpendicular to the x-axis. If we rotate by $T_{\theta,x}$ after applying T_P, and then apply T_P^{-1}, the composite will have the affect of rotating around ℓ by θ. The matrix of this composite is the same as the matrix $[T_{\theta,x}]_{\mathcal{B}}$ since it is the product of the three matrices representing these operators.

The point of view of basis change just presented is different than considered previously. It is useful in a number of applied settings and is explored in more detail in the final group project of this section.

Problems

1. For each of the following ordered bases, find the transition matrix from the basis $\mathcal{B}$ to the standard basis and find the $\mathcal{B}$-coordinates of the vectors listed.
   (a) $\mathcal{B} = \{(1,2),(2,1)\}$ for $\mathbf{R}^2$. Vectors: $(2,4)$, $(4,2)$, $(3,3)$.
   (b) $\mathcal{B} = \{(1,1),(0,1)\}$ for $\mathbf{R}^2$. Vectors: $(17,4)$, $(3,1)$, $(0,0)$.
   (c) $\mathcal{B} = \{(1,1,1),(0,1,1),(0,0,1)\}$ for $\mathbf{R}^3$. Vectors: $(3,1,1)$, $(1,0,0)$.
2. For each of the following ordered bases, find the transition matrix from the basis $\mathcal{C}$ to the standard basis and find the $\mathcal{C}$-coordinates of the vectors listed.
   (a) $\mathcal{C} = \{(r,s),(t,u)\}$ for $\mathbf{R}^2$. Vectors: $(2,4)$, $(4,2)$, (a,b).
   (b) $\mathcal{C} = \{(1,-1,0),(0,0,1),(1,1,0)\}$ for $\mathbf{R}^3$. Vectors: $(3,1,1)$, $(1,0,0)$.
   (c) $\mathcal{C} = \{(1,0,0,0),(1,1,0,0),(1,1,1,0),(1,1,1,1)\}$ for $\mathbf{R}^4$.
   Vector: $(1,0,2,1)$.
3. Show that $[T]_{\mathcal{B}}$ is the zero matrix for every ordered basis $\mathcal{B}$ of $\mathbf{R}^n$ if and only if $T(\vec{v}) = \vec{0}$ for all $\vec{v} \in \mathbf{R}^n$.
4. Suppose that T is the linear transformation with standard matrix

$$\begin{pmatrix} 1 & 2 & 1 \\ 1 & 3 & 1 \\ 0 & 1 & 0 \end{pmatrix}.$$

Find $[T]_{\mathcal{B}}$ if $\mathcal{B}$ is the basis

(a) $\mathcal{B} = \{(1,2,0),(2,1,0),(0,0,1)\}$;

(b) $\mathcal{B} = \{(1,1,0),(0,1,1),(1,0,1)\}$;

(c) $\mathcal{B} = \{(1,1,1),(1,1,0),(0,0,1)\}$.

5. Let $\mathcal{B} = \{\vec{v}_1, \vec{v}_2, \vec{v}_3\}$ be an ordered basis of $\mathbf{R}^3$. Assume that $T : \mathbf{R}^3 \to \mathbf{R}^3$ is a linear transformation for which $T(\vec{v}_1) = \vec{v}_1 + \vec{v}_2$, $T(\vec{v}_2) = \vec{v}_2 + \vec{v}_3$, and $T(\vec{v}_3) = \vec{v}_1 + \vec{v}_2$.

   (a) Find $[T]_{\mathcal{B}}$.

   (b) If $\mathcal{B}'$ is the ordered basis $\mathcal{B}' = \{\vec{v}_3, \vec{v}_2, \vec{v}_1\}$, find $[T]_{\mathcal{B}'}$.

6. Let $T : \mathbf{R}^3 \to \mathbf{R}^3$ be defined by $T(a,b,c) = (a+c, 0, b)$. If $\mathcal{B} = \{(1,1,0),(0,1,1),(1,0,1)\}$ is an ordered basis for $\mathbf{R}^3$,

   (a) find $[T]$;

   (b) find $[T]_{\mathcal{B}}$;

   (c) find an invertible matrix P so that $P[T]_{\mathcal{B}}P^{-1} = [T]_{\mathcal{S}}$.

Group Project: Matrix Representations from Transformation Properties

Suppose that $\mathcal{B} = \{\vec{v}_1, \vec{v}_2, \ldots, \vec{v}_n\}$ is an ordered basis of $\mathbf{R}^n$.

(a) Define $T : \mathbf{R}^n \to \mathbf{R}^n$ by $T(\vec{v}_1) = \vec{v}_2$, $T(\vec{v}_2) = \vec{v}_3$, $\ldots$, $T(\vec{v}_{n-1}) = \vec{v}_n$, and $T(\vec{v}_n) = \vec{0}$. Find $[T]_{\mathcal{B}}$. What is $[T^i]_{\mathcal{B}}$ for $i \geq 1$?

(b) Define $T : \mathbf{R}^n \to \mathbf{R}^n$ by $T(\vec{v}_1) = \vec{v}_2, T(\vec{v}_2) = \vec{v}_3, \ldots, T(\vec{v}_{n-1}) = \vec{v}_n$, and $T(\vec{v}_n) = a_1\vec{v}_1 + a_2\vec{v}_2 + \cdots + a_n\vec{v}_n$. Find $[T]_{\mathcal{B}}$. When is this T one-one? When is this T onto?

(c) Suppose that $k \in \mathbf{R}$, $T(\vec{v}_1) = k\vec{v}_1$, and $T(\vec{v}_i) = k\vec{v}_i + \vec{v}_{i-1}$ for $i > 1$. Find $[T]_{\mathcal{B}}$. What are the eigenvalues of T? Is T diagonalizable?

Group Project: Matrix Representations for Rotations and Reflections

In this project you will have to extend the ideas used in the last part Sec. 8.3.

(a) Begin by discussing the peculiar way the discussion is set up there. Our first viewpoint was that a change of basis enables you to find the matrix of the operator with respect to a new basis, while in this later discussion the viewpoint taken is that the matrix is a composition of three linear operators. Why is this done? What are the advantages of each point of view? How are they unified?

Using these ideas find the standard matrix representations for the following linear operators on $\mathbf{R}^3$:

(b) the reflection through the plane spanned by $(1,1,0)$ and the z-axis;

(c) the rotation by θ around the line through the origin and $(1,1,1)$;

(d) the rotation that moves the points $(1,0,0)$ and $(0,1,0)$ to the points $(\frac{1}{\sqrt{2}}, 0, \frac{1}{\sqrt{2}})$ and $(\frac{1}{\sqrt{2}}, 0, -\frac{1}{\sqrt{2}})$, respectively.

ORTHOGONALITY AND LEAST-SQUARES PROBLEMS

9.1 Orthogonality and the Gram-Schmidt Process

In this chapter we extend the geometric ideas developed in Chap. 6 from $\mathbf{R}^2$ and $\mathbf{R}^3$ to $\mathbf{R}^n$. As an application we will develop the ideas behind least-squares approximations to data.

The Dot Product in $\mathbf{R}^n$

Throughout this chapter we shall use the brackets $\langle\ ,\ \rangle$ to represent the dot product. As pointed out in Sec. 6.1, one reason for this is to eliminate possible confusion of the "·" with scalar or real number multiplication. In Chap. 6 we only defined the dot product for $\mathbf{R}^2$ and $\mathbf{R}^3$. For general $\mathbf{R}^n$ the idea is the same, but we record the definition for emphasis.

Definition. Suppose $\vec{u} = (a_1, a_2, \ldots, a_n)$ and $\vec{v} = (b_1, b_2, \ldots, b_n)$ are vectors in $\mathbf{R}^n$. Then we define the *dot product* (or *inner product*) of $\vec{u}$

and $\vec{v}$ by

$$\langle \vec{u}, \vec{v} \rangle = a_1 b_1 + a_2 b_2 + \cdots + a_n b_n .$$

Often the dot product of $\vec{u}$ and $\vec{v}$ is denoted by $\vec{u} \cdot \vec{v}$. For any $\vec{u} \in \mathbf{R}^n$ we define the *norm* of $\vec{u}$ by $\|\vec{u}\| = \sqrt{\langle \vec{u}, \vec{u} \rangle} = \sqrt{a_1^2 + a_2^2 + \cdots + a_n^2}$.

It turns out that all of the properties derived for the dot product in $\mathbf{R}^2$ and $\mathbf{R}^3$ remain true for the dot product in $\mathbf{R}^n$. In particular, the entire statement of Theorem 34 holds verbatim. We will not repeat it here. However, we do recall from Theorem 35 that $\vec{u} \cdot \vec{v} = \|\vec{u}\| \|\vec{v}\| \cos(\theta)$, where θ is the angle between the two vectors $\vec{u}$ and $\vec{v}$. This result enables us to define the angle between two vectors $\vec{u}$ and $\vec{v}$ in $\mathbf{R}^n$ to be the angle θ, where $0 \leq \theta < \pi$ (we use radians, not degrees) and

$$\cos(\theta) = \frac{\langle \vec{u}, \vec{v} \rangle}{\|\vec{u}\| \|\vec{v}\|} .$$

This is useful, particularly because we do not have a geometric model of $\mathbf{R}^n$ whenever $n \geq 4$. Furthermore, this definition enables us to extend our definition of orthogonal from $\mathbf{R}^2$ and $\mathbf{R}^3$ to $\mathbf{R}^n$. We say that two nonzero vectors $\vec{u}$ and $\vec{v}$ are *orthogonal* if $\langle \vec{u}, \vec{v} \rangle = 0$, and we see that this occurs only when the angle between them is $\frac{\pi}{2}$.

Orthogonal and Orthonormal Bases

We now assume that V is a subspace of $\mathbf{R}^n$. We studied bases for V in Sec. 5.3. Often one wants bases whose elements are orthogonal (as is the standard basis for $\mathbf{R}^n$). Intuitively, orthogonal vectors point in different directions, and because of this they are linearly independent. This is given next.

Theorem 62. *Suppose that $\vec{v}_1, \vec{v}_2, \ldots, \vec{v}_m$ are mutually orthogonal vectors in $\mathbf{R}^n$. Then $\{\vec{v}_1, \vec{v}_2, \ldots, \vec{v}_m\}$ is linearly independent.*

Proof. We use the properties of the dot product given in Theorem 34. Suppose that $a_1\vec{v}_1 + a_2\vec{v}_2 + \cdots + a_n\vec{v}_n = \vec{0}$. It then follows for each i that

$$\begin{aligned} 0 &= \langle \vec{v}_i, \vec{0} \rangle \\ &= \langle \vec{v}_i, a_1\vec{v}_1 + a_2\vec{v}_2 + \cdots + a_n\vec{v}_n \rangle \\ &= \langle \vec{v}_i, a_1\vec{v}_1 \rangle + \langle \vec{v}_i, a_2\vec{v}_2 \rangle + \cdots + \langle \vec{v}_i, a_n\vec{v}_n \rangle \\ &= a_1\langle \vec{v}_i, \vec{v}_1 \rangle + a_2\langle \vec{v}_i, \vec{v}_2 \rangle + \cdots + a_n\langle \vec{v}_i, \vec{v}_n \rangle = a_i\langle \vec{v}_i, \vec{v}_i \rangle . \end{aligned}$$

Since $\vec{v}_i \neq \vec{0}$, we find that $\langle \vec{v}_i, \vec{v}_i \rangle \neq 0$ and consequently that $a_i = 0$. This shows that $\{\vec{v}_1, \vec{v}_2, \ldots, \vec{v}_m\}$ is linearly independent. □

Since orthogonal vectors are linearly independent, it makes sense to give the next definition.

Definition. Suppose that V is a subspace of $\mathbf{R}^n$. A basis $\{\vec{u}_1, \vec{u}_2, \ldots, \vec{u}_n\}$ is called an *orthogonal* basis for V if the $\vec{u}_i$ are mutually orthogonal; that is, $\langle \vec{u}_i, \vec{u}_j \rangle = 0$ whenever $i \neq j$. In addition, if each $\|\vec{u}_i\| = 1$ (that is, each $\vec{u}_i$ is a *unit vector*), we say that the basis is *orthonormal*.

The Key Property of Orthonormal Bases

The standard basis $\{\vec{e}_1, \vec{e}_2, \ldots, \vec{e}_n\}$ of $\mathbf{R}^n$ is orthonormal. Also, if $\vec{v} = a_1\vec{e}_1 + a_2\vec{e}_2 + \cdots + a_n\vec{e}_n$, then we have

$$\begin{aligned} \vec{v} &= a_1\vec{e}_1 + a_2\vec{e}_2 + \cdots + a_n\vec{e}_n \\ &= \langle \vec{v}, \vec{e}_1 \rangle \vec{e}_1 + \langle \vec{v}, \vec{e}_2 \rangle \vec{e}_2 + \cdots + \langle \vec{v}, \vec{e}_n \rangle \vec{e}_n. \end{aligned}$$

This result is true more generally.

Theorem 63. *Suppose that $\mathcal{B} = \{\vec{u}_1, \vec{u}_2, \ldots, \vec{u}_m\}$ is an orthonormal basis of a subspace V of $\mathbf{R}^n$. Then for any $\vec{v} \in V$,*

$$\vec{v} = a_1\vec{u}_1 + a_2\vec{u}_2 + \cdots + a_m\vec{u}_m,$$

where $a_i = \langle \vec{v}, \vec{u}_i \rangle$. In other words,

$$(\vec{v})_{\mathcal{B}} = \begin{pmatrix} \langle \vec{v}, \vec{u}_1 \rangle \\ \vdots \\ \langle \vec{v}, \vec{u}_m \rangle \end{pmatrix}.$$

Proof. Since $\{\vec{u}_1, \vec{u}_2, \ldots, \vec{u}_m\}$ is a basis of V, we know that there is a unique expression of the form $\vec{v} = a_1\vec{u}_1 + a_2\vec{u}_2 + \cdots + a_m\vec{u}_m$. We must show that $a_i = \langle \vec{v}, \vec{u}_i \rangle$. We compute as follows:

$$\begin{aligned} \langle \vec{v}, \vec{u}_i \rangle &= \langle a_1\vec{u}_1 + a_2\vec{u}_2 + \cdots + a_m\vec{u}_m, \vec{u}_i \rangle \\ &= \langle a_1\vec{u}_1, \vec{u}_i \rangle + \cdots + \langle a_i\vec{u}_i, \vec{u}_i \rangle + \cdots + \langle a_m\vec{u}_m, \vec{u}_i \rangle \\ &= a_1\langle \vec{u}_1, \vec{u}_i \rangle + \cdots + a_i\langle \vec{u}_i, \vec{u}_i \rangle + \cdots + a_m\langle \vec{u}_m, \vec{u}_i \rangle \\ &= a_1 0 + \cdots + a_i 1 + \cdots + a_m 0 \\ &= a_i. \end{aligned}$$

This proves the theorem. □

As an example, we consider the orthonormal basis

$$\mathcal{B} = \left\{ \begin{pmatrix} \frac{1}{\sqrt{2}} \\ 0 \\ \frac{1}{\sqrt{2}} \end{pmatrix}, \begin{pmatrix} \frac{1}{\sqrt{6}} \\ \frac{2}{\sqrt{6}} \\ -\frac{1}{\sqrt{6}} \end{pmatrix}, \begin{pmatrix} \frac{1}{\sqrt{3}} \\ -\frac{1}{\sqrt{3}} \\ -\frac{1}{\sqrt{3}} \end{pmatrix} \right\}$$

of $\mathbf{R}^3$ and we consider the vector

$$\vec{v} = \begin{pmatrix} 1 \\ 1 \\ 1 \end{pmatrix}.$$

Theorem 63 says

$$\begin{pmatrix} 1 \\ 1 \\ 1 \end{pmatrix} = \frac{2}{\sqrt{2}} \begin{pmatrix} \frac{1}{\sqrt{2}} \\ 0 \\ \frac{1}{\sqrt{2}} \end{pmatrix} + \frac{2}{\sqrt{6}} \begin{pmatrix} \frac{1}{\sqrt{6}} \\ \frac{2}{\sqrt{6}} \\ -\frac{1}{\sqrt{6}} \end{pmatrix} + \frac{-1}{\sqrt{3}} \begin{pmatrix} \frac{1}{\sqrt{3}} \\ -\frac{1}{\sqrt{3}} \\ -\frac{1}{\sqrt{3}} \end{pmatrix},$$

and consequently,

$$\begin{pmatrix} 1 \\ 1 \\ 1 \end{pmatrix}_{\mathcal{B}} = \begin{pmatrix} \frac{2}{\sqrt{2}} \\ \frac{2}{\sqrt{6}} \\ -\frac{1}{\sqrt{3}} \end{pmatrix}.$$

The Gram-Schmidt Theorem

Suppose that V is the two-dimensional subspace of $\mathbf{R}^3$ that is spanned by two vectors $\vec{u}$ and $\vec{v}$. We saw in Sec. 6.2 that the vectors $\vec{v}$ and $\vec{u} - \text{proj}_{\vec{v}}(\vec{u})$ were orthogonal. Since this second pair of vectors also spans V, we have produced an orthogonal basis for V. This same process will work for any two-dimensional subspace of $\mathbf{R}^n$ (arbitrary n). This next theorem generalizes this idea and helps us find orthogonal bases for vector spaces.

Theorem 64 (Gram-Schmidt theorem). *Suppose that $\vec{w}_1, \ldots, \vec{w}_m$ are mutually orthogonal and nonzero vectors in $\mathbf{R}^n$. Suppose $\vec{v} \notin \text{span}\{\vec{w}_1, \ldots, \vec{w}_m\}$. We define*

$$\vec{w}_{m+1} = \vec{v} - \sum_{j=1}^{m} \frac{\langle \vec{v}, \vec{w}_j \rangle}{\langle \vec{w}_j, \vec{w}_j \rangle} \vec{w}_j = \vec{v} - \sum_{j=1}^{m} \text{proj}_{\vec{w}_j}(\vec{v}).$$

Then the vectors $\vec{w}_1, \vec{w}_2, \ldots, \vec{w}_m, \vec{w}_{m+1}$ are mutually orthogonal. Moreover, we have $\text{span}\{\vec{w}_1, \ldots, \vec{w}_m, \vec{v}\} = \text{span}\{\vec{w}_1, \ldots, \vec{w}_m, \vec{w}_{m+1}\}$.

Proof. According to the definition of $\vec{w}_{m+1}$, one sees immediately that $\vec{v} \in \text{span}\{\vec{w}_1, \ldots, \vec{w}_m, \vec{w}_{m+1}\}$ and $\vec{w}_{m+1} \in \text{span}\{\vec{w}_1, \ldots, \vec{w}_m, \vec{v}\}$. Therefore, the two spans $\text{span}\{\vec{w}_1, \ldots, \vec{w}_m, \vec{w}_{m+1}\}$ and $\text{span}\{\vec{w}_1, \ldots, \vec{w}_m, \vec{v}\}$ are equal. To prove the orthogonality of $\vec{w}_1, \ldots, \vec{w}_m, \vec{w}_{m+1}$, since $\vec{w}_1, \ldots, \vec{w}_m$ are mutually orthogonal, we must show that $\langle \vec{w}_{m+1}, \vec{w}_k \rangle = 0$ for each $k = 1, 2, \ldots, m$. For this, we compute

$$\begin{aligned}
\langle \vec{w}_{m+1}, \vec{w}_k \rangle &= \left\langle \vec{v} - \sum_{j=1}^{m} \frac{\langle \vec{v}, \vec{w}_j \rangle}{\langle \vec{w}_j, \vec{w}_j \rangle} \vec{w}_j, \vec{w}_k \right\rangle \\
&= \langle \vec{v}, \vec{w}_k \rangle - \sum_{j=1}^{m} \frac{\langle \vec{v}, \vec{w}_j \rangle}{\langle \vec{w}_j, \vec{w}_j \rangle} \langle \vec{w}_j, \vec{w}_k \rangle \\
&= \langle \vec{v}, \vec{w}_k \rangle - \frac{\langle \vec{v}, \vec{w}_k \rangle}{\langle \vec{w}_k, \vec{w}_k \rangle} \langle \vec{w}_k, \vec{w}_k \rangle \\
&= \langle \vec{v}, \vec{w}_k \rangle - \langle \vec{v}, \vec{w}_k \rangle \\
&= 0.
\end{aligned}$$

This proves the theorem. □

The Gram-Schmidt Process

Assume that $\{\vec{v}_1, \vec{v}_2, \ldots, \vec{v}_n\}$ is a basis for a vector space V. Suppose, however, that we desire an orthonormal basis for V. One can be constructed using $\{\vec{v}_1, \vec{v}_2, \ldots, \vec{v}_n\}$ and Theorem 64. This procedure is known as the *Gram-Schmidt process.*

In the Gram-Schmidt process we first set $\vec{w}_1 = \vec{v}_1$. Next we set

$$\vec{w}_2 = \vec{v}_2 - \frac{\langle \vec{v}_2, \vec{w}_1 \rangle}{\langle \vec{w}_1, \vec{w}_1 \rangle} \vec{w}_1 = \vec{v}_2 - \text{proj}_{\vec{w}_1}(\vec{v}_2).$$

The Gram-Schmidt theorem guarantees that $\vec{w}_1$ and $\vec{w}_2$ are orthogonal. As a third step we set

$$\vec{w}_3 = \vec{v}_3 - \frac{\langle \vec{v}_3, \vec{w}_1 \rangle}{\langle \vec{w}_1, \vec{w}_1 \rangle} \vec{w}_1 - \frac{\langle \vec{v}_3, \vec{w}_2 \rangle}{\langle \vec{w}_2, \vec{w}_2 \rangle} \vec{w}_2.$$

Again using the Gram-Schmidt theorem, we note that $\vec{w}_1, \vec{w}_2$, and $\vec{w}_3$ are mutually orthogonal. The idea of how to proceed is now clear. Suppose that $\vec{w}_1, \vec{w}_2, \ldots, \vec{w}_{k-1}$ have been chosen. Then we define $\vec{w}_k$ by

$$\vec{w}_k = \vec{v}_k - \sum_{j=1}^{k-1} \frac{\langle \vec{v}_k, \vec{w}_j \rangle}{\langle \vec{w}_j, \vec{w}_j \rangle} \vec{w}_j.$$

Inductively, this procedure gives us n mutually orthogonal vectors $\vec{w}_1, \vec{w}_2, \ldots, \vec{w}_n$. The Gram-Schmidt theorem also guarantees that both spans

$\operatorname{span}\{\vec{v}_1, \vec{v}_2, \ldots, \vec{v}_n\}$ and $\operatorname{span}\{\vec{w}_1, \vec{w}_2, \ldots, \vec{w}_n\}$ are equal. The new basis $\{\vec{w}_1, \vec{w}_2, \ldots, \vec{w}_n\}$ is called the *Gram-Schmidt orthogonalization* of the basis $\{\vec{v}_1, \vec{v}_2, \ldots, \vec{v}_n\}$.

It is often convenient to normalize $\vec{w}_k$ to a unit vector $\vec{u}_k$ immediately after its computation. When doing so, we can use the $\vec{u}_j$ for $1 \leq j < k$ in defining $\vec{w}_k$. This has the advantage that $\langle \vec{u}_j, \vec{u}_j \rangle = 1$ in this calculation. However, in some applications it is not convenient to normalize until the end.

For example, consider the subspace V of $\mathbf{R}^4$ whose basis is the set containing three vectors

$$\vec{v}_1 = \begin{pmatrix} 1 \\ 1 \\ 1 \\ 1 \end{pmatrix}, \quad \vec{v}_2 = \begin{pmatrix} 1 \\ 0 \\ 1 \\ 0 \end{pmatrix}, \quad \vec{v}_3 = \begin{pmatrix} 1 \\ 0 \\ 0 \\ 1 \end{pmatrix}.$$

We will apply the Gram-Schmidt process, normalizing our orthogonal vectors as we go along. Since $\|\vec{v}_1\| = 2$, we set $\vec{u}_1 = \frac{1}{2}\vec{v}_1$. The Gram-Schmidt process now gives

$$\vec{w}_2 = \begin{pmatrix} 1 \\ 0 \\ 1 \\ 0 \end{pmatrix} - \left(\frac{1}{2} + \frac{1}{2}\right) \begin{pmatrix} \frac{1}{2} \\ \frac{1}{2} \\ \frac{1}{2} \\ \frac{1}{2} \end{pmatrix} = \begin{pmatrix} \frac{1}{2} \\ -\frac{1}{2} \\ \frac{1}{2} \\ -\frac{1}{2} \end{pmatrix}.$$

Since $\|\vec{w}_2\| = 1$, we set $\vec{u}_2 = \vec{w}_2$. Finally, the Gram-Schmidt process gives

$$\vec{w}_3 = \begin{pmatrix} 1 \\ 0 \\ 0 \\ 1 \end{pmatrix} - \left(\frac{1}{2} + \frac{1}{2}\right) \begin{pmatrix} \frac{1}{2} \\ \frac{1}{2} \\ \frac{1}{2} \\ \frac{1}{2} \end{pmatrix} - \left(\frac{1}{2} - \frac{1}{2}\right) \begin{pmatrix} \frac{1}{2} \\ -\frac{1}{2} \\ \frac{1}{2} \\ -\frac{1}{2} \end{pmatrix} = \begin{pmatrix} \frac{1}{2} \\ -\frac{1}{2} \\ -\frac{1}{2} \\ \frac{1}{2} \end{pmatrix}.$$

Since $\|\vec{w}_3\| = 1$ we can set $\vec{u}_3 = \vec{w}_3$, and $\{\vec{u}_1, \vec{u}_2, \vec{u}_3\}$ is the desired orthonormal basis of V.

Orthogonal Matrices

Suppose we apply the Gram-Schmidt procedure to the basis

$$\left\{ \begin{pmatrix} 1 \\ 0 \\ 1 \end{pmatrix}, \begin{pmatrix} 1 \\ 1 \\ 0 \end{pmatrix}, \begin{pmatrix} 1 \\ 0 \\ -1 \end{pmatrix} \right\}$$

of $\mathbf{R}^3$. We obtain the orthogonal basis

$$\left\{ \begin{pmatrix} 1 \\ 0 \\ 1 \end{pmatrix}, \begin{pmatrix} \frac{1}{2} \\ 1 \\ -\frac{1}{2} \end{pmatrix}, \begin{pmatrix} \frac{2}{3} \\ -\frac{2}{3} \\ -\frac{2}{3} \end{pmatrix} \right\}.$$

If we normalize these vectors, we obtain the orthonormal basis

$$\mathcal{B} = \left\{ \begin{pmatrix} \frac{1}{\sqrt{2}} \\ 0 \\ \frac{1}{\sqrt{2}} \end{pmatrix}, \begin{pmatrix} \frac{1}{\sqrt{6}} \\ \frac{2}{\sqrt{6}} \\ -\frac{1}{\sqrt{6}} \end{pmatrix}, \begin{pmatrix} \frac{1}{\sqrt{3}} \\ -\frac{1}{\sqrt{3}} \\ -\frac{1}{\sqrt{3}} \end{pmatrix} \right\}.$$

Note that this is the basis we used earlier.

The transition matrix from the basis $\mathcal{B}$ to the standard basis is the matrix P whose columns are the vectors in $\mathcal{B}$. Since P has as columns an orthonormal basis for $\mathbf{R}^3$, we will see that $P^{-1} = P^t$. We check this here. $P^t \cdot P$ is

$$\begin{pmatrix} \frac{1}{\sqrt{2}} & 0 & \frac{1}{\sqrt{2}} \\ \frac{1}{\sqrt{6}} & \frac{2}{\sqrt{6}} & -\frac{1}{\sqrt{6}} \\ \frac{1}{\sqrt{3}} & -\frac{1}{\sqrt{3}} & -\frac{1}{\sqrt{3}} \end{pmatrix} \begin{pmatrix} \frac{1}{\sqrt{2}} & \frac{1}{\sqrt{6}} & \frac{1}{\sqrt{3}} \\ 0 & \frac{2}{\sqrt{6}} & -\frac{1}{\sqrt{3}} \\ \frac{1}{\sqrt{2}} & -\frac{1}{\sqrt{6}} & -\frac{1}{\sqrt{3}} \end{pmatrix} = \begin{pmatrix} 1 & 0 & 0 \\ 0 & 1 & 0 \\ 0 & 0 & 1 \end{pmatrix}.$$

In particular, we find that the transition matrix from the standard basis to the basis $\mathcal{B}$ is the matrix P^t. This motivates the following definition.

Definition. An $n \times n$ matrix P is called *orthogonal* if $P^{-1} = P^t$.

Whenever P is an $n \times n$ matrix and you compute the product P^tP, the ijth entry is the dot product of the ith row of P^t with the jth column of P. Since the jth row of P^t is the jth column of P, we see that P being orthogonal means that the dot product of the ith and jth columns of P is 0 if $i \neq j$ and is 1 if $i = j$. This is the same as saying P's columns form an orthonormal basis for $\mathbf{R}^n$. We record this as the next theorem.

Theorem 65. *An $n \times n$ matrix P is orthogonal if and only if its columns (or rows) form an orthonormal basis for* $\mathbf{R}^n$.

We are now in a position to give a proof of Theorem 49, which states that every symmetric matrix is diagonalizable. Suppose that S is a symmetric $n \times n$ matrix. According to Theorem 48, we know that S has an eigenvector $\vec{v}_1$. We can assume that $\|\vec{v}_1\| = 1$, and using the Gram-Schmidt procedure we can find an orthonormal basis $\mathcal{B} = \{\vec{v}_1, \vec{v}_2, \dots, \vec{v}_n\}$ with the eigenvector $\vec{v}_1$ as first element.

Let P be the orthogonal matrix with columns $\vec{v}_1, \vec{v}_2, \dots, \vec{v}_n$. What can we say about the matrix $S_1 = P^{-1}SP = P^tSP$? First, we recall from Sec. 8.3 that S_1 is the matrix that represents the operator T_S in the orthonormal basis $\mathcal{B}$. Since $\vec{v}_1$ is an eigenvector of T_S with eigenvalue say k_1, we find that the first column of S_1 must be the $\mathcal{B}$-coordinates of $T_S(\vec{v}_1)$, namely

$$(T_S(\vec{v}_1))_{\mathcal{B}} = \begin{pmatrix} k_1 \\ 0 \\ \vdots \\ 0 \end{pmatrix}$$

is the first column of S_1.

Recall from Theorem 12 that $(AB)^t = B^t A^t$ whenever the matrix product AB is defined. Using this we find $(S_1)^t = (P^t SP)^t = P^t S^t (P^t)^t = P^t SP = S_1$. So, S_1 is symmetric. This, together with our calculation of the first column of S_1, shows that in fact

$$S_1 = \begin{pmatrix} k_1 & \vec{0} \\ \vec{0} & S' \end{pmatrix},$$

where S' is an $(n-1)\times(n-1)$ symmetric matrix. Since S' is a smaller symmetric matrix, we can use mathematical induction to say that $T_{S'} : \mathbf{R}^{n-1} \to \mathbf{R}^{n-1}$ is diagonalizable. This means that S' has $n-1$ linearly independent eigenvectors in $\mathbf{R}^{n-1}$. If we take these eigenvectors and "put them" in $\mathbf{R}^n$ by giving them 0 as a first coordinate, we see that they are eigenvectors for S_1. Since $\vec{e}_1$ is also an eigenvector for S_1, we see that S_1 has n linearly independent eigenvectors. This also means that S has n linearly independent eigenvectors and consequently is diagonalizable.

Our proof that any $n \times n$ symmetric matrix S is diagonalizable gives, in fact, more information. Observe that in the last stage, if we assumed that the $n-1$ eigenvectors for S' were orthonormal, then the collection of n eigenvectors we obtained for S_1 would also be orthonormal. We noticed in the example in the Coordinate-Change Matrix subsection in Sec. 8.3 that the eigenvectors of the matrix considered there were orthogonal. What we have just observed shows that whenever a symmetric matrix has distinct eigenvalues, any collection of the associated eigenvectors must be orthogonal. Summarizing this, we have the following result.

Theorem 66. *If S is an $n \times n$ symmetric matrix, then S has an orthonormal basis of eigenvectors. In particular, there is a orthogonal matrix P such that $P^t SP$ is diagonal.*

The Q-R Factorization

Suppose that we apply the Gram-Schmidt procedure to a collection of columns of an $n \times m$ matrix, where $m < n$. Of course, we won't obtain an orthogonal matrix, because the matrix isn't square. However, this procedure does have an important interpretation. For example, consider

$$A = \begin{pmatrix} 1 & 1 & 1 \\ 1 & 0 & 0 \\ 1 & 1 & 0 \\ 1 & 0 & 1 \end{pmatrix}.$$

Earlier we applied the Gram-Schmidt process to the columns of A. The columns of A were denoted $\vec{v}_1, \vec{v}_2, \vec{v}_3$, and the Gram-Schmidt procedure gave us

$$\begin{aligned} \vec{u}_1 &= \frac{1}{2}\vec{v}_1, \\ \vec{u}_2 &= \vec{v}_2 - \vec{u}_1 = \vec{v}_2 - \frac{1}{2}\vec{v}_1, \\ \vec{u}_3 &= \vec{v}_3 - \vec{u}_1 - 0\vec{u}_2 = \vec{v}_3 - \frac{1}{2}\vec{v}_1. \end{aligned}$$

If we view the three parts of our Gram-Schmidt process as column operations on A, we see that this process can be interpreted as multiplication on the right by an upper triangular matrix. We have

$$\begin{pmatrix} 1 & 1 & 1 \\ 1 & 0 & 0 \\ 1 & 1 & 0 \\ 1 & 0 & 1 \end{pmatrix} \begin{pmatrix} \frac{1}{2} & -\frac{1}{2} & -\frac{1}{2} \\ 0 & 1 & 0 \\ 0 & 0 & 1 \end{pmatrix} = \begin{pmatrix} \frac{1}{2} & \frac{1}{2} & \frac{1}{2} \\ \frac{1}{2} & -\frac{1}{2} & -\frac{1}{2} \\ \frac{1}{2} & \frac{1}{2} & -\frac{1}{2} \\ \frac{1}{2} & -\frac{1}{2} & \frac{1}{2} \end{pmatrix}.$$

The columns of the right-hand matrix are the orthonormal vectors that resulted from applying the Gram-Schmidt procedure.

Inverting the square matrix in the above equation, and multiplying this equation on the right by this inverse, gives the expression

$$A = \begin{pmatrix} 1 & 1 & 1 \\ 1 & 0 & 0 \\ 1 & 1 & 0 \\ 1 & 0 & 1 \end{pmatrix} = \begin{pmatrix} \frac{1}{2} & \frac{1}{2} & \frac{1}{2} \\ \frac{1}{2} & -\frac{1}{2} & -\frac{1}{2} \\ \frac{1}{2} & \frac{1}{2} & -\frac{1}{2} \\ \frac{1}{2} & -\frac{1}{2} & \frac{1}{2} \end{pmatrix} \begin{pmatrix} 2 & 1 & 1 \\ 0 & 1 & 0 \\ 0 & 0 & 1 \end{pmatrix}.$$

This factorization of A is known as a Q-R *factorization*, and we denote the factors according to the equation $A = QR$. The matrix Q has orthonormal columns, and the square matrix R is upper triangular. The Gram-Schmidt process shows that any $n \times m$ matrix of rank m has a Q-R factorization. The Q-R factorization turns out to be quite useful when solving least-squares problems. This is discussed in Sec. 9.3.

Problems

1. (a) Apply the Gram-Schmidt process to the vectors $(3, 4)$, $(5, 12)$ in $\mathbf{R}^2$ to find an orthonormal basis.

   (b) Apply the Gram-Schmidt process to the vectors $(2, 2, 1)$, $(0, 4, 1)$, $(8, 3, 5)$ in $\mathbf{R}^3$ to find an orthogonal basis.

2. (a) Apply the Gram-Schmidt process to the vectors $(1, 0, 1)$ and $(1, 1, 1)$ in $\mathbf{R}^3$ to find an orthonormal basis for their span.

   (b) Use part (a) to find a Q-R factorization of

$$A = \begin{pmatrix} 1 & 1 \\ 0 & 1 \\ 1 & 1 \end{pmatrix}.$$

3. Suppose that V is a subspace of $\mathbf{R}^n$ and $\vec{u}, \vec{v}, \vec{w} \in V$. If $\vec{u} \perp \vec{v}$, and $\vec{v} \perp \vec{w}$, is it also true that $\vec{u} \perp \vec{w}$? Give a proof or a counterexample.

4. Suppose that $\{\vec{u}_1, \vec{u}_2, \ldots, \vec{u}_n\}$ is an orthonormal basis of $\mathbf{R}^n$. Show for any $\vec{v} \in \mathbf{R}^n$ that $\|\vec{v}\|^2 = \langle \vec{v}, \vec{u}_1 \rangle^2 + \langle \vec{v}, \vec{u}_2 \rangle^2 + \cdots + \langle \vec{v}, \vec{u}_n \rangle^2$.

5. Find an orthogonal matrix P such that $P^t SP$ is diagonal if S is the symmetric matrix below.

   (a) $\begin{pmatrix} 1 & 1 \\ 1 & 1 \end{pmatrix}$ (b) $\begin{pmatrix} 1 & 0 & 1 \\ 0 & 1 & 1 \\ 1 & 1 & 2 \end{pmatrix}$ (c) $\begin{pmatrix} 1 & 1 & 1 \\ 1 & 1 & 1 \\ 1 & 1 & 1 \end{pmatrix}$

6. Find a Q-R factorization of the following matrix.

$$\begin{pmatrix} 1 & -1 & 1 \\ 0 & 0 & 1 \\ 1 & -1 & 0 \\ 0 & 1 & -1 \end{pmatrix}$$

7. Let $\vec{v} = (1, 1, 0, 0)$. Find an orthonormal basis for the subspace $W = \{\vec{w} \in \mathbf{R}^4 \mid \langle w, \vec{v} \rangle = 0\}$.

8. Let $\vec{v}_1 = (1, 1, 1, 1, 1)$, $\vec{v}_2 = (1, 0, 1, 0, 1) \in \mathbf{R}^5$. Find an orthogonal basis for the subspace $U = \{\vec{u} \in \mathbf{R}^5 \mid \langle \vec{v}, \vec{v}_1 \rangle = 0 \text{ and } \langle \vec{v}, \vec{v}_2 \rangle = 0\}$.

Group Project: The Distance Between a Point and a Subspace

In this problem you will use the ideas behind the Gram-Schmidt procedure to compute the distance between a point and a subspace in $\mathbf{R}^n$.

(a) Consider a line $\mathcal{L} = \text{span}\{\vec{v}\}$ in $\mathbf{R}^n$, where $\vec{v} \neq \vec{0}$. For any $\vec{w} \in \mathbf{R}^n$ explain why the distance between $\vec{w}$ (viewed as a point) and $\mathcal{L}$ is $\|\vec{w} - \text{proj}_{\vec{v}}(\vec{w})\|$.

(b) Using the ideas from (a), describe how the Gram-Schmidt theorem can help you find the distance between the subspace $\text{span}\{\vec{v}_1, \vec{v}_2, \ldots, \vec{v}_s\}$ and a point in $\mathbf{R}^n$.

(c) Find the distance between the subspace $V = \text{span}\{(1,1,1,1),(1,1,0,1)\} \subset \mathbf{R}^4$ and the point $(0,1,-2,0)$.

9.2 Orthogonal Projections

This section deals with orthogonal projections. An orthogonal projection of a three-dimensional object is like its shadow when the sun is directly overhead. In order to compute the matrix representations of orthogonal projections, we need to begin with the study of orthogonal complements.

Orthogonal Complements

Whenever S is a subspace of $\mathbf{R}^n$, the set of vectors perpendicular to S form another subspace of $\mathbf{R}^n$, known as the orthogonal compliment of S.

Definition. Suppose that S is a subspace of $\mathbf{R}^n$. The *orthogonal complement* of S in $\mathbf{R}^n$ is defined to be the subspace $S^\perp$ (read "S perp") defined by $S^\perp = \{\vec{v} \in \mathbf{R}^n \mid \langle \vec{v}, \vec{w} \rangle = 0 \text{ for all } \vec{w} \in S\}$.

For example, if $S \subset \mathbf{R}^3$ is the xy-plane, then $S^\perp$ is the z-axis. Observe that the bilinearity property of the dot product (Theorem 34) guarantees that $S^\perp$ is a vector space. See Prob. 1 at the end of this section for more details.

Orthogonal Decomposition

This next result is an important consequence of the Gram-Schmidt theorem. It says that if $V \subseteq \mathbf{R}^n$ is a subspace, then $\mathbf{R}^n$ is spanned by V together with its orthogonal complement $V^\perp$. More precisely, it shows that any such V determines a unique decomposition of any vector as a sum of two vectors, one in V and the other orthogonal to V.

Theorem 67. *Let V be a subspace of $\mathbf{R}^n$. Then*

(i) *Every $\vec{w} \in \mathbf{R}^n$ can be expressed as $\vec{w} = \vec{v} + \vec{v}\,'$ with $\vec{v} \in V$ and $\vec{v}\,' \in V^\perp$.*

(ii) $V \cap V^\perp = \{\vec{0}\}$.

(iii) *The expression $\vec{w} = \vec{v} + \vec{v}\,'$ with $\vec{v} \in V$ and $\vec{v}\,' \in V^\perp$ in part* (i) *is unique.*

Proof. (i) Let $\{\vec{u}_1, \vec{u}_2, \ldots, \vec{u}_s\}$ be an orthonormal basis for V, and let $\vec{w} \in \mathbf{R}^n$. In case $\vec{w} \in V$, then $\vec{w} = \vec{w} + \vec{0}$ is the desired expression. In case $\vec{w} \notin \text{span}\{\vec{u}_1, \vec{u}_2, \ldots, \vec{u}_s\}$, then by the Gram-Schmidt theorem we find that

$$\vec{v}\,' = \vec{w} - \sum_{j=1}^{s} \langle \vec{w}, \vec{u}_j \rangle \, \vec{u}_j$$

is orthogonal to each of $\vec{u}_1, \vec{u}_2, \ldots, \vec{u}_s$. Since $\vec{u}_1, \vec{u}_2, \ldots, \vec{u}_s$ span V, any element $\vec{z} \in V$ is a linear combination of $\vec{u}_1, \vec{u}_2, \ldots, \vec{u}_s$, and consequently $\langle \vec{v}\,', \vec{z} \rangle = 0$ for all $\vec{z} \in V$. This shows that $\vec{v}\,' \in V^{\perp}$. Setting $\vec{v} = \sum_{j=1}^{s} \langle \vec{w}, \vec{u}_j \rangle \, \vec{u}_j$, we find that $\vec{w} = \vec{v} + \vec{v}\,'$ with $\vec{v} \in V$ and $\vec{v}\,' \in V^{\perp}$. This gives (i).

For (ii) we suppose that $\vec{u} \in V \cap V^{\perp}$. Applying part (i) we express $\vec{u} = \vec{v} + \vec{v}\,'$ with $\vec{v} \in V$ and $\vec{v}\,' \in V^{\perp}$. Since $\vec{u} \in V$ we have $\langle \vec{u}, \vec{v} \rangle = 0$, and since $\vec{u} \in V^{\perp}$ we have $\langle \vec{u}, \vec{v}\,' \rangle = 0$. This shows $\langle \vec{u}, \vec{u} \rangle = \langle \vec{u}, \vec{v} + \vec{v}\,' \rangle = \langle \vec{u}, \vec{v} \rangle + \langle \vec{u}, \vec{v}\,' \rangle = 0$. So $\vec{u} = \vec{0}$, giving (ii).

(iii) We now suppose that $\vec{w} = \vec{v} + \vec{v}\,' = \vec{z} + \vec{z}\,'$, where $\vec{v}, \vec{z} \in V$ and $\vec{v}\,', \vec{z}\,' \in V^{\perp}$. Then $\vec{v} - \vec{z} = \vec{v}\,' - \vec{z}\,' \in V \cap V^{\perp}$, so by (ii) we find that $\vec{v} - \vec{z} = \vec{v}\,' - \vec{z}\,' = \vec{0}$. We conclude that $\vec{v} = \vec{z}$ and $\vec{v}\,' = \vec{z}\,'$, establishing the uniqueness assertion. □

The geometric meaning of the theorem can be understood as follows. Imagine a plane $\mathcal{P}$ in $\mathbf{R}^3$ passing through the origin O. Let P be a point in $\mathbf{R}^3$ and let $Q \in \mathcal{P}$ be chosen so that $\overline{PQ}$ is perpendicular to $\mathcal{P}$. Then Q is the *orthogonal projection* of P onto $\mathcal{P}$. The unique expression $\vec{w} = \vec{v} + \vec{v}\,'$ in part (iii) of Theorem 67 corresponds to the addition of geometric vectors $\overrightarrow{OP} = \overrightarrow{OQ} + \overrightarrow{QP}$ (see Fig. 9.1).

Orthogonal Projection

The uniqueness assertion of Theorem 67 (iii) enables us to give the following definition.

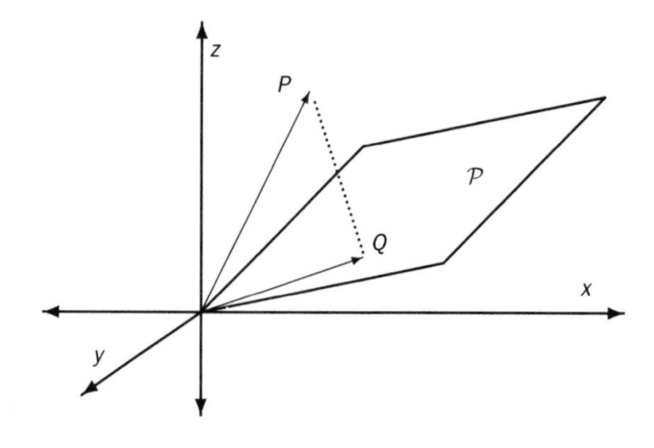

Fig. 9.1. Orthogonal projection onto the plane $\mathcal{P}$

Definition. If V is a subspace of $\mathbf{R}^n$, the *orthogonal projection* $\text{Proj}_V : \mathbf{R}^n \to \mathbf{R}^n$ is the function defined by $\text{Proj}_V(\vec{w}) = \vec{v}$ whenever $\vec{w} = \vec{v} + \vec{v}\,'$ with $\vec{v} \in V$ and $\vec{v}\,' \in V^\perp$. We shall often refer to Proj_V as the *projection operator* onto V.

Using this definition, observe that whenever $\vec{w} = \vec{v} + \vec{v}\,'$ with $\vec{v} \in V$ and $\vec{v}\,' \in V^\perp$, we also have $\text{Proj}_{V^\perp}(\vec{w}) = \vec{v}\,'$. (This is because $(V^\perp)^\perp = V$; see Prob. 5(b).) In particular, for any $\vec{w} \in \mathbf{R}^n$ we have $\text{Proj}_V(\vec{w}) + \text{Proj}_{V^\perp}(\vec{w}) = \vec{v} + \vec{v}\,' = \vec{w}$. In other words, the operator $\text{Proj}_V + \text{Proj}_{V^\perp} = I$ is the identity operator on $\mathbf{R}^n$.

We shall see shortly how to express orthogonal projection by matrix multiplication. First, however, we note that the proof of Theorem 67 gives a description of the orthogonal projection. In the proof we computed $\vec{w}$ from $\vec{v}$ by projecting $\vec{v}$ onto an orthogonal basis of V and summed the results. We state this as a corollary.

Corollary. *Suppose that* V *is a subspace of* $\mathbf{R}^n$. *Let* $\vec{u}_1, \vec{u}_2, \ldots, \vec{u}_s$ *be an orthonormal basis of* V. *The orthogonal projection* Proj_V *can be computed by the formula*

$$\text{Proj}_V(\vec{w}) = \langle \vec{w}, \vec{u}_1 \rangle \vec{u}_1 + \langle \vec{w}, \vec{u}_2 \rangle \vec{u}_2 + \cdots + \langle \vec{w}, \vec{u}_s \rangle \vec{u}_s.$$

For example, the subspace V of $\mathbf{R}^4$ considered in Sec. 9.1 had the basis of orthonormal vectors

$$\left\{ \begin{pmatrix} \frac{1}{2} \\ \frac{1}{2} \\ \frac{1}{2} \\ \frac{1}{2} \end{pmatrix}, \begin{pmatrix} \frac{1}{2} \\ -\frac{1}{2} \\ \frac{1}{2} \\ -\frac{1}{2} \end{pmatrix}, \begin{pmatrix} \frac{1}{2} \\ -\frac{1}{2} \\ -\frac{1}{2} \\ \frac{1}{2} \end{pmatrix} \right\}.$$

The corollary tells us that the projection of a vector $(a, b, c, d) \in \mathbf{R}^n$ is given by

$$\text{Proj}_V \begin{pmatrix} a \\ b \\ c \\ d \end{pmatrix} = \frac{a+b+c+d}{2} \begin{pmatrix} \frac{1}{2} \\ \frac{1}{2} \\ \frac{1}{2} \\ \frac{1}{2} \end{pmatrix} + \frac{a-b+c-d}{2} \begin{pmatrix} \frac{1}{2} \\ -\frac{1}{2} \\ \frac{1}{2} \\ -\frac{1}{2} \end{pmatrix}$$

$$+ \frac{a-b-c+d}{2} \begin{pmatrix} \frac{1}{2} \\ -\frac{1}{2} \\ -\frac{1}{2} \\ \frac{1}{2} \end{pmatrix}$$

$$= \frac{1}{4} \begin{pmatrix} 3 & -1 & 1 & 1 \\ -1 & 3 & 1 & 1 \\ 1 & 1 & 3 & -1 \\ 1 & 1 & -1 & 3 \end{pmatrix} \begin{pmatrix} a \\ b \\ c \\ d \end{pmatrix}.$$

The 4×4 matrix in this expression is the standard matrix $[\mathrm{Proj}_V]$ of the projection onto V. Since $\mathrm{Proj}_V + \mathrm{Proj}_{V^\perp} = I$, we find that $[\mathrm{Proj}_{V^\perp}] = I_4 - [\mathrm{Proj}_V]$. Hence our calculations also show

$$[\mathrm{Proj}_{V^\perp}] = \frac{1}{4}\begin{pmatrix} 1 & 1 & -1 & -1 \\ 1 & 1 & -1 & -1 \\ -1 & -1 & 1 & 1 \\ -1 & -1 & 1 & 1 \end{pmatrix}.$$

Observe that $[\mathrm{Proj}_{V^\perp}]$ is a rank-1 matrix. This should be expected since $V^\perp$ is one-dimensional. In fact $V^\perp$ is the subspace spanned by a column of $[\mathrm{Proj}_{V^\perp}]$.

Matrix Applications

Recall that for any matrix A, $\ker(A)$ denotes the null space of A, $\mathrm{row}(A)$ denotes the row space of A, and $\mathrm{col}(A)$ denotes the column space of A.

Theorem 68. *Suppose that A is an $m \times n$ matrix. Then* $\ker(A) = (\mathrm{row}(A))^\perp \subseteq \mathbf{R}^n$.

Proof. Let R_i denote the ith row of A. The definition of matrix multiplication shows that $\vec{v} \in \ker(A)$ if and only if $R_i\vec{v} = 0$ for each i. However, viewing R_i and $\vec{v}$ as vectors in $\mathbf{R}^n$, the matrix product $R_i\vec{v}$ is precisely the dot product of R_i and $\vec{v}$. Since $\mathrm{row}(A)$ is the subspace of $\mathbf{R}^n$ spanned by the rows of A, we find that $\vec{v} \in (\mathrm{row}(A))^\perp$ if and only if $\vec{v} \in \ker(A)$. □

Earlier we studied the column space V of

$$A = \begin{pmatrix} 1 & 1 & 1 \\ 1 & 0 & 0 \\ 1 & 1 & 0 \\ 1 & 0 & 1 \end{pmatrix}.$$

According to Theorem 68 $V^\perp$ is $\ker(A^t)$. We readily see

$$\ker(A^t) = \ker\begin{pmatrix} 1 & 1 & 1 & 1 \\ 1 & 0 & 1 & 0 \\ 1 & 0 & 0 & 1 \end{pmatrix} = \mathrm{span}\left\{\begin{pmatrix} 1 \\ 1 \\ -1 \\ -1 \end{pmatrix}\right\},$$

which agrees with our previous determination that $V^\perp$ was the span of a column of $[\mathrm{Proj}_{V^\perp}]$.

If A is a rank-n, $m \times n$ matrix, then we can use Theorem 68 to obtain the following.

Corollary. *Suppose that A is a rank-n, $m \times n$ matrix. Then A^tA is an invertible $n \times n$ matrix.*

Proof. According to Theorem 15 it suffices to show that $\ker(A^tA) = \{\vec{0}\}$. If $A^tA\vec{v} = \vec{0}$, then we see that $A\vec{v} \in \ker(A^t)$. However, we also have that $A\vec{v} \in \mathrm{col}(A) = \mathrm{row}(A^t)$. Since $\mathrm{row}(A^t)^\perp = \ker(A^t)$, we conclude that $A\vec{v} \in \mathrm{row}(A^t) \cap \mathrm{row}(A^t)^\perp = \{\vec{0}\}$. We find that $A\vec{v} = \vec{0}$. Since $\mathrm{rk}(A) = n$, Theorem 15 shows $\ker(A) = \{\vec{0}\}$, and consequently $\vec{v} = \vec{0}$. □

We now have a new method for computing the matrices that give projection operators.

Theorem 69. *Suppose A is a rank-n $m\times n$ matrix. Then the standard matrix of the projection operator onto* $\mathrm{col}(A)$ *can be computed by the matrix equation* $[\mathrm{Proj}_{\mathrm{col}(A)}] = A(A^tA)^{-1}A^t$.

Proof. Since $\mathrm{rk}(A) = n$ and A is $m \times n$, A^tA is invertible by the previous corollary. Therefore, the product $\tilde{A} = A(A^tA)^{-1}A^t$ is defined. Theorem 68 shows that $\mathrm{col}(A)^\perp = \ker(A^t)$. Therefore, if $\vec{w}\,' \in \ker(A^t)$, then $\tilde{A}\vec{w}\,' = A(A^tA)^{-1}(A^t\vec{w}\,') = A(A^tA)^{-1}\vec{0} = \vec{0}$.

Next note that if C_i is the ith column of A, then we know that $\tilde{A}C_i$ is the ith column of $\tilde{A}A$. Observe that $\tilde{A}A = A(A^tA)^{-1}(A^tA) = A$. From this, $\tilde{A}C_i = C_i$ follows. Since any vector $\vec{w} \in \mathrm{col}(A)$ is a sum of such columns of A, we find that $\tilde{A}\vec{w} = \vec{w}$. Since $\tilde{A}\vec{w}\,' = \vec{0}$ when $\vec{w}\,' \in \ker(A^t)$, and $\tilde{A}\vec{w} = \vec{w}$ when $\vec{w} \in \mathrm{col}(A)$, we conclude that $[\mathrm{Proj}_{\mathrm{col}(A)}] = \tilde{A}$ as required. □

An Example

Note that whenever a subspace $V \subset \mathbf{R}^n$ has a basis of m vectors, Theorem 69 can be applied to the matrix whose columns are these vectors. For example, we consider the two-dimensional subspace $V = \mathrm{span}\{(1, 2, 1), (-1, 3, 0)\}$ of $\mathbf{R}^3$. We set A to be the matrix

$$A = \begin{pmatrix} 1 & -1 \\ 2 & 3 \\ 1 & 0 \end{pmatrix}$$

whose columns span V. Then direct calculation gives

$$A^tA = \begin{pmatrix} 1 & 2 & 1 \\ -1 & 3 & 0 \end{pmatrix}\begin{pmatrix} 1 & -1 \\ 2 & 3 \\ 1 & 0 \end{pmatrix} = \begin{pmatrix} 6 & 5 \\ 5 & 10 \end{pmatrix},$$

and consequently

$$(A^tA)^{-1} = \frac{1}{35}\begin{pmatrix} 10 & -5 \\ -5 & 6 \end{pmatrix}.$$

We may now compute that

$$\begin{aligned} A(A^tA)^{-1}A^t &= \frac{1}{35}\begin{pmatrix} 1 & -1 \\ 2 & 3 \\ 1 & 0 \end{pmatrix}\begin{pmatrix} 10 & -5 \\ -5 & 6 \end{pmatrix}\begin{pmatrix} 1 & 2 & 1 \\ -1 & 3 & 0 \end{pmatrix} \\ &= \frac{1}{35}\begin{pmatrix} 26 & -3 & 15 \\ -3 & 34 & 5 \\ 15 & 5 & 10 \end{pmatrix} = [\mathrm{Proj}_V]. \end{aligned}$$

The reader can readily check that

$$\frac{1}{35}\begin{pmatrix} 26 & -3 & 15 \\ -3 & 34 & 5 \\ 15 & 5 & 10 \end{pmatrix}\begin{pmatrix} 1 \\ 2 \\ 1 \end{pmatrix} = \begin{pmatrix} 1 \\ 2 \\ 1 \end{pmatrix} \quad \text{and}$$

$$\frac{1}{35}\begin{pmatrix} 26 & -3 & 15 \\ -3 & 34 & 5 \\ 15 & 5 & 10 \end{pmatrix}\begin{pmatrix} 1 \\ -3 \\ 0 \end{pmatrix} = \begin{pmatrix} 1 \\ -3 \\ 0 \end{pmatrix}.$$

This is as it should be, since vectors in V must project to themselves. If we compute

$$\begin{pmatrix} 1 & 0 & 0 \\ 0 & 1 & 0 \\ 0 & 0 & 1 \end{pmatrix} - \frac{1}{35}\begin{pmatrix} 26 & -3 & 15 \\ -3 & 34 & 5 \\ 15 & 5 & 10 \end{pmatrix} = \frac{1}{35}\begin{pmatrix} 9 & 3 & -15 \\ 3 & 1 & -5 \\ -15 & -5 & 25 \end{pmatrix},$$

we obtain the matrix of the projection operator $\mathrm{Proj}_{V^\perp}$. Observe that this latter matrix has rank 1. This is because $V^\perp$ is the one-dimensional subspace of $\mathbf{R}^3$ spanned by $(3, 1, -5)$ (or any column of $[\mathrm{Proj}_{V^\perp}]$).

Problems

1. Show that for any subset $S \subset \mathbf{R}^n$, the set $S^\perp$ is a subspace of $\mathbf{R}^n$. Use the properties of the dot product stated in Theorem 34 to show that $S^\perp$ contains $\vec{0}$ and is closed under scalar multiplication and vector addition. Note that you need to assume only that S is a subset, not a subspace.
2. Consider the subspace $W = \mathrm{span}\{(2, 1, 0, 0), (0, 0, 1, 2), (1, 1, 1, 1)\} \subseteq \mathbf{R}^4$. Show that

$$[\mathrm{Proj}_W] = \frac{1}{10}\begin{pmatrix} 9 & 2 & -2 & 1 \\ 2 & 6 & 4 & -2 \\ -2 & 4 & 6 & 2 \\ 1 & -2 & 2 & 9 \end{pmatrix},$$

   and find $[\mathrm{Proj}_{W^\perp}]$.

3. Find the standard matrices of the projection operators $\text{Proj}_V : \mathbf{R}^3 \to \mathbf{R}^3$ and $\text{Proj}_{V^\perp} : \mathbf{R}^3 \to \mathbf{R}^3$, where V is as given.
   (a) $V = \text{span}\{(1,4,0),(1,3,1)\}$
   (b) $V = \text{span}\{(2,4,-8)\}$
   (c) $V = \{(x,y,z) \mid 4x - y - z = 0\}$
4. Use the Theorem 69 to compute the projection operator $\text{Proj}_W : \mathbf{R}^4 \to \mathbf{R}^4$ for
   (a) $W = \text{span}\{(1,1,0,0),(0,0,1,1)\}$;
   (b) $W = \text{span}\{(2,4,3,1)\}$;
   (c) $W = \text{span}\{(1,1,0,0),(0,1,0,1),(0,0,1,1)\}$.
5. Suppose that U and V are subspaces of $\mathbf{R}^n$.
   (a) If $\dim(V) = d$, what is $\dim(V^\perp)$?
   (b) Use (a) to help show $V = (V^\perp)^\perp$.
   (c) If $U \subseteq V$, show $V^\perp \subseteq U^\perp$.
   (d) Show that $(U + V)^\perp = U^\perp \cap V^\perp$.
6. Suppose that U and V are subspaces of $\mathbf{R}^n$. Show that $U^\perp + V^\perp = (U \cap V)^\perp$.
7. If V is a subspace of $\mathbf{R}^n$, show that $\|\text{Proj}_V(\vec{w})\| \le \|\vec{w}\|$ for all $\vec{w} \in \mathbf{R}^n$. Why is this true intuitively?
8. Why are the matrices $[\text{Proj}_V]$ studied in this section always symmetric?
9. Give an alternative proof of the corollary following Theorem 68 by showing that $\ker(A^tA) = \ker(A)$, so that $\text{rk}(A^tA) = \text{rk}(A)$. (Hint: Compute $\vec{x}^{\,t}A^tA\vec{x}$.)
10. Suppose that B is an $n \times n$ invertible matrix. Let V_i be the subspace of $\mathbf{R}^n$ spanned by the ith row of B. Show that the orthogonal complement $V^\perp$ of V is spanned by the $(n-1)$ columns of B^{-1}, excluding the ith.

Group Project: Orthogonal Matrices

In this investigation you will study more properties of orthogonal matrices.

(a) Show that if A and B are both orthogonal matrices then AB is orthogonal.

(b) Suppose that $\vec{v}, \vec{w} \in \mathbf{R}^n$ and that A is an orthogonal $n \times n$ matrix. Show that the dot product $\langle \vec{v}, \vec{w} \rangle = \langle A\vec{v}, A\vec{w} \rangle$.

(c) What you just checked in (b) shows that the function $F : \mathbf{R}^n \to \mathbf{R}^n$ defined by $F(\vec{v}) = A\vec{v}$ is what is called an *isometry*. What does this mean geometrically in $\mathbf{R}^2$ or $\mathbf{R}^3$? Rotations and reflections are examples of isometries. Show that every 2×2 orthogonal matrix is either a rotation or a reflection matrix.

9.3 Least-Squares Approximations

In Sec. 1.2 we noted that the output of many functions studied in science looks very much like that of a linear function. Recall that such behavior is called *local linearity.* An important problem for scientists in many subjects is to take some data, find a good linear approximation, and use this to help understand and predict outcomes. But what is the best way to find these linear approximations?

The geometric ideas developed in the first two sections of this chapter turn out to be quite useful in helping us understand and devise methods for finding good linear approximations. We give a brief orientation to the ideas of one method in this section. We will study what are called *least-squares approximations.* Undoubtedly, many of you reading this text will see this process in much greater detail in later classes, perhaps in an econometrics class, where this process is known as *linear regression.*

Estimation of Cost

According to a study of accounting records dating 1935–1938,[1] the four-week cost of operating a certain leather belt shop can be accurately estimated by

$$C = -60.178 + .77b + 70.18w,$$

where b is the thousands of square feet of single-ply belting, w is the belt weight in pounds per square foot, and C is the total cost in thousands of dollars. This looks useful, but you must be wondering where the numbers in this equation came from. Our next task is to find out.

We begin by thinking about a possible matrix formulation of this situation. One has a collection of data showing various b_i and w_i values with their respective outputs C_i. We want to try to find coefficients β_1, β_2, and β_3 so that the equation $C = \beta_1 + \beta_2 b + \beta_3 w$ gives a good prediction of the behavior we want to describe. In other words, for each triple of leather shop data b_i, w_i, C_i, we should have $C_i \approx \beta_1 + \beta_2 b_i + \beta_3 w_i$. If we think of β_1, β_2, and β_3 as unkowns, this problem can be represented by the matrix equation

$$\begin{pmatrix} 1 & b_1 & w_1 \\ 1 & b_2 & w_2 \\ 1 & b_3 & w_3 \\ \vdots & \vdots & \vdots \\ 1 & b_n & w_n \end{pmatrix} \begin{pmatrix} \beta_1 \\ \beta_2 \\ \beta_3 \end{pmatrix} \approx \begin{pmatrix} C_1 \\ C_2 \\ C_3 \\ \vdots \\ C_n \end{pmatrix},$$

where we have n triples of data to approximate. This makes our problem look more like one of matrix theory. The economists studying the leather

[1] J. Dean, *The Relation of Cost to Output for a Leather Belt Shop*, National Bureau of Economic Research, New York (1941).

belt shop found that $\beta_1 = -60.178$, $\beta_2 = .77$, and $\beta_3 = 70.18$ give good approximate solutions.

The phrase "a good approximate solution" can be interpreted in many ways. One useful notion is the following.

Definition. Let A be an $m \times n$ matrix. A vector $\vec{\beta} \in \mathbf{R}^n$ is called a *least-squares solution* to the (possibly inconsistent) system $A\vec{X} = \vec{C}$ if the distance

$$\|A\vec{\beta} - \vec{C}\| = \sqrt{\langle A\vec{\beta} - \vec{C}, A\vec{\beta} - \vec{C}\rangle}$$

in $\mathbf{R}^n$ is a minimum among all possible choices for $\vec{\beta}$.

If we analyze our matrix equation above, we see that minimizing the length $\|A\vec{\beta} - \vec{C}\|$ is the same as minimizing its square, which expanded is

$$(\beta_1 + b_1\beta_2 + w_1\beta_3 - C_1)^2 + (\beta_1 + b_2\beta_2 + w_2\beta_3 - C_2)^2$$
$$+ \cdots + (\beta_1 + b_n\beta_2 + w_n\beta_3 - C_n)^2.$$

This accounts for the terminology "least squares."

Projections and Least Squares

Suppose that $\mathcal{P}$ is a plane in $\mathbf{R}^3$ and that Q is a point not on $\mathcal{P}$. Then the distance between Q and $\mathcal{P}$ is the distance of the segment from Q to $\mathcal{P}$ that is perpendicular to $\mathcal{P}$. This is the same as the distance between Q and $\mathrm{Proj}_{\mathcal{P}}(Q)$, the projection of Q on $\mathcal{P}$. (See the group project at the end of Sec. 9.1 for more details.) The same ideas extend to higher dimensions. The distance between a point and a subspace is the distance between the point and the projection of that point onto the subspace.

Now suppose that A is an $n \times m$ matrix and that we are trying to find a least-squares solution to $A\vec{X} = \vec{C}$. Recall that for any $\vec{\beta} \in \mathbf{R}^m$, the vector $A\vec{\beta}$ is an element of the column space of A. This means that the least-squares solutions $\vec{\beta}$ correspond to points $A\vec{\beta} \in \mathrm{col}(A)$ for which the distance between $\vec{C}$ and $A\vec{\beta}$ is minimal. What we just noted shows that this minimal distance occurs where $A\vec{\beta} = \mathrm{Proj}_{\mathrm{col}(A)}(\vec{C})$, that is, where $A\vec{\beta}$ is the projection of $\vec{C}$ onto $\mathrm{col}(A)$. This means that finding the least-squares solutions to $A\vec{X} = \vec{C}$ is the same as finding the actual solutions to $A\vec{X} = \mathrm{Proj}_{\mathrm{col}(A)}(\vec{C})$. This observation is our next theorem.

Theorem 70. *Consider the (possibly inconsistent) system $A\vec{X} = \vec{C}$. Then $A\vec{X} = \vec{C}$ has a least-squares solution, and all its least-squares solutions are the solutions to the consistent system $A\vec{X} = \mathrm{Proj}_{\mathrm{col}(A)}(\vec{C})$.*

If A is an $m \times n$ rank-n matrix, then Theorem 69 gives a computation of $\mathrm{Proj}_{\mathrm{col}(A)}$. Plugging this information into Theorem 70 gives our next result.

Theorem 71. *If A is an $m \times n$ rank n matrix, then the least-squares solutions to the system $A\vec{X} = \vec{C}$ are the solutions to the system $A\vec{X} = A(A^tA)^{-1}A^t\vec{C}$. This system has the unique solution $\vec{X} = (A^tA)^{-1}A^t\vec{C}$.*

Proof. Since the matrix $[\mathrm{Proj}_{\mathrm{col}(A)}] = A(A^tA)^{-1}A^t$ by Theorem 69, Theorem 70 shows that the least-squares solutions to $A\vec{X} = \vec{C}$ are the solutions to $A\vec{X} = A(A^tA)^{-1}A^t\vec{C}$. However, since $\mathrm{rk}(A) = n$, this system of equations has a unique solution. As $\vec{X} = (A^tA)^{-1}A^t\vec{C}$ is a solution to this system, it must be the desired least-squares solution. □

Linear Regression

Suppose that a collection of n data points $p_1 = (r_1, s_1), \ldots, p_n = (r_n, s_n) \in \mathbf{R}^2$ are given, where we assume that they all lie reasonably close to a line. Our problem is to find the equation of a line for which these points are close. This means we want to find an equation of the line $Y = mX + b$ which fits our data in the least-squares sense. To find our "best" m and b, we must find a least-squares solution the system

$$\begin{pmatrix} r_1 & 1 \\ r_2 & 1 \\ \vdots & \vdots \\ r_n & 1 \end{pmatrix} \begin{pmatrix} m \\ b \end{pmatrix} = \begin{pmatrix} s_1 \\ s_2 \\ \vdots \\ s_n \end{pmatrix}.$$

Since the coefficient matrix has rank 2, the least-squares solution can be computed using Theorem 71.

For example, let us consider the nine values considered in our storage battery capacity example in Sec. 1.2. The values were given in the following chart.

Discharge rate (amps)	1	5	10	15	20	25	30	40	50
Amp − hr Capacity	70	68	66	62	58	54	50	40	28

Suppose we wanted to find a line that approximated all of these points, instead of the local linear approximation illustrated in Fig. 1.3. Then we set A to be the 9×2 matrix where

$$A^t = \begin{pmatrix} 1 & 1 & 1 & 1 & 1 & 1 & 1 & 1 & 1 \\ 1 & 5 & 10 & 15 & 20 & 25 & 30 & 40 & 50 \end{pmatrix}$$

(we write A^t here to save space), and $\vec{C}$ is the 9×1 matrix where

$$\vec{C}^{\,t} = (70 \quad 68 \quad 66 \quad 62 \quad 58 \quad 54 \quad 50 \quad 40 \quad 28).$$

Entering these matrices into a calculator gives

$$(A^tA)^{-1}A^t\vec{C} \approx \begin{pmatrix} 73.626 \\ -.850 \end{pmatrix}.$$

This tells us that the line $C = 73.626 - .850r$ is a good approximation to the battery's amp-hr capacity C as a function of its discharge rate r. It is called the *least-squares line.* Recall that our locally linear approximation obtained in Sec. 1.2 was $C \approx 74 - \frac{4}{5}r$. Note how close these linear functions are, but also note the difference. Our equation in Sec. 1.2. was a line through two data points, that happened to pass through several others. Incidentally, most calculators will compute the equation of the least-squares line if you enter the data points and ask the machine to calculate a *linear regression,* which is usually found in the statistics menu. The least-squares line for our storage battery capacity data is shown in Fig. 9.2.

We can also use least squares to obtain a two-variable linear approximation to the values of ΔC_L given in Table 1.2, which is reproduced here.

	5°	10°	15°	20°
.1	.08	.13	.19	.24
.2	.12	.17	.23	.30
.3	.17	.22	.29	.42

Here we are looking for real numbers a_1, a_2, a_3 so that $\Delta C_L \approx a_1 + a_2\delta + a_3E$. Our discussion above shows that we are looking for a least-squares solution to the equation $A\vec{X} = \vec{C}$ where A is a 12×3 matrix whose transpose is

$$A^t = \begin{pmatrix} 1 & 1 & 1 & 1 & 1 & 1 & 1 & 1 & 1 & 1 & 1 & 1 \\ 5 & 10 & 15 & 20 & 5 & 10 & 15 & 20 & 5 & 10 & 15 & 20 \\ .1 & .1 & .1 & .1 & .2 & .2 & .2 & .2 & .3 & .3 & .3 & .3 \end{pmatrix},$$

and where $\vec{C}$ is the 12×1 vector whose transpose is

$$\vec{C}^t = (\,.08 \;\; .13 \;\; .19 \;\; .24 \;\; .12 \;\; .17 \;\; .23 \;\; .30 \;\; .17 \;\; .22 \;\; .29 \;\; .42\,).$$

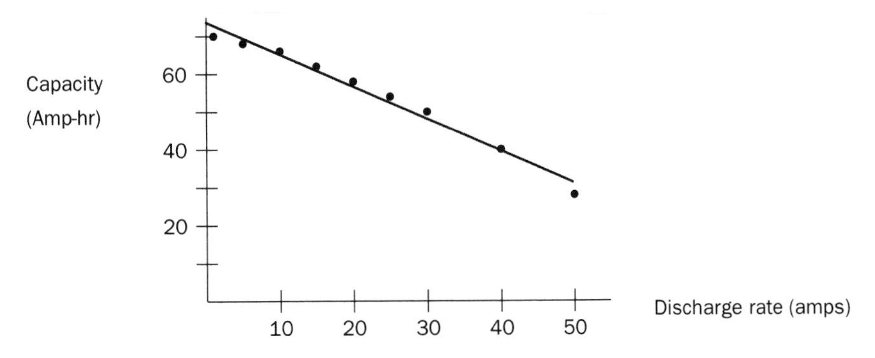

Fig. 9.2. Least-squares approximation to storage battery capacity data

Note that the reason we have 12 rows in our matrices is that we have 12 data values that we want our approximation to take into account. The second and third columns of A give the various input values (they are pictured as the lower two rows of A^t above), and the column $\vec{C}$ lists in corresponding order the output values from the table. We enter these matrices into a calculator and obtain the least-squares solution

$$(A^tA)^{-1}A^t\vec{C} \approx \begin{pmatrix} -.065 \\ .013 \\ .575 \end{pmatrix}.$$

This gives the linear regression (least-squares approximation)

$$\Delta C_L \approx -.065 + .013\delta + .575E,$$

which is close to the approximation

$$\Delta C_L \approx -.02 + .01\delta + .5E$$

obtained in Sec. 1.2. The latter approximation was based on estimates of rates of change of output where one variable was held fixed. We saw at the time that while the approximation was close for values of δ near 10° and E near .2 it was less accurate at values farther away. In contrast, our linear regression will be more accurate overall, but less accurate near the values of $\delta \approx 10°$ and $E \approx .2$.

The Least-Squares Quadratic

Since our data in the battery capacity problem are not exactly linear, we might decide to try to approximate the values using a nonlinear function. The simplest choice would be to try find a quadratic function, say $C = a + br + cr^2$. Solving this problem is essentially the same as finding a least-squares line, the only difference is that our matrix A becomes 9×3 instead of 9×2. The reason for this is that the constants a, b, and c arise as least-squares solutions to the system whose ith equation is $a + r_ib + r_i^2c = C_i$. In other words, A is the transpose of the matrix

$$A^t = \begin{pmatrix} 1 & 1 & 1 & 1 & 1 & 1 & 1 & 1 & 1 \\ 1 & 5 & 10 & 15 & 20 & 25 & 30 & 40 & 50 \\ 1 & 25 & 100 & 225 & 400 & 625 & 900 & 1600 & 2500 \end{pmatrix},$$

where the third column of A is obtained by taking the squares of the second column of A. Entering this A and the same $\vec{C}$ as before into a calculator gives

$$(A^tA)^{-1}A^t\vec{C} \approx \begin{pmatrix} 70.618 \\ -.454 \\ -.008 \end{pmatrix}.$$

This means that our battery capacity function can be approximated by the quadratic function $C \approx 70.618 - .454r - .008r^2$.

The Normal Equation

One may wonder what happens if the matrix A in a least-squares problem does not have full rank as required by Theorem 71. This may happen, for example, when an experiment is repeated a number of times and different outputs arise from the same input. The next result gives an alternative consistent system whose solutions are the least-squares solutions to the original system.

Theorem 72. *If A is an $m \times n$ matrix, then the least-squares solutions to the system $A\vec{X} = \vec{C}$ are the same as the solutions to the system $(A^t A)\vec{X} = A^t\vec{C}$.*

Proof. According to Theorem 70, if $\vec{\beta}$ is a least-squares solution to $A\vec{X} = \vec{C}$, then $A\vec{\beta} = \mathrm{Proj}_{\mathrm{col}(A)}(\vec{C})$. But one knows from the defining property of the projection that $(\vec{C} - \mathrm{Proj}_{\mathrm{col}(A)}(\vec{C})) \in \mathrm{col}(A)^{\perp}$. But Theorem 69 shows that $\mathrm{col}(A)^{\perp} = \ker(A^t)$, and therefore putting these equations together shows $(\vec{C} - A\vec{\beta}) \in \ker(A^t)$. This says that $A^t(\vec{C} - A\vec{\beta}) = \vec{0}$; in other words, $(A^t A)\vec{\beta} = A^t\vec{C}$. Hence all least-squares solutions are solutions to the system $(A^t A)\vec{X} = A^t\vec{C}$.

Conversely, if $(A^t A)\vec{\beta} = A^t\vec{C}$, then $A\vec{\beta} - \vec{C} \in \ker(A^t) = \mathrm{col}(A)^{\perp}$. Since $A\vec{\beta} \in \mathrm{col}(A)$ and $\vec{C} = (A\vec{\beta}) - (A\vec{\beta} - \vec{C})$, we find by definition that $\mathrm{Proj}_W(\vec{C}) = A\vec{\beta}$. Theorem 71 shows that $\vec{\beta}$ is a least-squares solution to $A\vec{X} = \vec{C}$. □

Definition. The system $(A^t A)\vec{X} = A^t\vec{C}$ in Theorem 72 is called the *normal equation* associated with the least-squares problem $A\vec{X} = \vec{C}$.

For example, Theorem 72 shows that the solutions to the least-squares problem

$$\begin{pmatrix} 1 & 2 & 4 \\ 1 & 2 & 4 \\ 1 & 2 & 4 \\ 1 & 2 & 4 \end{pmatrix} \begin{pmatrix} X \\ Y \\ Z \end{pmatrix} = \begin{pmatrix} 47 \\ 51 \\ 49 \\ 50 \end{pmatrix}$$

are the solutions to the system of equations

$$\begin{pmatrix} 1 & 1 & 1 & 1 \\ 2 & 2 & 2 & 2 \\ 4 & 4 & 4 & 4 \end{pmatrix} \begin{pmatrix} 1 & 2 & 4 \\ 1 & 2 & 4 \\ 1 & 2 & 4 \\ 1 & 2 & 4 \end{pmatrix} \begin{pmatrix} X \\ Y \\ Z \end{pmatrix} = \begin{pmatrix} 1 & 1 & 1 & 1 \\ 2 & 2 & 2 & 2 \\ 4 & 4 & 4 & 4 \end{pmatrix} \begin{pmatrix} 47 \\ 51 \\ 49 \\ 50 \end{pmatrix}.$$

This latter system is

$$\begin{pmatrix} 4 & 8 & 16 \\ 8 & 16 & 32 \\ 16 & 32 & 64 \end{pmatrix} \begin{pmatrix} X \\ Y \\ Z \end{pmatrix} = \begin{pmatrix} 197 \\ 394 \\ 788 \end{pmatrix}.$$

We obtain that the least-squares solutions to this problem are the set of solutions to $X + 2Y + 4Z = \frac{197}{4}$. Note that this means there are infinitely many least-squares solutions to the original system, but each least-squares solution has as output 197/4 (in this case the average value of the four outputs in the original system).

Least-Squares Solutions and the *Q*-*R* Factorization

The conclusion of Theorem 71 simplifies in case we have determined the Q-R factorization of the matrix A. If $A = QR$ is a Q-R factorization, then as the matrix Q has orthonormal columns we find $Q^tQ = I_m$. Therefore, we find $A^tA = (QR)^tQR = R^t(Q^tQ)R = R^tR$ and the least-squares solution becomes $\vec{X} = (A^tA)^{-1}A^t\vec{C} = (R^tR)^{-1}(QR)^t\vec{C} = R^{-1}(R^t)^{-1}R^tQ^t\vec{C} = R^{-1}Q^t\vec{C}$.

For example, in Sec. 9.1 we found the Q-R factorization

$$A = \begin{pmatrix} 1 & 1 & 1 \\ 1 & 0 & 0 \\ 1 & 1 & 0 \\ 1 & 0 & 1 \end{pmatrix} = \begin{pmatrix} \frac{1}{2} & \frac{1}{2} & \frac{1}{2} \\ \frac{1}{2} & -\frac{1}{2} & -\frac{1}{2} \\ \frac{1}{2} & \frac{1}{2} & -\frac{1}{2} \\ \frac{1}{2} & -\frac{1}{2} & \frac{1}{2} \end{pmatrix} \begin{pmatrix} 2 & 1 & 1 \\ 0 & 1 & 0 \\ 0 & 0 & 1 \end{pmatrix}.$$

This shows that the least-squares solution to

$$\begin{pmatrix} 1 & 1 & 1 \\ 1 & 0 & 0 \\ 1 & 1 & 0 \\ 1 & 0 & 1 \end{pmatrix} \begin{pmatrix} X \\ Y \\ Z \end{pmatrix} = \begin{pmatrix} 8 \\ 2 \\ 5 \\ 6 \end{pmatrix}$$

is given by

$$\begin{pmatrix} X \\ Y \\ Z \end{pmatrix} = \begin{pmatrix} \frac{1}{2} & -\frac{1}{2} & -\frac{1}{2} \\ 0 & 1 & 0 \\ 0 & 0 & 1 \end{pmatrix} \begin{pmatrix} \frac{1}{2} & \frac{1}{2} & \frac{1}{2} & \frac{1}{2} \\ \frac{1}{2} & -\frac{1}{2} & \frac{1}{2} & -\frac{1}{2} \\ \frac{1}{2} & -\frac{1}{2} & -\frac{1}{2} & \frac{1}{2} \end{pmatrix} \begin{pmatrix} 8 \\ 2 \\ 5 \\ 6 \end{pmatrix},$$

which after calculation shows $X = \frac{9}{4}$, $Y = \frac{5}{2}$, and $Z = \frac{7}{2}$.

Problems

1. Find the least-squares solution of the following. You may use your calculator or computer for the matrix computations.

(a)
$$\begin{aligned} X + Y &= 2 \\ X + 3Y &= 5 \\ X + 5Y &= 9 \\ X + 11Y &= 14 \end{aligned}$$

(b)
$$\begin{aligned} X - Y + 2Z &= 3 \\ X + Y - Z &= 6 \\ 2X - 2Y + Z &= 4 \\ 2X - Y + 6Z &= 9 \end{aligned}$$

2. (a) Show by an algebraic calculation that $\tilde{A} = A(A^tA)^{-1}A^t$ is symmetric with $\tilde{A}^2 = \tilde{A}$ whenever A is $m \times n$ and has rank n.

   (b) Give a geometric explanation of why the result in part (a) is true. (You will need to interpret the geometric meaning of Theorem 69.)

3. A matrix P is called a *projection matrix* if it is symmetric and satisfies $P^2 = P$. (This was checked for $\tilde{A}$ in Prob. 2.) If P is an $n \times n$ projection matrix, show that $I_n - P$ is also a projection matrix. What do these properties mean geometrically if P is the matrix of an orthogonal projection?

4. (a) Use the method of linear regression to find linear approximations to the one-variable functions in the rows of Table 1.4. (These rows correspond to the fixed values of $E = .1, .2, .3$.) How do these three linear functions compare to the three linear approximations for ΔC_D given in Sec. 1.2?

   (b) Find the least-squares, two-variable linear approximation for ΔC_D using the 12 values given in Table 1.4. How does your approximation compare with the nonlinear approximation to ΔC_D given in Table 1.5? Which approximation would you trust if you were designing an airplane?

5. Suppose that P is the 2×2 matrix of an orthogonal projection onto a line in $\mathbf{R}^2$. The matrix $R = I_2 - 2P$ is called a *reflection* matrix. Explain, in geometric terms, what the transformation given by R is doing. Show algebraically that $R^2 = I_2$. Why does this make sense geometrically?

6. (a) Find the least-squares line for the data points $(2,3)$, $(3,7)$, $(5,18)$, $(6,24)$, $(7,31)$, and $(9,43)$.

   (b) Find the least-squares quadratic approximation to the points in part (a).

   (c) Find the least-squares cubic approximation to the points in part (a).

7. Use the Q-R factorization

$$\begin{pmatrix} 1 & 1 \\ 0 & 1 \\ 1 & 1 \end{pmatrix} = \begin{pmatrix} \frac{1}{\sqrt{2}} & 0 \\ 0 & -1 \\ \frac{1}{\sqrt{2}} & 0 \end{pmatrix} \begin{pmatrix} \sqrt{2} & \sqrt{2} \\ 0 & -1 \end{pmatrix}$$

   to find the least-squares solution to

$$\begin{pmatrix} 1 & 1 \\ 0 & 1 \\ 1 & 1 \end{pmatrix} \begin{pmatrix} x \\ y \end{pmatrix} = \begin{pmatrix} 51 \\ 17 \\ 53 \end{pmatrix}.$$

8. Consider the following table.

	2	4	6	8	10
1	3	6	9	12	14
3	10	14	16	18	22
5	24	28	32	35	38

Using a calculator or a computer to find a linear approximation to the function represented by this table of data.

Group Project: Weighted Least Squares

Suppose that you desire to fit a line to some data in which some numbers are more reliable (or important) than others. Then you may want to use a variation of the least-squares line in which the reliability of the data is taken into account. Suppose, for example, you are considering $P_1 = (r_1, s_2)$, $P_2 = (r_2, s_2), \ldots, P_n = (r_n, s_n)$ in $\mathbf{R}^2$. Then you could try to find a weighted least-squares line $Y = mX + b$ by minimizing the sum

$$\sum_{i=1}^{n} w_i(s_i - mr_i - b)^2,$$

where the real numbers w_i are "weights."

(a) Explain why you might do this. How would you determine the weights w_i?

(b) How would you find the values of m and b that minimize the sum? What least-squares problem must you solve? Express this in matrix form.

(c) Rework Prob. 5(a), but this time weight the points $(5, 18)$, $(6, 24)$, and $(7, 31)$ as twice as important as first three. Use your calculator or computer as needed.

Group Project: Quadratic Approximations in Several Variables

(a) As a group, compare your solutions to Prob. 8, and discuss how satisfied you are with the approximations obtained.

(b) Next, figure out how to use least squares to find a quadratic approximation to this function. Your quadratic approximation will look like $Q(X, Y) = a_1 + a_2X + a_3Y + a_4X^2 + a_5XY + a_6Y^2$. Write the least-squares problem in matrix form.

(c) Using a calculator or a computer, obtain your least-squares quadratic approximation. Compare its values to those you studied in part (a). Which approximation is better?

ANSWERS TO ODD-NUMBERED PROBLEMS

Section 1.1

1. Since you can travel 220 miles in 5 hours at a constant speed, you are traveling at $220 \div 5 = 44$ miles per hour. At this speed your distance traveled at time t is $D(t) = 44t$ miles.

3. (a) $P_A = 5 \cdot 25 + .25 \cdot M$ and $P_B = 5 \cdot 35 + .10 \cdot M$ for M miles in five days. Setting these functions equal shows $5 \cdot 25 + .25 \cdot M = 5 \cdot 35 + .10 \cdot M$, and so $50 = .15 \cdot M$. We find that $M = \frac{50}{.15} = 333\frac{1}{3}$ miles for the same price in five days. For seven days the answer is $466\frac{2}{3}$ miles, and for eight days the answer is $533\frac{1}{3}$ miles.
(b) $M = \frac{10D}{.15} = 66\frac{2}{3} \cdot D$.
(c) Substituting $P_A = 25 \cdot D + .25 \cdot 66\frac{2}{3} \cdot D = 41\frac{2}{3} \cdot D$ and $P_B = 35 \cdot D + .10 \cdot 66\frac{2}{3} \cdot D = 41\frac{2}{3} \cdot D$, checking that the costs are the same.

5. (a) $Y = -2X_1 - 2X_2 + 2X_3$ (b) $Y = X_2 + X_3 + 2$

7. (a) $X = \frac{1}{2}Y + 2$ (b) $X = -\frac{1}{3}Y + \frac{7}{3}$ (c) $X = \frac{1}{100}Y - \frac{99}{100}$
(d) This means that the expressions are those of inverse functions.
(e) For (b) we have $-3(\frac{-1}{3}Y + \frac{7}{3}) + 7 = Y - 7 + 7$ and for (c) $Y = 100(\frac{1}{100}Y - \frac{99}{100}) + 99 = Y - 99 + 99$.
(f) This linear function doesn't have an inverse!

9. (a) In order to determine this linear function uniquely, one needs another input-output value. For example, it would be useful to know some values where the input of X_1 is zero. One can see from the data given that the coefficient of X_2 must be 0.
(b) Since the coefficient of X_2 is 0 in any linear function satisfying this input-output chart, these linear functions must all look like $Y = kX_1 + (3 - k)$, where k is some real constant.

Section 1.2

1. For example, the first column drops by \$417, \$139, \$70, and \$41 for each successive 12-month increase (starting at 12 months). This is not characteristic of a linear function. However, one could try a slope of $-\frac{100}{12}$, where the -100 is close to the average of the second and third drops listed. Then, since our linear approximation should pass through $(M, n) = (295, 36)$, we obtain $M - 295 \approx -\frac{100}{12}(n - 36)$ or $M \approx -8.3n + 593$. This linear approximation is not accurate when n is 12 or 60 (but is more reasonable when n is 35, 36, or 37). Many other answers are possible.

3. If the elevation function is locally linear in some region, then the contours representing equal elevation changes should look like equally spaced parallel lines.

5. (a) The increase from 10 to 20 mph was 65 amps, and from 20 to 30 mph was 85 amps. This isn't too far from the behavior of a linear function, so it's reasonable to estimate that the current drain at 15 mph would be halfway between the 10 and 20 currents, say about 117 amps. If we assume that the rate of increase near 30 mph is about $\frac{85}{10} = 8.5$ amps/mph, then at 32 mph we would approximate $235 + 17 = 252$ amps.
(b) The increase between 30 and 40 mph is 105 amps, and the increase between 40 and 50 mph is 130 amps. So the increase between 50 and 60 mph would probably be 160 or more, so a rough guess is $470 + 160 = 630$ amps.

Section 1.3

1. (a) Nonsense. (b) $\begin{pmatrix} 5 & -9 \\ 8 & 18 \end{pmatrix}$ (c) Nonsense. (d) $\begin{pmatrix} -2 & 0 \\ 2 & 4 \\ 0 & 0 \end{pmatrix}$

(e) $\begin{pmatrix} 7 & 9 \\ 4 & 8 \end{pmatrix}$ (f) $\begin{pmatrix} 1 & -3 \\ 2 & 14 \end{pmatrix}$ (g) Nonsense. (h) $\begin{pmatrix} -2 & 3 \\ 12 & -18 \\ -8 & 10 \\ -12 & 18 \end{pmatrix}$

(i) Nonsense. (j) $\begin{pmatrix} 0 & -3 \\ 0 & 5 \\ 0 & 0 \end{pmatrix}$ (k) $\begin{pmatrix} -35 & -45 \\ 30 & 10 \end{pmatrix}$ (l) Nonsense.

3. (a) $\begin{pmatrix} P_1 \\ P_2 \end{pmatrix} = \begin{pmatrix} 2 & -3 \\ 1 & 1 \end{pmatrix} \begin{pmatrix} X \\ Y \end{pmatrix}$

(b) $\begin{pmatrix} Q_1 \\ Q_2 \\ Q_3 \end{pmatrix} = \begin{pmatrix} 3 & -1 & -1 \\ 1 & 1 & 1 \\ 2 & 0 & 1 \end{pmatrix} \begin{pmatrix} X \\ Y \\ Z \end{pmatrix}$

(c) $\begin{pmatrix} R_1 \\ R_2 \end{pmatrix} = \begin{pmatrix} 1 & -1 & -1 & -1 \\ 1 & -1 & -1 & -1 \end{pmatrix} \begin{pmatrix} X \\ Y \\ Z \\ W \end{pmatrix}$

(d) $\begin{pmatrix} S_1 \\ S_2 \\ S_3 \\ S_4 \end{pmatrix} = \begin{pmatrix} 0 & 2 & 1 \\ 4 & -5 & -7 \\ 1 & -1 & 2 \\ 1 & 0 & 1 \end{pmatrix} \begin{pmatrix} X \\ Y \\ Z \end{pmatrix}$

5. Assume that

$$A = \begin{pmatrix} x & y \\ z & w \end{pmatrix} \quad \text{and} \quad B = \begin{pmatrix} s & t \\ u & v \end{pmatrix}.$$

Then by the definition of matrix multiplication we have

$$AB = \begin{pmatrix} sx + uy & * \\ * & tz + vw \end{pmatrix} \quad \text{and} \quad BA = \begin{pmatrix} sx + tz & * \\ * & uy + vw \end{pmatrix},$$

where the values in the $*$ places are omitted. Subtracting, we find that

$$AB - BA = \begin{pmatrix} uy - tz & * \\ * & tz - uy \end{pmatrix} = \begin{pmatrix} a & b \\ c & d \end{pmatrix}.$$

Hence, we find $a + d = (uy - tz) + (tz - uy) = 0$.

7. (a) The initial path matrix for this graph and its powers are

$$M = \begin{pmatrix} 0 & 2 & 1 \\ 2 & 0 & 1 \\ 1 & 1 & 0 \end{pmatrix}, \qquad M^2 = \begin{pmatrix} 5 & 1 & 2 \\ 1 & 5 & 2 \\ 2 & 2 & 2 \end{pmatrix},$$

$$M^3 = \begin{pmatrix} 4 & 12 & 6 \\ 12 & 4 & 6 \\ 6 & 6 & 4 \end{pmatrix}, \quad M^4 = \begin{pmatrix} 30 & 14 & 16 \\ 14 & 30 & 16 \\ 16 & 16 & 12 \end{pmatrix},$$

and adding these powers gives

$$M + M^2 + M^3 + M^4 = \begin{pmatrix} 39 & 29 & 25 \\ 29 & 39 & 25 \\ 25 & 25 & 18 \end{pmatrix}.$$

We find that between points A and C there are 6 paths of length 3, 16 paths of length 4, and 25 paths of length at most 4.

(b) In this graph one finds that there are 15 paths between A and C of length at most 3, 16 paths between A and A of length at most 3, and 8 paths between C and C of length at most 3.

Section 1.4

1. Excluding wheels, the parts for a car total \$0.80 and the parts for a truck total \$1.00. Since we have to spend \$25 for a box of wheels, the new constraint in this problem is $.8 \cdot C + T + 25 \leq 100$. Graphing this new inequality on top of our previous feasible region gives a new feasible

region, which is the triangle with corners $(25, 25)$, $(25, 55)$, and $(62.5, 25)$. Evaluating our profit function on these three points gives \$10.00, \$34.00, and \$32.50, respectively. So making 25 cars and the rest trucks is the best strategy this time.

3. The vertices of the feasible region turn out to be $(0, 0)$, $(0, 6)$, $(2, 4)$, and $(4, 0)$. The maximum value of $2X + 5Y + 1$ on these inputs is attained at $(0, 6)$ and is 31.

5. (a) The vertices of the feasible region are $(0, 0)$, $(0, 3)$, $(3, 0)$, $(3, \frac{3}{2})$, and $(1, 3)$. The maximum value of $Z = 3X + 4Y$ is 15 and the minimum value is 0.

(b) The maximum value is attained at both vertices $(3, \frac{3}{2})$ and $(1, 3)$. This occurred because one of the lines of constant value of the function $X = 3X + 4Y$ happened to contain one of the sides of our feasible region. (Namely, the side with vertices $(3, \frac{3}{2})$ and $(1, 3)$ lies on the line $3X + 4Y = 15$.) Because of this, every point of the feasible region on the segment with vertices $(3, \frac{3}{2})$ and $(1, 3)$ attains the maximum value 15 for our function.

7. If A and B denote the number of birds of species A and B respectively, then we have the territorial constraint $100\frac{A}{2} + 300\frac{B}{2} \leq 100{,}000$. The food constraint equation is $A + .5 \cdot B \leq 500$ since there is enough food for 500 species A birds and the B birds eat half as much as the A birds. The corners of the feasible region for this problem are $(0, 0)$, $(500, 0)$, $(0, 666\frac{2}{3})$, and $(200, 600)$. The maximum of the function $A + B$ occurs when $A = 200$ and $B = 600$, and there are 800 birds total.

9. (a) 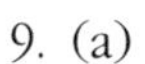

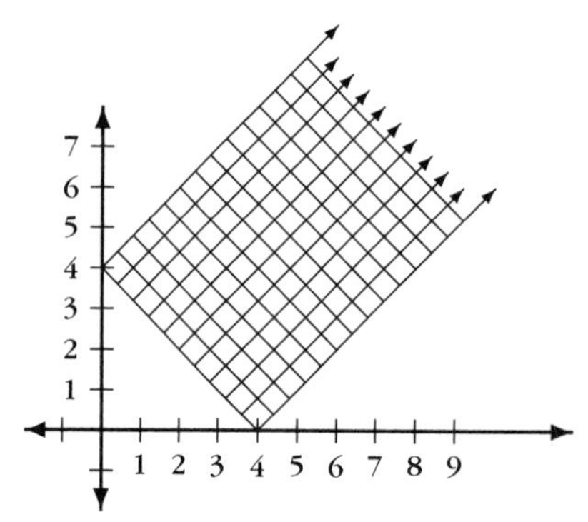

(b) $X + Y$ has minimum of 4, but no maximum.

(c) $-X - Y$ has a maximum of -4, but no minimum.

(d) No, this is not possible.

(e) This example shows that some unbounded regions can have the property that for any linear function, there always is a maximum or minimum.

Section 2.1

1. (a) $\begin{pmatrix} 2 \\ 10 \end{pmatrix}$ (b) $\begin{pmatrix} -10 \\ -30 \end{pmatrix}$ (c) $\begin{pmatrix} -2\pi \\ 4 - 6\pi \end{pmatrix}$

3. The vectors on the unit circle representing successive 72° rotations to three decimal places are $(1, 0)$, $(.309, .951)$, $(-.809, .588)$, $(-.809, -.588)$, and $(.309, -.951)$. Adding these vectors in succession gives the five vertices of a regular pentagon with sides of length 1 as $(1, 0)$, $(1.309, .951)$, $(.5, 1.539)$, $(-.309, .951)$, and $(0, 0)$.

5. If $\vec{v} = \begin{pmatrix} a \\ b \end{pmatrix}$ and $\left\| \begin{pmatrix} a \\ b \end{pmatrix} + \begin{pmatrix} 1 \\ 1 \end{pmatrix} \right\| = 2$, then $(a+1)^2 + (b+1)^2 = 4$. This means that $\vec{v}$ is the position vector of points on the circle of radius 2 centered at $(-1, -1)$.

Section 2.2

1. (a) The direction vector for this line is $\begin{pmatrix} 2 \\ 3 \end{pmatrix}$, which shows that the line has slope $\frac{3}{2}$. Consequently, a parametric description of the line is given by $\{(2t+1, 3t-1) \mid t \in \mathbf{R}\}$. A point-slope representation is given by $y+1 = \frac{3}{2}(x-1)$ and the slope-intercept equation of the line is $y = \frac{3}{2}x - \frac{5}{2}$.
(b) $\{(2t, t) \mid t \in \mathbf{R}\}$; $(y-0) = \frac{1}{2}(x-0)$; $y = \frac{1}{2}x$
(c) $\{(t, 3t) \mid t \in \mathbf{R}\}$; $(y-0) = 3(x-0)$; $y = 3x$
(d) $\{(t, -t-6) \mid t \in \mathbf{R}\}$; $(y-0) = -(x+6)$; $y = -x-6$

3. (a) The line through the points $(1, 1)$ and $(0, 0)$ is given by the equation $y = x$. The second line is given by the equation $y = 2x - 1$. Therefore, the x-coordinate of their intersection is given by $x = 2x - 1$. Hence $x = 1$, and we see that the point of intersection is $(1, 1)$.
(b) The point of intersection is $(-3, -2)$.
(c) The point of intersection is $(1, 1)$.

5. The perpendicular line is $\{(2+2t, 3+t) \mid t \in \mathbf{R}\}$.

Section 2.3

1. (a) A direction vector for the line $\overleftrightarrow{PQ}$ can be found subtracting the coordinates of P from Q and is

$$\begin{pmatrix} 1 \\ 3 \\ -3 \end{pmatrix}.$$

Since $P = (1, 1, 3)$, is a point of $\overleftrightarrow{PQ}$, we find that this line is described parametrically by $\overleftrightarrow{PQ} = \{(t+1, 3t+1, -3t+3) \mid t \in \mathbf{R}\}$.
(b) A parameterization of $\overleftrightarrow{RS}$ is $\{(t, 2t-1, -t+2) \mid t \in \mathbf{R}\}$.
(c) A parameterization is $\{(t+1, 3t+1, -3t+1) \mid t \in \mathbf{R}\}$.

3. There are many possible correct answers to these problems. These are possible answers.

(a) Since the plane contains the origin $(0, 0, 0)$, the points P and Q give two direction vectors for the plane. Therefore, a parameterization is $\{(t + 2u, t + 4u, t + u) \mid t, u \in \mathbf{R}\}$.

(b) Subtracting the coordinates of A from B and C gives the direction vectors

$$\begin{pmatrix} 0 \\ -3 \\ -3 \end{pmatrix} \quad \text{and} \quad \begin{pmatrix} -1 \\ -2 \\ -3 \end{pmatrix}.$$

Since $A = (1, 2, 4)$ lies on the plane, we obtain the parameterization $\{(-u+1, -3t - 2u + 2, -3t - 3u + 4) \mid t, u \in \mathbf{R}\}$.

(c) Solving $z = -x + y + 4$ shows a parameterization is $\{(t, u, -t + u + 4) \mid t, u \in \mathbf{R}\}$.

(d) This plane contains the origin and the points $(1, 2, 0)$ and $(0, 1, -1)$ (obtained by setting $t = 0$ and $t = -1$ in the parameterization, respectively). Therefore, a parameterization is $\{(t, 2t + u, -u) \mid t, u \in \mathbf{R}\}$.

5. (a) We know that the points on this plane are given by $x = t + u$, $y = t - u + 1$, and $z = t$, where t and u range over $\mathbf{R}$. We must use this information to find the relationship between x, y, and z. Inspecting these equations, we see that adding x and y gives $x + y = 2t + 1$. Since $z = t$, we make this substitution to obtain $x + y = 2z + 1$ or $x + y - 2z - 1 = 0$, either of which is a possible answer.

(b) Since O is the origin, this plane can be parameterized by $\{(t - u, 3t, 5t + 2u) \mid t, u \in \mathbf{R}\}$. So we manipulate the equations $x = t - u$, $y = 3t$, and $z = 5t + 2u$ to obtain an equation for the plane. If we add $2x$ and z, we find that $2x + z = 7t$. Since $y = 3t$, we find that $2x + z = \frac{7}{3}y$, which can be rewritten if you like as $6x + 3z - 7y = 0$.

7. (a) Both of these planes contain the point $(0, 0, 0)$. We need to find another point. Setting $x = 1$, we must have $z = -1$ by the first equation. Then $y = -2$ by the second equation. Since the line passes through the origin and contains $(1, -2, -1)$, it is parameterized by $\{(t, -2t, -t) \mid t \in \mathbf{R}\}$.

(b) Substituting $x = 0$ into these equations gives $z = 6$ and $y = 5$. Hence, $(0, 5, 6)$ lies on this line. Similarly, $(6, 5, 0)$ lies on this line, and therefore a direction vector for the line is $(6, 0, -6)$. We find that a parameterization is $\{(6t, 5, -6t + 6) \mid t \in \mathbf{R}\}$.

Section 2.4

1. Here are some answers, but there are more! Draw some pictures of these.

(a) Triangles usually look like other triangles when viewed in perspective, although their shape can change. However, if a vertex of a triangle lies on the horizon, then under the perspective correspondence the triangle would correspond to a segment with two parallel rays emerging. (Often, the reverse of this is seen when a railroad track vanishing at the horizon is

drawn as a triangle with a vertex as the infinite point.) What happens if the horizon passes through a side of the triangle?
(b) Squares can perspectively correspond to other quadrilaterals that need not be squares or even rectangles.
(c) Angles can perspectively correspond to any other angle.
(d) Lengths can change drastically in a perspective correspondence.

Section 3.1

1. (a) $X = \frac{9}{2}$, $Y = -\frac{1}{2}$
(b) $X = 0$, $Y = 0$
(c) $X = t$, $Y = \frac{1}{2}t + 1$, where $t \in \mathbf{R}$ is a parameter
(d) $\emptyset$, that is, there are no solutions.

3. (a) $X = -\frac{1}{2}t + \frac{5}{2}$, $Y = -\frac{5}{2}t + \frac{5}{2}$, $Z = t$, where t is a parameter.
(b) $X = 0, Y = 0, Z = 0$

5. (a) $(7 - t - u - v, t, u, v)$, where t, u, v are parameters
(b) $(3 + t, 2 - u, t, u)$, where t and u are parameters

7. Subtracting twice the first equation from the second gives the equation $0 = v - 2u$. Therefore, this system of equations has a solution only if $v = 2u$. In this case it has infinitely many solutions. In particular, the system will never have a unique solution.

9. If the polarity of the 8-volt battery is reversed, we have to study the system of equations:

$$\begin{aligned} 4I_{CB} + 2I_{BA} \qquad\qquad &= -8 \\ 4I_{CB} \qquad\qquad + I_{BD} &= \ \ 6 \\ I_{CB} - \quad I_{BA} - I_{BD} &= \ \ 0. \end{aligned}$$

This system is the same as the system considered in the section *except* that the 8 is replaced by -8. The solution is $I_{CB} = \frac{2}{7}$, $I_{BA} = \frac{-32}{7}$, and $I_{BD} = \frac{34}{7}$. In this circuit, the 6 volt battery is discharging at $\frac{34}{7}$ amps, a higher rate than the $\frac{32}{7}$ amp discharge of the 8 volt battery. However, since polarities are opposite, the net current flow between B and C is only $\frac{2}{7}$ amp.

Section 3.2

1. (a) Subtract twice the second equation from the first.
(b) Add twice the second equation to the first equation and then add the (new) first equation to the second. (Other methods are possible.)

3. (a) $\{(2 - \frac{1}{2}t, 1 - \frac{1}{2}t, t) \mid t \in \mathbf{R}\}$
(b) $\emptyset$; this system of equations is inconsistent.

(c) $W = t$, $Z = u$, $Y = u - 2t + 8$, and $X = 2u - 3t + 13$, where t and u are parameters
(d) $\{(\frac{2}{3}u - v, \frac{5}{6}u - v, u, v) \mid u, v \in \mathbf{R}\}$.

5. This sequence is invalid because too much was done at once! After one row operation is applied, a further operation must be applied to the *new* system. Here, two operations were applied to the original system. Note that the resulting systems cannot be row equivalent since $(0, -\frac{4}{3}, 0)$ is a solution to the second but not the first.

Remark: The author included this problem because he has seen this type of mistake on student exams. There is nothing wrong with saving some space and writing down a few operations at a time—but be careful and make sure that what you've done really arises from a sequence of operations.

7. Consider the inconsistent systems given by $X + 0Y = 0$ and $X + 0Y = 2$, and the inconsistent system given by $X + Y = 0$ and $X + Y = 2$. Any elementary operations applied to the first system will have zero coefficient in front of Y. Any elementary operations applied to the second system that has zero coefficient for Y will also produce a zero coefficient for X. So the systems are not equivalent.

Section 3.3

1. (a) $\{(t + 2, u - 1, u, t) \mid u, t \in \mathbf{R}\}$
(b) $\varnothing$; this system of equations is inconsistent.

3. Applying Gauss-Jordan elimination, we can assume that this system is in reduced echelon form. (Row operations do not change the solution set.) If the system is inconsistent, it has no solutions, so in particular it cannot have a unique solution. Thus we can assume the system is consistent. Since there are at most three equations in this reduced echelon form, there are at most three determined variables. In particular, there must be at least one free variable (since there are four variables). If we use the process of back-substitution, it follows that there is a solution to the system with this free variable taking any scalar value. This shows the system has more than one solution, which is what we want.

5. After Gaussian elimination we obtain a third equation, which reads $0 = c - b - a$. Therefore, the system will be consistent only if $a + b = c$ to begin with. One expects a result like this because row operations will give linear expressions involving a, b, and c on the right-hand side of the equal sign. Of course, after Gaussian elimination the third equation could have read $Z = \cdots$, in which case the system would have had a solution for all possible a, b, c.

7. Since these systems are consistent, they can be reduced to one of the following types of reduced echelon form systems:

$$\begin{aligned} X \quad &= a \\ Y &= b, \end{aligned} \qquad \begin{aligned} X + aY &= b \\ 0Y &= 0, \end{aligned} \qquad \begin{aligned} Y &= a \\ 0 &= 0. \end{aligned}$$

In the first case, the system has a unique solution, completely determined by the echelon form. In the second case, the solution set is precisely $\{(b-at, t) \mid t \in \mathbf{R}\}$, and in the third case, the solution set is $\{(t, a) \mid t \in \mathbf{R}\}$. In each case, the set of solutions is completely determined by the echelon form, and moreover, if the echelon forms are different, the solution sets are different. Thus, two consistent systems that have precisely the same solution set must have the same reduced echelon form. It follows that they are row equivalent.

9. Suppose that the system of equations

$$\begin{aligned} a_{11}X + a_{12}Y + a_{13}Z &= b_1 \\ a_{21}X + a_{22}Y + a_{23}Z &= b_2 \\ a_{31}X + a_{32}Y + a_{33}Z &= b_3 \end{aligned}$$

has the two different solutions (r, s, t) and (r', s', t'). Then $(r-r', s-s', t-t')$ is a solution to the system of equations

$$\begin{aligned} a_{11}X + a_{12}Y + a_{13}Z &= 0 \\ a_{21}X + a_{22}Y + a_{23}Z &= 0 \\ a_{31}X + a_{32}Y + a_{33}Z &= 0. \end{aligned}$$

Since the two solutions (r, s, t) and (r', s', t') of the original system are distinct, we know that $(r - r', s - s', t - t') \neq (0, 0, 0)$. Therefore, whenever $k \in \mathbf{R}$, $(r + k(r - r'), s + k(s - s'), t + k(t - t'))$ is a solution to the original system. This shows that the original system has infinitely many solutions.

11. Gaussian elimination applied to the augmented matrix gives

$$\left(\begin{array}{cc|c} 9 & 8 & 2 \\ 1 & 2 & 2 \\ 2 & 2 & 2 \end{array}\right) \mapsto \left(\begin{array}{cc|c} 1 & 2 & 2 \\ 0 & -10 & -16 \\ 0 & -2 & -2 \end{array}\right),$$

which is inconsistent. It follows that there are no solutions to the original equation.

13. (a) No inverse (b) $\begin{pmatrix} 3 & 5 \\ -1 & -2 \end{pmatrix}$

(c) $\begin{pmatrix} -\frac{1}{2} & \frac{1}{2} & \frac{1}{2} \\ \frac{3}{2} & -\frac{1}{2} & -\frac{1}{2} \\ 1 & 0 & -1 \end{pmatrix}$ (d) $\begin{pmatrix} \frac{10}{9} & -\frac{8}{9} & -\frac{1}{9} \\ -\frac{4}{9} & \frac{5}{9} & \frac{4}{9} \\ \frac{1}{9} & \frac{1}{9} & -\frac{1}{9} \end{pmatrix}$

15. A matrix A is invertible if and only if there is a matrix B such that $AB = I$. In this case, we claim that A^n is invertible with inverse B^n. To see this we repeatedly apply the associative law for matrix multiplication

and check: $A^nB^n = (AA\cdots AA)(BB\cdots BB) = (AA\cdots A)(AB)(B\cdots BB) = (AA\cdots A)I_n(B\cdots BB) = (AA\cdots A)\cdot(B\cdots BB) = (AA\cdots)(AB)(\cdots BB) = \cdots = (AA)(BB) = A(AB)B = AIB = AB = I$, as required.

17. Observe that a triangular matrix A is row equivalent to a row-echelon matrix with diagonal entries 1 whenever the diagonal entries of A are all nonzero. It follows from this that for any column of variables $\vec{X}$ and constants $\vec{v}$ the system of equations $A\vec{X} = \vec{v}$ can always be solved using back-substitution. As a consequence we see that the matrix equation $AM = I$ can be solved. This shows that A has an inverse.

19. We know that $AB = BA$. We multiply both equations on both sides by A^{-1} to find that $A^{-1}(AB)A^{-1} = A^{-1}(BA)A^{-1}$. Since $AA^{-1} = I = A^{-1}A$, we can use associativity to find that $BA^{-1} = A^{-1}B$. We next multiply these expressions on both sides by B^{-1} to find that $B^{-1}(A^{-1}B)B^{-1} = B^{-1}(BA^{-1})B^{-1}$. Using associativity and $BB^{-1} = I$ shows that $B^{-1}A^{-1} = A^{-1}B^{-1}$, as required.

Section 3.4

1. (a) $\begin{pmatrix} 1 & 0 & \frac{3}{4} \\ 0 & 1 & \frac{3}{2} \\ 0 & 0 & 0 \end{pmatrix}$, rank 2. (b) $\begin{pmatrix} 1 & 0 & 0 \\ 0 & 1 & \frac{3}{2} \\ 0 & 0 & 0 \\ 0 & 0 & 0 \end{pmatrix}$, rank 2.

(c) $\begin{pmatrix} 1 & 1 & 1 & 0 \\ 0 & 0 & 0 & 1 \\ 0 & 0 & 0 & 0 \end{pmatrix}$, rank 2. (d) $\begin{pmatrix} 1 & 0 & 1 \\ 0 & 1 & 0 \\ 0 & 0 & 0 \end{pmatrix}$, rank 2.

3. The reduced row-echelon form of this matrix is $\begin{pmatrix} 1 & 0 & 4 \\ 0 & 1 & -\frac{1}{2} \\ 0 & 0 & 0 \end{pmatrix}$. All of its row-echelon forms can be obtained by adding an arbitrary multiple of the second row to the first. They are $\begin{pmatrix} 1 & t & 4-\frac{1}{2}t \\ 0 & 1 & -\frac{1}{2} \\ 0 & 0 & 0 \end{pmatrix}$, where t is a real number.

5. Applying Gaussian elimination to this system gives

$$\left(\begin{array}{cccc|c} 1 & 2 & 3 & 0 & 6 \\ 0 & 1 & 4 & 1 & 6 \end{array}\right) \mapsto \left(\begin{array}{cccc|c} 1 & 0 & -5 & -2 & -6 \\ 0 & 1 & 4 & 1 & 6 \end{array}\right).$$

The solutions to the associated homogeneous system can be found by replacing the right-hand columns by zeros and using back-substitution. They are $\{(5u + 2v, -4u - v, u, v) \mid u, v \in \mathbf{R}\}$. Theorem 5 guarantees that this solution set is closed under vector addition and scalar multiplication. Using back-substitution with $Z = W = 0$, we see that the point $(-6, 6, 0, 0)$ is a solution to the original system. Theorem 6 now shows that the solutions to the original system can be obtained by adding this solution to all the homogeneous solutions. They are $\{(-6+5u+2v, 6-4u-v, u, v) \mid u, v \in \mathbf{R}\}$. Finally, Theorem 7 guarantees that the system is consistent since the rank of the coefficient matrix equals the rank of the augmented matrix.

7. (a) This system is inconsistent. The associated homogeneous system has as its only solution $(0, 0)$.
(b) The ranks of the coefficient matrix and the augmented matrix for this system are both 3. The corollary shows that this system and its associated homogeneous system both require one parameter to describe their solution sets. The set of solutions to the system is $\{(-\frac{1}{4}t, 1+\frac{1}{4}t, -\frac{1}{2}t, t) \mid t \in \mathbf{R}\}$.
(c) The ranks of the coefficient matrix and the augmented matrix for this system are both 3. Therefore, both systems have unique solutions. The solution to the original system is $(-\frac{1}{2}, -\frac{1}{2}, \frac{3}{2})$.
(d) The ranks of the coefficient matrix and the augmented matrix for this system are both 1. Therefore, two parameters are needed to describe each solution set. The set of solutions to the original system are $\{(1-u-7v, u, v) \mid u, v \in \mathbf{R}\}$.

9. For example, consider the system described by the augmented matrix

$$\left(\begin{array}{cccc|c} 1 & 0 & 0 & 0 & 0 \\ 0 & 1 & 0 & 0 & 0 \\ 1 & 0 & 0 & 0 & 1 \\ 0 & 1 & 0 & 0 & 1 \end{array}\right).$$

11. Since B is a 5×2 matrix, the matrix equation $MX = B$ is equivalent to two systems of five equations in seven unknowns. Since M has rank-5, any system of equations with coefficient matrix M has a solution because the rank of the augmented matrix must also be 5. This shows that $MX = B$ can always be solved for X (which will be a 7×2 matrix).

Section 3.5

1. (a) The maximum value of $f(X, Y)$ is 31 and occurs where $X = 0$ and $Y = 6$.
(b) The maximum value of $g(X, Y)$ is 18 and occurs where $X = 2$ and $Y = 4$.

3. The maximum value of $h(X, Y)$ is $\frac{11}{3}$ and occurs where $X = \frac{8}{3}$ and $Y = 0$. The simplex algorithm worked quickly because of the negative coefficient in the function $h(X, Y)$. Only one coefficient in the bottom row of the tableau had to be made positive.

5. The maximum value of $T(X, Y, Z)$ is 48 and occurs where $X = \frac{22}{3}$, $Y = 0$, and $Z = \frac{4}{3}$.

7. The maximum value of 92 occurs where $X = 2$, $Y = 1$, and $Z = 0$.

Section 4.1

1. (a) no (b) yes (c) no

3. (a) Multiplication on the left by P has the effect of reordering the rows of A; thus PA has the third row of A as its first row, the first row of A as its second row, and the second row of A as its third row.
(b) The matrix QA has as its first row the sum of the rows of A, as its second row the sum of the second and third rows of A, and the same third row as A.

5. Suppose that $A = \begin{pmatrix} a & b \\ c & d \end{pmatrix}$. Since $AB = BA$ for all matrices B, we can in particular choose $B = \begin{pmatrix} 0 & 1 \\ 0 & 0 \end{pmatrix}$. We compute

$$\begin{pmatrix} a & b \\ c & d \end{pmatrix}\begin{pmatrix} 0 & 1 \\ 0 & 0 \end{pmatrix} = \begin{pmatrix} 0 & a \\ 0 & c \end{pmatrix} = \begin{pmatrix} 0 & 1 \\ 0 & 0 \end{pmatrix}\begin{pmatrix} a & b \\ c & d \end{pmatrix} = \begin{pmatrix} c & d \\ 0 & 0 \end{pmatrix}.$$

We conclude that $a = d$ and that $c = 0$. The analogous computation, using $B = \begin{pmatrix} 0 & 0 \\ 1 & 0 \end{pmatrix}$, shows that $b = 0$. This shows that $A = \begin{pmatrix} a & 0 \\ 0 & a \end{pmatrix}$.

7. (a) We use the associative law repeatedly. Then $(AB)(B^{-1}A^{-1}) = A(BB^{-1})A^{-1} = AI_nA^{-1} = AA^{-1} = I_n$ and $(B^{-1}A^{-1})(AB) = B^{-1}(A^{-1}A)B = B^{-1}I_nB = B^{-1}B = I_n$. This shows that AB is invertible with inverse $B^{-1}A^{-1}$.
(b) Theorem 12 shows that $A^t(A^{-1})^t = (A^{-1}A)^t = (I_n)^t = I_n$ and that $(A^{-1})^tA^t = (AA^{-1})^t = (I_n)^t = I_n$. This shows that A^t is invertible with inverse $(A^{-1})^t$.

9. Suppose that $T = (t_{ij})$ is an upper triangular $n \times n$ matrix and $T^n = 0$. The definition of matrix multiplication shows that $T^n(i, i) = t_{ii}^n$ for each i. Hence $t_{ii}^n = 0$; this shows that $t_{ii} = 0$ for each i. Yes, the converse is also true. Hint: Show that $T^k(i, j) = 0$ whenever $i \leq j + k$ by induction on k. The case of $k = n$ gives $T^n = 0$.

11. (a) In order for $A = A^t$, the number of rows of A must equal the number of columns of A^t. So, A must be square.
(b) See that $(A + A^t)(i, j) = A(i, j) + A^t(i, j) = A(i, j) + A(j, i)$. So $(A + A^t)(i, j) = (A + A^t)(j, i)$, which shows that $A + A^t$ is symmetric.
(c) Suppose that AB is symmetric. Then $AB = (AB)^t = B^tA^t = BA$, since both A and B are symmetric. This shows that A and B commute. Conversely, if A and B commute, then $(AB)^t = B^tA^t = BA = AB$, so AB is symmetric.

13. There are many possible solutions to these problems. Some possible solutions are

(a) $\begin{pmatrix} 1 & 4 \\ 3 & 2 \end{pmatrix} = \begin{pmatrix} 1 & 0 \\ 3 & 1 \end{pmatrix} \begin{pmatrix} 1 & 0 \\ 0 & -10 \end{pmatrix} \begin{pmatrix} 1 & 4 \\ 0 & 1 \end{pmatrix}$;

(b) $\begin{pmatrix} 1 & 0 & 1 \\ 1 & 1 & 0 \\ 0 & 0 & 2 \end{pmatrix} = \begin{pmatrix} 1 & 0 & 0 \\ 1 & 1 & 0 \\ 0 & 0 & 1 \end{pmatrix} \begin{pmatrix} 1 & 0 & 0 \\ 0 & 1 & 0 \\ 0 & 0 & 2 \end{pmatrix} \begin{pmatrix} 1 & 0 & 1 \\ 0 & 1 & 0 \\ 0 & 0 & 1 \end{pmatrix}$.

15. An upper triangular $n \times n$ matrix has rank n precisely when all its diagonal entries are nonzero. This is exactly the situation when it can be row-reduced to the identity matrix.

Section 4.2

1. This is a calculator exercise, and the solution won't be given. But when looking at high powers of a matrix you should be concerned with three possibilities, namely, when the entries all become close to zero, when they stabilize, and when they get larger and larger without bound. In looking at these examples, try to decide why these different possibilities occur.

3. Each month this bank adds to your deposit 1% of your previous balance. This means that if $D(n)$ represents the deposit after n months, then $D(n) = D(n-1) + 0.01D(n-1)$. This is a difference equation. Iterating this we find that $D(n) = (1.01)^n D(0)$, which is the usual formula for compounding 1% interest per month for n months.

Section 4.3

1. (a) The determinant is 11.
(b) The determinant is -8.
(c) The determinant is 0.
(d) The determinant is -2.
(e) The determinant is 2.
(f) The determinant is 5.

3. An $n \times n$ matrix A has n^2 entries. If more than $n^2 - n$ entries of A are zero, then A has at most $n - 1$ nonzero entries. Since A has n rows, this means that there is some row with all zero entries. The cofactor expansion of the determinant along this row must be zero. Hence the determinant of A is zero.

5. $\begin{pmatrix} a & b \\ c & d \end{pmatrix} \begin{pmatrix} e & f \\ g & h \end{pmatrix} = \begin{pmatrix} ae + bg & af + bh \\ ce + dg & cf + dh \end{pmatrix}$, which has determinant $(ae + bg)(cf + dh) - (af + bh)(ce + dg) = [acef + adeh + bcfg + bdgh] - [acef + adfg + bceh + bdgh] = adeh + bcfg - adfg - bceh = ad(eh - fg) + bc(fg - eh) = (ad - bc)(eh - fg)$, as required.

7. The matrix $-A$ is obtained from A by multiplying each of the rows of A by -1. If n is odd, Theorem 14(i) shows $\det(-A) = (-1)^n \det(A) = -\det(A)$. Now, if A is skew symmetric, as $A^t = -A$, the corollary following

Theorem 13 gives $\det(A) = \det(A^t) = \det(-A) = -\det(A)$. This shows that $\det(A) = 0$. In case n is even, the result is not true. For example, the skew-symmetric matrix $\begin{pmatrix} 0 & -1 \\ 1 & 0 \end{pmatrix}$ has determinant 1.

Section 4.4

1. (a) $\frac{1}{9}\begin{pmatrix} -3 & -2 & 1 \\ -3 & 1 & 4 \\ -6 & 2 & -1 \end{pmatrix}$ (b) $\frac{1}{3}\begin{pmatrix} 4 & -2 & 1 \\ -7 & 5 & -1 \\ -8 & 7 & -2 \end{pmatrix}$

3. (a) 6 (b) 0 (c) 143 (d) -1

5. Theorem 16 shows that $\det(A^n) = \det(A)^n$. Therefore, if $\det(A^n) = 1$, then $\det(A)^n = 1$. It follows that $\det(A) = 1$ if n is odd, and $\det(A) = \pm 1$ if n is even.

Section 4.5

1. (a) $\begin{pmatrix} 1 & 3 \\ 2 & 7 \end{pmatrix} = \begin{pmatrix} 1 & 0 \\ 2 & 1 \end{pmatrix}\begin{pmatrix} 1 & 3 \\ 0 & 1 \end{pmatrix}$

(b) $\begin{pmatrix} 1 & 1 & 0 \\ 0 & 1 & 1 \\ 1 & 1 & 1 \end{pmatrix} = \begin{pmatrix} 1 & 0 & 0 \\ 0 & 1 & 0 \\ 1 & 0 & 1 \end{pmatrix}\begin{pmatrix} 1 & 1 & 0 \\ 0 & 1 & 1 \\ 0 & 0 & 1 \end{pmatrix}$

(c) $\begin{pmatrix} 1 & 1 & 3 \\ 1 & 0 & 1 \\ 1 & 0 & 0 \end{pmatrix} = \begin{pmatrix} 1 & 0 & 0 \\ 1 & 1 & 0 \\ 1 & 1 & 1 \end{pmatrix}\begin{pmatrix} 1 & 1 & 3 \\ 0 & -1 & -2 \\ 0 & 0 & -1 \end{pmatrix}$

3. (a) $\begin{pmatrix} 0 & 1 \\ 1 & 0 \end{pmatrix}\begin{pmatrix} 0 & 3 \\ 2 & 7 \end{pmatrix} = \begin{pmatrix} 1 & 0 \\ 0 & 1 \end{pmatrix}\begin{pmatrix} 2 & 7 \\ 0 & 3 \end{pmatrix}$

(b) $\begin{pmatrix} 0 & 0 & 1 \\ 1 & 0 & 0 \\ 0 & 1 & 0 \end{pmatrix}\begin{pmatrix} 0 & 1 & 0 \\ 0 & 1 & 1 \\ 1 & 1 & 0 \end{pmatrix} = \begin{pmatrix} 1 & 0 & 0 \\ 0 & 1 & 0 \\ 0 & 1 & 1 \end{pmatrix}\begin{pmatrix} 1 & 1 & 0 \\ 0 & 1 & 0 \\ 0 & 0 & 1 \end{pmatrix}$

(c) $\begin{pmatrix} 0 & 0 & 1 \\ 1 & 0 & 0 \\ 0 & 1 & 0 \end{pmatrix}\begin{pmatrix} 0 & 1 & 3 \\ 1 & 0 & 1 \\ 1 & 0 & 0 \end{pmatrix} = \begin{pmatrix} 1 & 0 & 0 \\ 0 & 1 & 0 \\ 1 & 0 & 1 \end{pmatrix}\begin{pmatrix} 1 & 0 & 0 \\ 0 & 1 & 3 \\ 0 & 0 & 1 \end{pmatrix}$

Section 5.1

1. No.

3. (a) $\left\{\begin{pmatrix} 2t \\ t \\ 2t \end{pmatrix} \mid t \in \mathbf{R}\right\}$ (b) $\mathbf{R}^3$ (c) $\left\{\begin{pmatrix} r \\ 2t+u \\ t+3u \\ u \\ t \end{pmatrix} \mid t, u \in \mathbf{R}\right\}$ (d) $\mathbf{R}^2$

5. It is clear that $\text{span}\{\vec{v}_1, \vec{v}_2, \ldots, \vec{v}_n\} \subseteq \text{span}\{\vec{v}, \vec{v}_1, \vec{v}_2, \ldots, \vec{v}_n\}$, since any linear combination of $\vec{v}_1, \vec{v}_2, \ldots, \vec{v}_n$ is also a linear combination of $\vec{v}, \vec{v}_1, \vec{v}_2, \ldots,$

$\vec{v}_n$. For the reverse inclusion, we may suppose that $\vec{v} = a_1\vec{v}_1 + a_2\vec{v}_2 + \cdots + a_n\vec{v}_n$ for some real numbers $a_1, a_2, \ldots, a_n$. Consider $\vec{w} \in \text{span}\{\vec{v}, \vec{v}_1, \vec{v}_2, \ldots, \vec{v}_n\}$. If $\vec{w} = c\vec{v} + b_1\vec{v}_1 + b_2\vec{v}_2 + \cdots + b_n\vec{v}_n$, then $\vec{w} = c(a_1\vec{v}_1 + a_2\vec{v}_2 + \cdots + a_n\vec{v}_n) + b_1\vec{v}_1 + b_2\vec{v}_2 + \cdots + b_n\vec{v}_n = (ca_1 + b_1)\vec{v}_1 + (ca_2 + b_2)\vec{v}_2 + \cdots + (ca_n + b_n)\vec{v}_n \in \text{span}\{\vec{v}_1, \vec{v}_2, \ldots, \vec{v}_n\}$. This shows what we wanted.

7. $V \cap W$ will be always be a subspace. Since each of V and W is closed under vector addition and scalar multiplication, so is $V \cap W$.

However, $V \cup W$ need *not* be a subspace. The only time it is a subspace is when either $V \subseteq W$ or $W \subseteq V$, in which case the union $V \cup W$ will be either V or W.

Section 5.2

1. (a) A spanning set for the row space is $\{(1, 2, 1)\}$, a spanning set for the column space is $\{(1, 3, 2)\}$, and a spanning set for the null space is $\{(-1, 0, 1), (-2, 1, 0)\}$.
(b) A spanning set for the row space is $\{(1, 3, 4), (1, 3, 5)\}$, a spanning set for the column space is $\{(1, 0), (0, 1)\}$, and a spanning set for the null space is $\{(-3, 1, 0)\}$.
(c) A spanning set for the row space is $\{(2, 1, 4), (1, 0, 5)\}$, a spanning set for the column space is $\{(2, 1, 3), (1, 0, 1)\}$, and a spanning set for the null space is $\{(-5, 6, 1)\}$.

3. (a) Not linearly independent (b) Not linearly independent
(c) Not linearly independent (d) Not linearly independent

5. (a) Any pair is a minimal spanning set.
(b) Any pair except the last two is a minimal spanning set.
(c) All three form a minimal spanning set.
(d) Any pair excluding the zero vector forms a minimal spanning set.

7. Suppose that $\{\vec{v}_1, \vec{v}_2, \ldots, \vec{v}_s\} \subseteq \{\vec{v}_1, \vec{v}_2, \ldots, \vec{v}_n\}$ is our subset, where $s \leq n$. Assume that $a_1\vec{v}_1 + a_2\vec{v}_2 + \cdots + a_s\vec{v}_s = \vec{0}$ for some real numbers $a_1, a_2, \ldots, a_s$. Then we have $a_1\vec{v}_1 + a_2\vec{v}_2 + \cdots + a_s\vec{v}_s + 0\vec{v}_{s+1} + \cdots 0\vec{v}_n = \vec{0}$. Since $\{\vec{v}_1, \vec{v}_2, \ldots, \vec{v}_n\}$ is linearly independent, we find that $a_1 = 0$, $a_2 = 0$, $\ldots$, $a_s = 0$. This shows that $\{\vec{v}_1, \vec{v}_2, \ldots, \vec{v}_s\}$ is linearly independent.

9. Suppose that $a_1\vec{v}_1 + a_2\vec{v}_2 + \cdots + a_{n-1}\vec{v}_{n-1} + a_n\vec{v}_n = \vec{0}$, where the $a_i \in \mathbf{R}$. Suppose first that $a_n \neq 0$. Then we can write $\vec{v}_n = -\frac{1}{a_n}(a_1\vec{v}_1 + a_2\vec{v}_2 + \cdots + a_{n-1}\vec{v}_{n-1})$. But this contradicts the assumption that $\vec{v}_n \notin \text{span}\{\vec{v}_1, \vec{v}_2, \ldots, \vec{v}_{n-1}\}$. Consequently, $a_n = 0$. This means that $a_1\vec{v}_1 + a_2\vec{v}_2 + \cdots + a_{n-1}\vec{v}_{n-1} = \vec{0}$. We are now done, by using induction on n, for we have exactly the same problem except with $n - 1$ vectors instead of n.

11. Suppose that $\{\vec{v}_1, \vec{v}_2, \ldots, \vec{v}_n\}$ is linearly independent and A is invertible. Since we want to show that $\{A\vec{v}_1, A\vec{v}_2, \ldots, A\vec{v}_n\}$ is linearly independent, we assume $\vec{0} = a_1A\vec{v}_1 + a_2A\vec{v}_2 + \cdots + a_nA\vec{v}_n$, where $a_i \in \mathbf{R}$. Multiplying by

A^{-1}, we find $\vec{0} = A^{-1}\vec{0} = A^{-1}(a_1 A\vec{v}_1 + \cdots + a_n A\vec{v}_n) = a_1\vec{v}_1 + \cdots + a_n\vec{v}_n$. The linear independence of $\{\vec{v}_1, \vec{v}_2, \ldots, \vec{v}_n\}$ gives $a_1 = 0, a_2 = 0, \ldots, a_n = 0$ and shows that $\{A\vec{v}_1, A\vec{v}_2, \ldots, A\vec{v}_n\}$ is linearly independent.

Section 5.3

1. $\left\{\begin{pmatrix} -3 \\ 5 \\ -\frac{13}{5} \\ 0 \\ 1 \end{pmatrix}, \begin{pmatrix} 0 \\ 0 \\ -\frac{2}{5} \\ 1 \\ 0 \end{pmatrix}\right\}$ is an answer (others possible, bases are not unique).

3. (a) The matrix has rank 2. A basis for its row space is $\{(1,1,3),(3,0,2)\}$, a basis for its column space (which is $\mathbf{R}^2$) is $\{(1,0),(0,1)\}$, and a basis for the null space is $\{(-2,-7,3)\}$.
(b) The matrix has rank 2. A basis for its row space is $\{(1,0),(0,1)\}$, a basis for its column space is $\{(1,0,2,1),(1,1,2,3)\}$, and a basis for the null space is the empty set, $\varnothing$ (that is, the null space is $\{\vec{0}\}$).
(c) Row-reducing shows the matrix has rank 3. Therefore, its row and column spaces are $\mathbf{R}^3$ and the standard basis works for each. A basis for the null space is $\varnothing$.
(d) The matrix has rank 3. The standard basis for $\mathbf{R}^3$ works for its row space, the three columns of the matrix give a basis for its column space, and a basis for the null space is $\varnothing$.

5. (a) $\left\{\begin{pmatrix} 1 \\ 2 \\ 3 \end{pmatrix}, \begin{pmatrix} 4 \\ 5 \\ 6 \end{pmatrix}\right\}$ (b) $\left\{\begin{pmatrix} 1 \\ 0 \\ 1 \end{pmatrix}, \begin{pmatrix} 2 \\ 4 \\ 8 \end{pmatrix}\right\}$

7. Observe that

$$\begin{pmatrix} 1 & 0 & 0 \\ 0 & 1 & 0 \end{pmatrix} \quad \text{and} \quad \begin{pmatrix} 1 & 0 & 1 \\ 0 & 1 & 0 \end{pmatrix}$$

are both rank-2, 2×3 matrices. In fact, both are reduced row-echelon matrices. Observe that their row spaces are different since the row space of the first matrix cannot have an element whose third coordinate is nonzero, while the second matrix has such an element in its row space.

9. We can write $\vec{e}_1 = (1,1,1) - (0,1,1)$, $\vec{e}_2 = (1,1,1) - (1,0,1)$, and $\vec{e}_3 = (1,0,1) + (0,1,1) - (1,1,1)$. It follows that every vector in $\mathbf{R}^3$ is a linear combination of these three vectors. Since $\dim(\mathbf{R}^3) = 3$, we see that they must be a basis for $\mathbf{R}^3$.

Section 6.1

1. (a) $\|\vec{u}\| = \sqrt{2}$ (b) $\|\vec{v} - \vec{w}\| = \left\| \begin{pmatrix} 2 \\ 1 \end{pmatrix} \right\| = \sqrt{5}$

(c) $\vec{w} \cdot \vec{v} = 12$ (d) $\begin{pmatrix} 1 \\ 2 \end{pmatrix} \cdot \begin{pmatrix} 1 \\ -4 \end{pmatrix} = -7$

3. (a) $\|(1, -7)\| = \sqrt{50} = 5\sqrt{2}$ (b) 1

(c) $\cos^{-1}\left(\frac{(1,-7)\cdot(0,-1)}{5\sqrt{2}\cdot 1}\right) = \cos^{-1}\left(\frac{7}{5\sqrt{2}}\right) = \cos^{-1}(.9899) = .141$ radians

5. We consider $\vec{v}, \vec{w} \in \mathbf{R}^2$ or $\mathbf{R}^3$. We expand according to properties given in Theorem 34:

$$
\begin{aligned}
\|\vec{v} + \vec{w}\|^2 + \|\vec{v} - \vec{w}\|^2 &= \langle \vec{v} + \vec{w}, \vec{v} + \vec{w} \rangle + \langle \vec{v} - \vec{w}, \vec{v} - \vec{w} \rangle \\
&= \langle \vec{v}, \vec{v} \rangle + \langle \vec{v}, \vec{w} \rangle + \langle \vec{w}, \vec{v} \rangle + \langle \vec{w}, \vec{w} \rangle \\
&\quad + \langle \vec{v}, \vec{v} \rangle - \langle \vec{v}, \vec{w} \rangle - \langle \vec{w}, \vec{v} \rangle + \langle \vec{w}, \vec{w} \rangle \\
&= 2\langle \vec{v}, \vec{v} \rangle + 2\langle \vec{w}, \vec{w} \rangle \\
&= 2\|\vec{v}\|^2 + 2\|\vec{w}\|^2.
\end{aligned}
$$

Section 6.2

1. (a) $\begin{pmatrix} -\frac{2}{\sqrt{5}} \\ -\frac{4}{\sqrt{5}} \end{pmatrix}$ (b) $\begin{pmatrix} -\sqrt{2} \\ \sqrt{2} \end{pmatrix}$ (c) $\begin{pmatrix} -\frac{12}{\sqrt{5}} \\ 3 - \frac{24}{\sqrt{5}} \end{pmatrix} = \begin{pmatrix} -\frac{12\sqrt{5}}{5} \\ \frac{15-24\sqrt{5}}{5} \end{pmatrix}$

3. (a) $\ell = \{(7t, t) \mid t \in \mathbf{R}\}$ (b) $\ell = \{(1 + t, 1) \mid t \in \mathbf{R}\}$

5. The distance between P_1 and both P_2 and P_6 is 2, because we set up the model that way. The distance between P_1 and both P_3 and P_5 is $\frac{4\sqrt{6}}{3} \approx 3.27$, and the distance between P_1 and P_4 is $\frac{2\sqrt{33}}{3} \approx 3.83$.

To make the comparable calculations for benzene, we have to find the distances between the vertices of a regular hexagon whose side also has length 2. Doubling the coordinates for the hexagon given in Sec. 2.1 and computing the distances, we find the distance between P_1 and both P_3 and P_5 is $2\sqrt{3} \approx 3.46$ and the distance between P_1 and P_4 is 4. Note that the distances between the carbon atoms in benzene is greater than cyclohexane, mostly because the angle (120°) between the carbons in benzene is greater than in cyclohexane.

Section 6.3

1. (a) $\begin{pmatrix} 1 \\ -3 \\ 1 \end{pmatrix}$ (b) $\begin{pmatrix} 8 \\ -5 \\ -11 \end{pmatrix}$ (c) 19 (d) 0

3. (a) $3\sqrt{2}$ (b) $\frac{5\sqrt{6}}{2}$

5. (a) $\frac{\sqrt{11}}{11}$ (b) 1

7. (a) $\{(3-2t, 2-2t, 1+t) \mid t \in \mathbf{R}\}$ (b) $\{(1+t, t, 1+t) \mid t \in \mathbf{R}\}$

9. $x - y - 2(z-1) = 0$

11. (a) $\theta = \frac{\pi}{3}$ radians $= 60°$ (b) $\theta = \frac{\pi}{4}$ radians $= 45°$

13. For this we use the properties of the cross product given in Theorem 40. Since $\vec{u} + \vec{v} + \vec{w} = \vec{0}$, we find that $\vec{0} = (\vec{u} + \vec{v} + \vec{w}) \times \vec{v} = \vec{u} \times \vec{v} + \vec{v} \times \vec{v} + \vec{w} \times \vec{v} = \vec{u} \times \vec{v} + \vec{w} \times \vec{v}$. Hence $\vec{u} \times \vec{v} = -\vec{w} \times \vec{v} = \vec{v} \times \vec{w}$. The other equality is established similarly.

Section 7.1

1. (a) $X^2 - 4X - 5 = (X+1)(X-5)$, $E_{-1} = \text{span}\left\{\begin{pmatrix} -1 \\ 1 \end{pmatrix}\right\}$, $E_5 = \text{span}\left\{\begin{pmatrix} 1 \\ 1 \end{pmatrix}\right\}$

(b) $X^2 - 1$, $E_1 = \text{span}\left\{\begin{pmatrix} 1 \\ 1 \end{pmatrix}\right\}$, $E_{-1} = \text{span}\left\{\begin{pmatrix} -1 \\ 1 \end{pmatrix}\right\}$

(c) $X^2 + 1$; there are no real eigenvalues

3. To solve this, we need a matrix $\begin{pmatrix} a & b \\ c & d \end{pmatrix}$ such that

$$\ker\begin{pmatrix} a-1 & b \\ c & d-1 \end{pmatrix} = \text{span}\left\{\begin{pmatrix} 1 \\ -2 \end{pmatrix}\right\}$$

and

$$\ker\begin{pmatrix} a+1 & b \\ c & d+1 \end{pmatrix} = \text{span}\left\{\begin{pmatrix} 2 \\ 1 \end{pmatrix}\right\}.$$

The first condition means that $(a-1) - 2b = 0$ and $c - 2(d-1) = 0$, and the second condition means that $2(a+1) + b = 0$ and $2c + (d+1) = 0$. Solving these systems for a, b, c, and d give the matrix $\frac{1}{5}\begin{pmatrix} -3 & -4 \\ -4 & 3 \end{pmatrix}$. Left multiplication by this matrix fixes all points on the line given by E_1 and takes all points on line E_{-1} to their opposite. Since E_1 and E_{-1} are perpendicular, this is what a reflection in the line E_1 does.

5. $E_1 = \text{span}\left\{\begin{pmatrix} 1 \\ -1 \\ 0 \end{pmatrix}\right\}$ and $E_2 = \text{span}\left\{\begin{pmatrix} 2 \\ 0 \\ -1 \end{pmatrix}\right\}$

7. By a direct calculation

$$\det\begin{pmatrix} X-2 & 0 & -1 \\ -1 & X-1 & 0 \\ -1 & 0 & X-1 \end{pmatrix} = (X-2)(X-1)(X-1) - (X-1)$$

$$= (X-1)[X^2 - 3X + 1].$$

Therefore, the given numbers are the eigenvalues since (using the quadratic formula to factor the quadratic) they are the roots of the characteristic polynomial.

9. If A is a real 3×3 matrix, then its characteristic polynomial $C_A(X)$ is a real polynomial of degree 3. However, every-odd degree real polynomial has a real root. Since the roots of $C_A(X)$ are the eigenvalues of A, A has a real eigenvalue c. By Theorem 43 this shows that $rk(A - cI_3) \leq 2$.

Section 7.2

1. (a) The characteristic polynomial is $X^3 - 12X - 4 = (X - 4)(X + 2)^2$. The eigenspaces are

$$E_4 = \text{span}\left\{\begin{pmatrix}1\\1\\1\end{pmatrix}\right\} \quad \text{and} \quad E_{-2} = \text{span}\left\{\begin{pmatrix}-1\\0\\1\end{pmatrix}, \begin{pmatrix}0\\-1\\1\end{pmatrix}\right\}.$$

If P is the matrix

$$\begin{pmatrix}1 & -1 & 0\\1 & 0 & -1\\1 & 1 & 1\end{pmatrix}, \quad \text{then} \quad P^{-1}AP = \begin{pmatrix}4 & 0 & 0\\0 & -2 & 0\\0 & 0 & -2\end{pmatrix}.$$

(b) The characteristic polynomial is $X^3 - 3X^2 + 3X - 1 = (X - 1)^3$. Hence A has the single eigenvalue 1. The eigenspace

$$E_1 = \text{span}\left\{\begin{pmatrix}1\\1\\1\end{pmatrix}\right\}$$

is one-dimensional and it follows that A is not diagonalizable.

(c) The characteristic polynomial is $X^2 + X + 1$, and thus this matrix has no real eigenvalues.

3. (a) First note that $\vec{v}$ and $\vec{w}$ are in the column space of A since we have $A\vec{v} = \vec{v}$ and $\frac{1}{2}A\vec{w} = \vec{w}$. Therefore, $\text{rk}(A) \geq 2$ since $\{\vec{v}, \vec{w}\}$ is linearly independent. But as $A\vec{u} = \vec{0}$ we see $\vec{u} \in \ker(A)$ so that $\text{null}(A) \geq 1$. Since $\text{rk}(A) + \text{null}(A) = 3$ (by the dimension theorem), we find $\text{rk}(A) = 2$ and $\text{null}(A) = 1$. Therefore a basis for $\text{col}(A)$ is $\{\vec{v}, \vec{w}\}$.

(b) Since $\text{null}(A) = 1$, we see that a basis for $\ker(A)$ is $\{\vec{u}\}$.

(c) The equations $A\vec{X} = \vec{v}$ and $A\vec{X} = \vec{v} + \vec{w}$ can both be solved for $\vec{X}$ since these vectors lie in the column space of A. However, $A\vec{X} = \vec{u}$ cannot be solved since $\vec{u}$ is not in the column space of A.

5. Since $\ker(A)$ is $(n - 1)$-dimensional, we see that A has $n - 1$ linearly independent eigenvectors $\{\vec{u}_1, \vec{u}_2, \ldots, \vec{u}_{n-1}\}$ of eigenvalue 0. Let $\vec{u}_n$ be another eigenvector of A with nonzero eigenvalue. Then we claim that $\{\vec{u}_1, \vec{u}_2, \ldots, \vec{u}_{n-1}, \vec{u}_n\}$ is linearly independent. For if $a_1\vec{u}_1 + \cdots + a_{n-1}\vec{u}_{n-1} + a_n\vec{u}_n = \vec{0}$, where $a_n \neq 0$, then the set of *two* eigenvectors $\{a_1\vec{u}_1 + \cdots +$

$a_{n-1}\vec{u}_{n-1}, \vec{u}_n\}$ is linearly dependent. This contradicts Theorem 45. It now follows from Theorem 44 that A is diagonalizable.

7. Suppose that $\vec{v}$ is an eigenvector of A with eigenvalue k. Then, $A\vec{v} = k\vec{v}$. But since $A^2 = A$, we also see that $A\vec{v} = A^2\vec{v} = A(k\vec{v}) = kA\vec{v} = k^2\vec{v}$. This shows that $k^2 = k$ and we see that either $k = 0$ or $k = 1$.

9. (a) If $A = P^{-1}BP$ and if $B = Q^{-1}CQ$, then substituting and applying the associative law gives $A = P^{-1}(Q^{-1}CQ)P = (P^{-1}Q^{-1})C(QP)$. Since $(P^{-1}Q^{-1}) = (QP)^{-1}$, we see that A is similar to C.

(b) Only the third pair is similar.

11. Suppose that $\vec{v}$ is an eigenvector of A with associated eigenvalue k. Then, as $A\vec{v} = k\vec{v}$, we find that $A^2\vec{v} = A(A\vec{v}) = A(k\vec{v}) = k^2\vec{v}$ and similarly $A^m\vec{v} = k^m\vec{v}$ for all m. Thus $\vec{v}$ is an eigenvector for A^m. Since A is an $n \times n$ diagonalizable matrix, Theorem 44 shows there is a set of n linearly independent eigenvectors of A. What we have just observed shows that this set is a set of n linearly independent eigenvectors for A^m. Consequently, by Theorem 44 again, A^m is diagonalizable.

As another proof, note that if $P^{-1}AP$ is a diagonal matrix, then so is $(P^{-1}AP)^m = P^{-1}A^mP$. So we see that A^m is diagonalizable. The two proofs given illustrate the geometric and symbolic viewpoints of diagonalizability.

Section 7.3

1. There are many possible solutions. The different diagonalizations can differ by squares on the diagonal and the order in which the diagonal entries are listed. Possible solutions are as follows.

(a) $\begin{pmatrix} 1 & 0 \\ -3 & 2 \end{pmatrix}\begin{pmatrix} 2 & 3 \\ 3 & 1 \end{pmatrix}\begin{pmatrix} 1 & -3 \\ 0 & 2 \end{pmatrix} = \begin{pmatrix} 2 & 0 \\ 0 & -14 \end{pmatrix}$

(b) $\begin{pmatrix} 1 & 1 & 0 \\ 1 & -1 & 0 \\ 1 & 0 & -1 \end{pmatrix}\begin{pmatrix} 0 & 2 & 0 \\ 2 & 0 & 2 \\ 0 & 2 & 0 \end{pmatrix}\begin{pmatrix} 1 & 1 & 1 \\ 1 & -1 & 0 \\ 0 & 0 & -1 \end{pmatrix} = \begin{pmatrix} 4 & 0 & 0 \\ 0 & -4 & 0 \\ 0 & 0 & 0 \end{pmatrix}$

3. (a) $\begin{pmatrix} 2 & -1 & \frac{1}{2} \\ -1 & 2 & 0 \\ \frac{1}{2} & 0 & -1 \end{pmatrix}$ (b) $\begin{pmatrix} 1 & 0 & 0 \\ 0 & 0 & \frac{1}{2} \\ 0 & \frac{1}{2} & 0 \end{pmatrix}$

5. We complete the square and find that $aX^2 + 2bXY + cY^2 = a(X + \frac{b}{a}Y)^2 + (c - \frac{b^2}{a})Y^2$. To be positive definite we need both of the diagonal coefficients $a > 0$ and $c - \frac{b^2}{a} > 0$. Multiplying by $a > 0$, the latter equation is equivalent to $ac - b^2 > 0$, as required.

7. (a) $\begin{pmatrix} \frac{1}{2} & \frac{1}{2} \\ \frac{1}{2} & \frac{1}{2} \end{pmatrix}$ (b) $\begin{pmatrix} \frac{1}{2} & \frac{3}{8} & \frac{1}{8} \\ \frac{1}{2} & \frac{3}{8} & \frac{1}{8} \\ \frac{1}{2} & \frac{3}{8} & \frac{1}{8} \end{pmatrix}$

9. Just as with the usual Fibonacci sequence, the limit is the golden ratio $\frac{1+\sqrt{5}}{2}$.

Section 8.1

1. (a) $\begin{pmatrix} 0 & 1 \\ 1 & 0 \end{pmatrix}$ (b) $\begin{pmatrix} 1 & 0 \\ 0 & 0 \end{pmatrix}$

3. (a) Suppose that $A = \begin{pmatrix} 0 & 1 \\ 1 & 0 \end{pmatrix}$ and $B = \begin{pmatrix} 2 & 0 \\ 0 & 2 \end{pmatrix}$. Then $AB = BA$. It follows from Theorem 54 that if $T_1 = T_A$ and $T_2 = T_B$, then $T_1 \circ T_B = T_2 \circ T_1$.
(b) Suppose that $A = \begin{pmatrix} 0 & 1 \\ 1 & 0 \end{pmatrix}$ and $B = \begin{pmatrix} 2 & 0 \\ 0 & 1 \end{pmatrix}$. Then $AB \neq BA$. It now follows from Theorem 54 that if $T_1 = T_A$ and $T_2 = T_B$, then $T_1 \circ T_B \neq T_2 \circ T_1$.

5. (a) We have that $[T_1] = \begin{pmatrix} 1 & 1 & 0 \\ 0 & 0 & 0 \end{pmatrix}$ and $[T_2] = \begin{pmatrix} 1 & 0 \\ 0 & 1 \\ 1 & 0 \end{pmatrix}$. Theorem 54 shows that $[T_1 \circ T_2] = \begin{pmatrix} 1 & 1 & 0 \\ 0 & 0 & 0 \end{pmatrix} \begin{pmatrix} 1 & 0 \\ 0 & 1 \\ 1 & 0 \end{pmatrix} = \begin{pmatrix} 1 & 1 \\ 0 & 0 \end{pmatrix}$. Similarly, $[T_2 \circ T_1] = \begin{pmatrix} 1 & 1 & 0 \\ 0 & 0 & 0 \\ 1 & 1 & 0 \end{pmatrix}$.

(b) $[T_1 \circ T_2] = \begin{pmatrix} 0 & 0 & 0 \\ 0 & 0 & 0 \end{pmatrix} \begin{pmatrix} ? & ? \\ ? & ? \\ ? & ? \end{pmatrix} = \begin{pmatrix} 0 & 0 \\ 0 & 0 \end{pmatrix}$. Similarly, $[T_2 \circ T_1] = \begin{pmatrix} 0 & 0 & 0 \\ 0 & 0 & 0 \\ 0 & 0 & 0 \end{pmatrix}$.

(c) $[T_1 \circ T_2] = \begin{pmatrix} 1 & 0 & 0 \\ 1 & 0 & 0 \end{pmatrix} \begin{pmatrix} 0 & 1 \\ 1 & 0 \\ 1 & 1 \end{pmatrix} = \begin{pmatrix} 0 & 1 \\ 0 & 1 \end{pmatrix}$. Similarly, $[T_2 \circ T_1] = \begin{pmatrix} 1 & 0 & 0 \\ 1 & 0 & 0 \\ 2 & 0 & 0 \end{pmatrix}$.

7. (a) To show that the set $\{\vec{v}_1, \vec{v}_2, \ldots, \vec{v}_n\}$ is linearly independent, we assume that $a_1\vec{v}_1 + a_2\vec{v}_2 + \cdots + a_n\vec{v}_n = \vec{0}$. Applying the linear transformation T, we find $\vec{0} = T(a_1\vec{v}_1 + a_2\vec{v}_2 + \cdots + a_n\vec{v}_n) = a_1T(\vec{v}_1) + a_2T(\vec{v}_2) + \cdots + a_nT(\vec{v}_n)$. Since the set $\{T(\vec{v}_1), T(\vec{v}_2), \ldots, T(\vec{v}_n)\}$ is linearly independent, we must have $a_1 = 0$, $a_2 = 0, \ldots, a_n = 0$. This shows what we wanted.
(b) The converse to part (a) is *not* true. For example, T could be the zero linear transformation, and then $\{T(\vec{v}_1), T(\vec{v}_2), \ldots, T(\vec{v}_n)\}$ could never be linearly independent!

Section 8.2

1. The determinant of a cascade parameter matrix is, by direct calculation,

$$z_{11}z_{22}/z_{21}^2 - (z_{11}z_{22}/z_{21}^2 - z_{21}/z_{21}) = - -1 = 1.$$

3. (a) $[T_1] = \begin{pmatrix} 1 & 0 & 1 \\ 3 & 0 & 3 \\ 2 & 0 & 2 \\ -1 & 0 & -1 \end{pmatrix}$ has rank 1 and nullity 2.

(b) $[T_2] = (0 \quad 1 \quad -1 \quad -1)$ has rank 1 and nullity 3.

(c) $[T_3] = I_4$ has rank 4 and nullity 0.

(d) $[T_4] = (7)$ has rank 1 and nullity 0.

(e) $[T_5]$ is the 5×5 matrix with all entries 0. It has rank 0 and nullity 5.

5. One possibility is T_A where A is $\begin{pmatrix} 1 & 0 & 0 \\ 1 & 0 & 0 \\ 1 & 0 & 0 \end{pmatrix}$. Since the rank of any such example must be 1, we see that the nullity must be 2. In this example, a basis for $\ker(T_A)$ is $\{\vec{e}_2, \vec{e}_3\}$.

7. We need that T is onto in order that T is invertible. Hence, if $T(0,0,1) = (a,b,c)$, we need that $a \neq 0$.

9. (a) We know by the dimension theorem that if $T : \mathbf{R}^3 \to \mathbf{R}^4$, then $\text{rk}(T) \leq 3$. But in order for T to be onto, Theorem 57 shows we must have $\text{rk}(T) = 4$. Hence T cannot be onto.

(b) We know by the dimension theorem that if $T : \mathbf{R}^4 \to \mathbf{R}^3$, then $\text{rk}(T) + \text{null}(T) = 4$. But since the range space of T is $\mathbf{R}^3$, we know $\text{rk}(T) \leq 3$. Therefore, $\text{null}(T) \geq 1$ and Theorem 56 shows that T cannot be one-one.

(c) Since $\text{im}(T)$ is a plane, we have $\text{rk}(T) = 2$. Consequently, $\text{null}(T) = 2$.

11. It is not necessarily true that $T = O$. For example, we could have $T = T_A$, where

$$A = \begin{pmatrix} 0 & 0 & 0 \\ 1 & 0 & 0 \\ 0 & 0 & 0 \end{pmatrix}.$$

Since A^2 is zero, we have $T_A \circ T_A = O$, but $T_A \neq O$.

13. Whenever $1 \leq i < n$, we have that $\{\vec{v}_{i+1}, \vec{v}_{i+2}, \ldots, \vec{v}_n\}$ is a basis for $\text{im}(T^i)$. Whenever $i \geq n$, then $\text{im}(T^i) = \{\vec{0}\}$. Hence, $\text{rk}(T^i) = n - i$ and $\text{null}(T^i) = i$ whenever $1 \leq i \leq n$ and whenever $i > n$, $\text{rk}(T^i) = 0$ and $\text{null}(T^i) = n$.

Section 8.3

1. (a) $\begin{pmatrix} 1 & 2 \\ 2 & 1 \end{pmatrix}$, $\begin{pmatrix} 2 \\ 4 \end{pmatrix}_{\mathcal{B}} = \begin{pmatrix} 2 \\ 0 \end{pmatrix}$, $\begin{pmatrix} 4 \\ 2 \end{pmatrix}_{\mathcal{B}} = \begin{pmatrix} 0 \\ 2 \end{pmatrix}$, $\begin{pmatrix} 3 \\ 3 \end{pmatrix}_{\mathcal{B}} = \begin{pmatrix} 1 \\ 1 \end{pmatrix}$

(b) $\begin{pmatrix} 1 & 0 \\ 1 & 1 \end{pmatrix}$, $\begin{pmatrix} 17 \\ 4 \end{pmatrix}_{\mathcal{B}} = \begin{pmatrix} 17 \\ -13 \end{pmatrix}$, $\begin{pmatrix} 3 \\ 1 \end{pmatrix}_{\mathcal{B}} = \begin{pmatrix} 3 \\ -2 \end{pmatrix}$, $\begin{pmatrix} 0 \\ 0 \end{pmatrix}_{\mathcal{B}} = \begin{pmatrix} 0 \\ 0 \end{pmatrix}$

(c) $\begin{pmatrix} 1 & 0 & 0 \\ 1 & 1 & 0 \\ 1 & 1 & 1 \end{pmatrix}$, $\begin{pmatrix} 3 \\ 1 \\ 1 \end{pmatrix}_{\mathcal{B}} = \begin{pmatrix} 3 \\ -2 \\ 0 \end{pmatrix}$, $\begin{pmatrix} 1 \\ 0 \\ 0 \end{pmatrix}_{\mathcal{B}} = \begin{pmatrix} 1 \\ -1 \\ 0 \end{pmatrix}$

3. If $[T]_\mathcal{B}$ is zero, then for $\vec{v} \in \mathbf{R}^n$, $(T(\vec{v}))_\mathcal{B} = [T]_\mathcal{B}(\vec{v})_\mathcal{B} = \vec{0}$ for all $\vec{v} \in \mathbf{R}^n$. This shows that $T(\vec{v}) = \vec{0}$ for all $\vec{v} \in \mathbf{R}^n$. Conversely, if $T(\vec{v}) = \vec{0}$ for all $\vec{v} \in \mathbf{R}^n$, then for any ordered basis $\mathcal{B} = \{\vec{v}_1, \vec{v}_2, \ldots, \vec{v}_n\}$, $(T(\vec{v}_i))_\mathcal{B} = (\vec{0})_\mathcal{B} = \vec{0}$. But $(T(\vec{v}_i))_\mathcal{B}$ is the ith column of $[T]_\mathcal{B}$, so $[T]_\mathcal{B}$ is zero.

5. (a) $[T]_\mathcal{B} = \begin{pmatrix} 1 & 0 & 1 \\ 1 & 1 & 1 \\ 0 & 1 & 0 \end{pmatrix}$ (b) $[T]_{\mathcal{B}'} = \begin{pmatrix} 0 & 1 & 0 \\ 1 & 1 & 1 \\ 1 & 0 & 1 \end{pmatrix}$

Section 9.1

1. (a) $\{(\frac{3}{5}, \frac{4}{5}), (\frac{4}{5}, -\frac{3}{5})\}$

(b) $\left\{(\frac{2}{3}, \frac{2}{3}, \frac{1}{3}), \left(-\frac{1}{\sqrt{2}}, \frac{1}{\sqrt{2}}, 0\right), \left(-\frac{1}{3\sqrt{2}}, -\frac{1}{3\sqrt{2}}, \frac{4}{3\sqrt{2}}\right)\right\}$

3. False; for example, take $\vec{u} = (1, 0, 0)$, $\vec{v} = (0, 1, 0)$, and $\vec{w} = (1, 0, 1)$ in $\mathbf{R}^3$.

5. (a) $P = \begin{pmatrix} \frac{1}{\sqrt{2}} & -\frac{1}{\sqrt{2}} \\ \frac{1}{\sqrt{2}} & \frac{1}{\sqrt{2}} \end{pmatrix}$ (b) $P = \begin{pmatrix} \frac{1}{\sqrt{2}} & -\frac{1}{\sqrt{3}} & -\frac{1}{\sqrt{6}} \\ -\frac{1}{\sqrt{2}} & -\frac{1}{\sqrt{3}} & -\frac{1}{\sqrt{6}} \\ 0 & \frac{1}{\sqrt{3}} & \frac{2}{\sqrt{6}} \end{pmatrix}$

(c) $P = \begin{pmatrix} -\frac{1}{\sqrt{2}} & \frac{1}{\sqrt{3}} & \frac{1}{\sqrt{6}} \\ \frac{1}{\sqrt{2}} & \frac{1}{\sqrt{3}} & \frac{1}{\sqrt{6}} \\ 0 & \frac{1}{\sqrt{3}} & -\frac{2}{\sqrt{6}} \end{pmatrix}$

7. An orthonormal basis is $\left\{\left(\frac{1}{\sqrt{2}}, -\frac{1}{\sqrt{2}}, 0, 0\right), (0, 0, 1, 0), (0, 0, 0, 1)\right\}$.

Section 9.2

1. If $S \subset \mathbf{R}^n$ is a subset, then $S^\perp = \{\vec{v} \in \mathbf{R}^n \mid \langle \vec{s}, \vec{v} \rangle = 0 \text{ for all } \vec{s} \in S\}$. To see that $S^\perp$ is a subspace, we suppose that $\vec{v}_1, \vec{v}_2 \in S^\perp$ and that k is a real number. Then for all $\vec{s} \in S$ we have $\vec{0} = \langle \vec{s}, \vec{v}_1 \rangle + \langle \vec{s}, \vec{v}_2 \rangle = \langle \vec{s}, \vec{v}_1 + \vec{v}_2 \rangle$ and $\vec{0} = k\vec{0} = k\langle \vec{s}, \vec{v}_1 \rangle = \langle \vec{s}, k\vec{v}_1 \rangle$. This shows that $\vec{v}_1 + \vec{v}_2, k\vec{v}_1 \in S^\perp$ as well. Hence $S^\perp$ is a subspace of $\mathbf{R}^n$.

3. (a) $\text{Proj}_V = \frac{1}{18}\begin{pmatrix} 2 & 4 & 4 \\ 4 & 17 & -1 \\ 4 & -1 & 17 \end{pmatrix}$ and $\text{Proj}_{V^\perp} = \frac{1}{18}\begin{pmatrix} 16 & -4 & -4 \\ -4 & 1 & 1 \\ -4 & 1 & 1 \end{pmatrix}$

(b) $\text{Proj}_V = \frac{1}{21}\begin{pmatrix} 1 & 2 & -4 \\ 2 & 4 & -8 \\ -4 & -8 & 16 \end{pmatrix}$ and $\text{Proj}_{V^\perp} = \frac{1}{21}\begin{pmatrix} 20 & -2 & 4 \\ -2 & 17 & 8 \\ 4 & 8 & 5 \end{pmatrix}$

(c) $\text{Proj}_V = \frac{1}{18}\begin{pmatrix} 2 & 4 & 4 \\ 4 & 17 & -1 \\ 4 & -1 & 17 \end{pmatrix}$ and $\text{Proj}_{V^\perp} = \frac{1}{18}\begin{pmatrix} 16 & -4 & -4 \\ -4 & 1 & 1 \\ -4 & 1 & 1 \end{pmatrix}$

5. (a) If $\dim(V) = d$, then $\dim(V^\perp) = n - d$. To see this, one must check that bases for V and $V^\perp$ taken together give a basis for $\mathbf{R}^n$.

(b) If $\vec{w} \in V$, then for all $\vec{u} \in V^{\perp}$ we have $\langle \vec{w}, \vec{u} \rangle = 0$. This shows that $\vec{w} \in (V^{\perp})^{\perp}$. If $\dim(V) = d$, then using (a) that $\dim(V^{\perp}) = n - d$, and by (a) again, $\dim((V^{\perp})^{\perp}) = n - (n - d) = d$. As $V \subseteq (V^{\perp})^{\perp}$, we have $V = (V^{\perp})^{\perp}$.

(c) If $U \subseteq V$ and if $\vec{v} \in V^{\perp}$, then $\langle \vec{v}, \vec{u} \rangle = 0$ for all $u \in V$. So, in particular, as $U \subseteq V$ we have $\langle \vec{v}, \vec{u} \rangle = 0$ for all $\vec{u} \in U$. This shows that $\vec{v} \in U^{\perp}$, and we have $V^{\perp} \subseteq U^{\perp}$.

(d) Suppose that $\vec{v} \in (U + V)^{\perp}$. Since $U, V \subseteq (U + V)$, we have for all $\vec{u} \in U, \vec{w} \in V$ that $\langle \vec{v}, \vec{u} \rangle = 0 = \langle \vec{v}, \vec{w} \rangle$ so that $\vec{v} \in U^{\perp} \cap V^{\perp}$. Conversely, suppose that $\vec{v} \in U^{\perp} \cap V^{\perp}$. For $\vec{u} \in U$ and $\vec{w} \in V$ we have $0 = \langle \vec{v}, \vec{u} \rangle = \langle \vec{v}, \vec{w} \rangle$ so that $0 = \langle \vec{v}, \vec{u} + \vec{w} \rangle$ for all $\vec{u} = \vec{w} \in U + V$. This shows $\vec{v} \in (U + V)^{\perp}$, as required.

7. Consider the orthogonal projection $\text{Proj}_V : \mathbf{R}^n \to \mathbf{R}^n$. Then for any $\vec{v} \in \mathbf{R}^n$, $\text{Proj}_V(\vec{v})$ and $\vec{v} - \text{Proj}_V(\vec{v})$ are orthogonal. Hence,

$$\|\text{Proj}_V(\vec{v})\|^2 + \|\vec{v} - \text{Proj}_V(\vec{v})\|^2 = \|\vec{v}\|^2,$$

and consequently $\|\text{Proj}_V(\vec{v})\| \leq \|\vec{v}\|$. (Note that $\|\text{Proj}_V(\vec{v})\| < \|\vec{v}\|$ whenever $\vec{v} \neq \text{Proj}_V(\vec{v})$.) An intuitive explanation for this is that projections should be shorter. For example, if the sun is directly overhead, the length of a shadow should not be more than the length of the object.

9. Suppose that A is a rank-n $m \times n$ matrix. Let $\vec{v} \in \ker(A^tA) \subseteq \mathbf{R}^n$; that is, assume that $(A^tA)\vec{v} = \vec{0}$. The dot product $(A\vec{v}) \cdot (A\vec{v})$ is given by the matrix multiplication $(A\vec{v})^t(A\vec{v}) = \vec{v}^tA^tA\vec{v} = \vec{v}^t(A^tA\vec{v}) = \vec{0}$. Consequently, by the positive-definite property of the dot product on $\mathbf{R}^n$, we see that $A\vec{v} = \vec{0}$. This shows that $\vec{v} \in \ker(A)$ and therefore $\vec{v} = \vec{0}$ since A has rank n. But this shows that $\ker(A^tA) = \{\vec{0}\}$; in other words $\text{rk}(A^tA) = n$ also. This proves that A^tA is invertible, as required.

Section 9.3

1. (a) The least-squares solution is approximately $X = 1.607$, $Y = 1.179$.

(b) The least-squares solution is approximately $X = 4.133$, $Y = 2.368$, and $Z = .525$.

3. Suppose that $P^2 = P$. Then by a direct calculation, $(I_n - P)^2 = (I_n - P)(I_n - P) = I_n - P - P + P^2 = I_n - P$, using $P^2 = P$ in the last step. This shows that $I_n - P$ is also a projection matrix. If P is the matrix of a projection operator onto a subspace $V \subset \mathbf{R}^n$, then $I_n - P$ is the matrix of the projection onto the orthogonal complement of V.

5. Suppose that P is the standard matrix of the projection onto a line ℓ in $\mathbf{R}^2$, and let $R = I_2 - 2P$. First, consider a point $Q \in \ell$. Since P is the matrix of the projection onto ℓ, we see that $PQ = Q$. It follows that $RQ = Q - 2Q = -Q$. Next suppose that $Q' \in \ell^{\perp}$. Then, $PQ' = \vec{0}$, and consequently, $RQ' = Q'$. This shows that R is the matrix of the reflection through $\ell^{\perp}$.

7. $X = 35$ and $Y = 17$

INDEX

LIBRARY
OF
COLLEGE
EMMITSBURG, MARYLAND